Advances in Acoustic Micrc

Volume 1

Advances in Acoustic Microscopy
Volume 1

Edited by

Andrew Briggs
University of Oxford
Oxford, United Kingdom

Plenum Press • New York and London

Library of Congress Cataloging-in-Publication Data

Advances in acoustic microscopy / edited by Andrew Briggs.
p. cm.
Includes bibliographical references and index.
ISBN 0-306-44798-3 (v. 1)
1. Materials--Microscopy. 2. Acoustic microscopy. I. Briggs, Andrew.
TA417.23.A38 1994
620.1'1274--dc20 95-3646
CIP

ISBN 0-306-44798-3

A Division of Plenum Publishing Corporation
233 Spring Street, New York, N. Y. 10013

10 9 8 7 6 5 4 3 2 1

Printed in the United States of America

For Diana, Felicity, and Lizzie

Ἀκούσας τὴν καθ᾽ ὑμᾶς πίστιν
καὶ τὴν ἀγάπην, οὐ παύομαι
εὐχαριστῶν ὑπὲρ ὑμῶν.

Having heard of your faith and love,
I never cease to give thanks for you.

Contributors

Jan D. Achenbach, Center for Quality Engineering and Failure Prevention, Northwestern University, Evanston, Illinois 60208

Abdullah Atalar, Electrical and Electronics Engineering Department, Bilkent University, Ankara, Turkey

M. Beghi, Dipartimento di Ingegneria Nucleare, Politecnico di Milano, Milano, Italy and Consorzio Interuniversitario Nazionale per la Fisica della Materia, Unitá di Ricerca Milano Politecnico, Milano, Italy

Jürgen Bereiter–Hahn, Cinematic Cell Research Group, Zoological Institute, Johann Wolfgang Goethe University, Frankfurt am Maim, Germany

C. E. Bottani, Dipartimento di Ingegneria Nucleare, Politecnico di Milano, Milano, Italy and Consorzio Interuniversitario Nazionale per la Fisica della Materia, Unitá di Ricerca Milano Politecnico, Milano, Italy

Ayhan Bozkurt, Electrical and Electronics Engineering Department, Bilkent University, Ankara, Turkey

G. A. D. Briggs, Department of Materials, University of Oxford, Oxford, England

Gabriel M. Crean, National Microelectronics Research Centre, University College, Lee Maltings, Cork, Ireland

Colm M. Flannery, National Microelectronics Research Centre, University College, Lee Maltings, Cork, Ireland

G. Ghislotti, Dipartimento di Ingegneria Nucleare, Politecnico di Milano, Milano, Italy

Jin O. Kim, Center for Quality Engineering and Failure Prevention, Northwestern University, Evanston, Illinois 60208

Dieter Knauss, Department of Materials, University of Oxford, Oxford, England

Hayrettin Köymen, Electrical and Electronics Engineering Department, Bilkent University, Ankara, Turkey

Yung-Chun Lee, Center for Quality Engineering and Failure Prevention, Northwestern University, Evanston, Illinois 60208

J. W. Martin, Department of Materials, University of Oxford, Oxford, England

Paolo Mutti, Department of Materials, University of Oxford, Oxford, England; *present address*: Dipartimento di Ingegneria Nucleare, Politecnico di Milano, Milano, Italy

Séan Cian Ó Mathúna, Power Electronics Ireland, National Microelectronics Research Centre, UCC, Ireland

J. R. Sandercock, JRS, Zurich, Switzerland

Zenon Sklar, Department of Materials, University of Oxford, Oxford, England

N. C. Stoodley, Department of Materials, University of Oxford, Oxford, England

Kazushi Yamanaka, Nanotechnology Division, Medical Engineering Laboratory, Ministry of International Trade and Industry, Namiki 1-2, Tsukaba, Ibaraki, Japan

Göksenin Yaralioğlu, Electrical and Electronics Engineering Department, Bilkent University, Ankara, Turkey

T. Zhai, Department of Materials, University of Oxford, Oxford, England

Preface

In 1992 *Acoustic Microscopy* was published by Oxford University Press, in the series of *Monographs on the Physics and Chemistry of Materials.* Reviews appeared in the *Journal of Microscopy* [**169** (1), 91] and in *Contemporary Physics* [**33** (4), 296]. At the time of going to press, it seemed that the field of acoustic microscopy had settled down from the wonderful developments in resolution that had been seen in the late seventies and the early eighties and from the no less exciting developments in quantitative elastic measurements that had followed. One reviewer wrote, "The time is ripe for such a book, now that the expansion of the subject has perceptively slowed after it was detonated by Lemons and Quate." [A. Howie, *Proc. RMS* **27** (4), 280]. In many ways, this remains true. The basic design for both imaging and quantitative instruments is well-established; the upper frequency for routine imaging is the 2 GHz established by the Ernst Leitz scanning acoustic microscope (ELSAM) in 1984. For the most accurate $V(z)$ measurements, the 225-MHz line-focus-beam lens, developed at Tohoku University a little before then, remains standard. The principles of the contrast theory have been confirmed by abundant experience; in particular the role of surface acoustic waves, such as Rayleigh waves, dominates the contrast in most high-resolution studies of many materials. But in other ways, it has been delightful to observe the astonishing advances that have taken place in acoustic microscopy since the monograph was written. The purpose of this volume is to give an account of some of the key developments since then.

Perhaps the fastest growing field of applications of acoustic microscopy is inspection using interior imaging at relatively modest frequencies. There is an analogy here with what is happening in scanning probe microscopy, where although many of the most scientifically dramatic pictures come from scanning tunneling microscopy (STM) imaging with atomic resolution in ultra high vacuum

(UHV), nevertheless the greater quantity of industrial applicaton is at a lower resolution in air with atomic force microscopy. Similarly there is a large volume of industrial inspection by acoustic microscopy that does not require gigahertz frequencies and where Rayleigh waves are not excited. Such inspection uses focused probes and frequencies intermediate between high-resolution acoustic microscopy and conventional nondestructive testing. Indeed in many ways, it forms a bridge joining those two kinds of examination. Because of the widespread importance of this kind of inspection, Chapter 1 is written by members of the National Microelectronics Research Centre in Cork, which has extensive experience of using acoustic microscopy for industrial inspection of electronic packaging; Chapter 1 is extensively illustrated with examples of their own work.

It has long been established that acoustic microscopy is a powerful technique for studying surface cracks. Enhanced contrast arises from the scattering of Rayleigh waves, which can strike a crack broadside and therefore be strongly scattered even when the crack opening is much less than the resolution of the microscope. The contrast theory for this has been worked out in detail and fairly thoroughly tested. It has also been known for some time that Rayleigh waves can be diffracted by the tip of a crack and this can be detected in time-resolved measurements. At much lower nondestructive testing (NDT) frequencies (typically 5 MHz), the depth of crack tips can be measured from the time of flight of the signal diffracted from the crack tip. There has been strong motivation to be able to make the same kind of measurement in an acoustic microscope to study the behavior of cracks at the earliest stages of fatigue. Chapter 2 describes how this has been achieved at Oxford University using nanosecond time resolution to measure the subsurface geometry of cracks in metals with a resolution of a few micrometres.

The use of acoustic microscopy in biology demands perseverance, but rewards are great because of the possibility of directly imaging and measuring the elastic properties of cells and tissue. The Cinematic Cell Research Group at Frankfurt has been active in this field for many years, and Professor Bereiter–Hahn has produced an authoritative account, richly illustrated from his own experiments. Chapter 3 is written from a cytologists point of view, with the particular hope that life scientists will find it expressed in their own language and addressing questions relevant to them. There are healthy warnings about constraints of sample preparation, and problems that remain to be solved to make wider applications possible, but there is also a great deal that has been thoroughly established and is now available for use in biological applications of acoustic microscopy.

The applications to short cracks and biological cells both involve quantitative measurements, and the capability for quantitative elastic measurements has become increasingly important in acoustic microscopy. The next three chapters are concerned with measuring various kinds of surface waves. Chapter 4, devoted to a critical account of various lens geometries that can be used in acoustic microscopy, is written by Professor Atalar, one of the original pioneers of acoustic microscopy with Professor Quate at Stanford University. Professor Atalar produced one of the original formulations of $V(z)$ theory and an early physical explanation of oscillations in terms of Rayleigh wave excitation. But $V(z)$ measurements are not the only way of measuring surface waves if they are dispersive, and the chapter includes a definitive account of a Lamb wave lens for making swept-frequency $V(f)$ measurements.

Chapters 5 and 6 give two highly professional accounts of how to analyze $V(z)$ data from line focus-beam lenses to determine elastic properties of layered specimens. Chapter 5, which comes from the Center for Quality Engineering and Failure Prevention at Northwestern University, contains examples from their own work there on analyzing the elastic properties of coatings. There is a subtle difference between the methods of inversion in Chapters 5 and 6. In Chapter 5 inversion is performed by calculating $V(z)$ for trial parameters of the sample, analyzing mode velocities in this calculated curve, and comparing to analyzed mode velocities in a $V(z)$ curve measured experimentally from the sample. Trial parameters are then improved by iteration until good agreement is obtained between mode velocities from calculated and measured $V(z)$ curves. A great strength of this procedure is that it explicitly focuses on modes that are most strongly excited in the acoustic microscope and therefore convey most information about the elastic structure of the sample.

In the method of inversion in Chapter 6, velocities of acoustic modes in the surface are calculated directly from the trial elastic parameters of the specimen without calculating a $V(z)$ curve as an intermediate step. In the theoretical model for doing this, elastic stress is explicitly included, so that it is possible to include stress among the parameters to be deduced from experimental $V(z)$ data. The chapter contains a number of examples from the authors' laboratory at Oxford University. These include elastic constants of hydroxyapatite and fluoroapatite, stresses and nonlinear elasticity in silicon, and changes in elastic constants of amorphous hydrogenated carbon coatings as a function of processing parameters.

In many studies of elastic properties of surfaces, it is desirable to sample a layer considerably thinner than the 10 micrometres or so (depending on the Rayleigh velocity) probed in a standard line-focus-beam measurement at 225 MHz. This applies for example to measuring thin coatings and also fine subsurface damage. To address this, surface Brillouin-scattering spectroscopy has been developed at the Politecnico di Milano to be able to measure surface waves at higher frequencies. In surface Brillouin spectroscopy, light waves (photons) are scattered by elastic waves (phonons) in the surface of the sample. Two quantities are conserved: energy, so that the difference in frequency between the incident and scattered photons is equal to the frequency of a phonon created or annihilated in the scattering process; and momentum, so that the change in the tangential component of the wave vector of the light is equal to the wave vector of the surface phonon. In measuring materials the frequency of surface acoustic waves with which the light interacts is typically in the range of 10–20 GHz. Therefore, this extends the range that can be measured by acoustic microscopy by two orders of magnitude.

The final chapter also begins with a method of exciting surface acoustic waves using light, developed by Dr. Yamanaka and his colleagues. In this case, excitation is coherent, and it arises from the interference between two beams of light. One of the beams is subject to a frequency shift that causes the interference pattern to move across the surface. A surface wave is excited when its velocity matches the speed of the moving pattern. An advantage of this method is that it eliminates the coupling fluid needed in standard acoustic microscopy. The second part of Chapter 8 is devoted to near-field acoustic microscopy, in which spatial resolution comes from a scanned probe in contact with the surface. The sample is mounted on a plane wave transducer that propagates an acoustic wave into the sample. Acoustic displacements on the opposite surface are detected by an atomic force microscope. Contact between the surface and the AFM tip is nonlinear, and so as the amplitude of the acoustic wave is increased, the tip experiences a deflection. If the acoustic amplitude of the transducer is modulated at a low frequency, this modulation can be detected via the AFM probe. The signal at the modulation frequency contains information about the response of the system to the acoustic frequency even when the probe is unable to respond significantly at the acoustic frequency itself. Also by applying shear stresses that vary at the acoustic frequency due to simultaneously vibrating the sample in a lateral direction, a new imaging method was developed to reveal subsurface delaminations and dislocations.

Five of the chapters arose from a European Industrial Workshop (EIW-11) in Oxford in August 1993, organized with Professor Walter Arnold of the Fraunhofer Institute, Saarbrücken, sponsored by the European Physical Society, and supported by the Commission of the European Communities—DG XII. We also wish to express our thanks to Calcul Scientifique Appliqué à la Mécanique pour l'Industrie for sponsorship to help with the preparaton of this and subsequent volumes.

It is our intention that the monograph *Acoustic Microscopy* remain the definitive introductory text on the subject. To avoid repetition, and to ensure a measure of continuity, the authors of *Advances in Acoustic Microscopy* assumed that the reader is familiar with *Acoustic Microscopy*. With that in mind, we avoided repeating or reviewing what is already available in that book and followed the same mathematical notation wherever we can. Readers of the present volume may therefore find it helpful to have a copy of the earlier book for relevant background information. It is our hope that the new book will help to fill lacunae left by the earlier one and will convey some of the very exciting advances that are being made in acoustic microscopy.

Andrew Briggs

Oxford

Contents

Symbols and Abbreviations

Chapters where symbol or abbreviation is principally used are indicated

$\boldsymbol{A}$	laser absorption coefficient at surface	8
$\boldsymbol{A}$	tip sample contact radius	8
$\boldsymbol{a}$	UFV (ultrasonic frequency vibration) amplitude of sample	8
A	illuminated area of sample surface	7
A_0	lowest asymmetric mode	8
a_0	lens aperture radius	4
a_0	pupil radius	4
a_1, a_2	elastooptic constants	7
$\boldsymbol{a}^*$	amplitude threshold of $\boldsymbol{a}$ for additional cantilever deflection	8
AES	Auger electron spectroscopy	5
AFM	atomic force microscopy	8
APPCHIP	advanced packaging for high performance	1
a_T	transducer radius	4
B	inclination angle of crack	2
BAW	bulk acoustic wave	8
C	contrast of interferometer	7
C	magnitude of leaky component	4
C	normalized constant	5
c	sound velocity in cyloplasm	3
c	speed of light in vacuum	7
$\boldsymbol{c}$	bulk wave velocity	8
$\boldsymbol{c}_l$	longitudinal wave velocity	8
$\boldsymbol{c}_t$	shear wave velocity	8

$\bar{c}$	ratio of longitudinal wave speed in lens to wave speed in coupling fluid	5
$c'ijkl$	elastic constants related to crystalline axes	5
c.w.	continuous wave	7
$c/n(\omega_i)$	phase velocity of light in medium	7
Carrier Add	conventional detection system with added carrier	4
CE	electronic echo	2
CERDIP	ceramic dual-in-line package	1
CERQUAD	ceramic quad	1
c_{IJ}	second-order elastic constants in reduced notation	5, 6
C_{ijkl}	effective second-order elastic constants	6
c_{ijkl}	elastic constants in tensor notation	5
c_{ijklmn}	third-order elastic constants	6
c_l	longitudinal wave speed in lens	5
CLSM	confocal laser scanning microscope	3
c_w	wave speed in coupling fluid	5
D	diameter of collection lens	7
D	mirror spacing of the Fabry–Perot interferometer	7
$\boldsymbol{D}$	plate thickness	8
d	grating spacing	7
$\boldsymbol{d}$	width of laser beam	8
$\boldsymbol{d}$	tip sample distance (indentation depth)	8
$\boldsymbol{d}_c$	tip sample distance (indentation depth) due to a static force	8
d_{face}	depth of crack segment	2
Diff. Phase	differential phase detection system	4
DIP	dual in line	1
DLC	diamondlike carbon	1
DM	working distance in material	1
$\mathrm{d_N}$	thickness of NbN layer in TiN/NbN superlattice	5
d_p	average diameter of precipitates	7
DPA	destructive physical analysis	1
$\delta \mathbf{P}_i$	fluctuating part of polarization vector	7
$\mathrm{d_T}$	thickness of TiN layer in TiN/NbN superlattice	5
d_{tip}	depth of crack tip	2
DW	working distance in water	1
E	Young modulus	1, 5
$\boldsymbol{E}_1$, $\boldsymbol{E}_2$	Young modulus of Tip (1) and Sample (2)	8
$\boldsymbol{e}$	phonon unit polarization vector	7

$\mathbf{E}^{\mathrm{I}}$	incident electric field	7
E_{i}	component of electric field	5
e_i	components of the polarization unit vector of incident electromagnetic wave	7
e_{ijk}	alternating tensor	5
$\boldsymbol{E}^*$	effective elastic modulus of tip sample combination	8
Env. Det.	conventional envelope detection system	4
Err.	absolute error	4
E^{S}	complex amplitude of scattered field	7
$\mathbf{E}^{\mathrm{S}}$	scattered electric field	7
e^{e}_{ij}	elongation tensor	6
F	finesse of interferometer	7
F	focal length of collection lens	7
$\boldsymbol{F}$	frequency of acoustic waves	8
f	frequency	3, 4, 5
f	volume fraction of precipitates	7
f	center frequency of lens	2
$\boldsymbol{f}$	sample vibration frequency	8
$\boldsymbol{F}_{att}$	long-range attraction force	8
$\boldsymbol{F}_c$	tip sample static force corresponding to z_c	8
$\boldsymbol{F}_F$	friction force	8
$\boldsymbol{F}_N$	normal force	8
f_0	cantilever resonance frequency	8
FFM	friction force microscopy	8
FFT	fast Fourier transform	4
f_{L}	focal length	5
FMM	force modulation mode	8
FM_θ	figure of merit	4
FR	crack face reflection	2
FR_{L}	longitudinal reflection at crack face	2
FR_{M}	mixed-mode face reflection	2
FR_{S}	shear wave reflection at crack face	2
FSR	free spectral range	7
FT	Fourier transform	2
FWHM	full-width half-maximum	7
G	shear modulus	5
$\mathbf{G}$	vector-scattering integral	7
g	focal length	4
h	sample thickness	7
h	thickness of thin film	5
$\bar{h}$:	normalized film thickness	5

$H(F)$	filter function	2
h/a	ratio of roughness to transverse correlation length	7
HFPSW	high-frequency pseudosurface wave (*alias* LR)	7
H_i	component of magnetic field	5
HOP	hot isostatic pressing	1
$\boldsymbol{I}$	amplitude of the laser beam	8
i	square root of -1	5
$I(\omega)$	instrumental response function of laser spectrometer system	7
IC	integrated circuit	1
ICD	diffraction from crack kink	2
IDT	interdigital transducer	8
ILB	inner lead bonds	1
INO	nitrosoreduction product	3
INO_2	1-methyl-2-nitroimidazole	3
I_s	scattered light intensity	7
$\boldsymbol{J}$	laser power density	8
k	wave number	4, 5, 7
$\boldsymbol{k}$	cantilever spring constant	8
$\boldsymbol{K = k / s}$	relative sample stiffness	8
k_0	wave vector of incident light	7
$\boldsymbol{k}_1$, $\boldsymbol{k}_2$, $\boldsymbol{K}$	magnitudes of the wave number of laser	8
k_B	Boltzmann constant	7
$\boldsymbol{k}_f$	wave number of scanning interference fringe (SIF)	8
$\mathbf{k}_i$	wave vector of incident photon	7
k_{ix}	component of indent photon wave vector parallel to surface	7
k_l	wave vector of longitudinal wave	5
$\mathbf{k}_s$	wave vector of scattered photon	7
k_{sx}	component of scattered wave vector parallel to surface	7
k_t	wave vector of transverse wave	5
k_w	wave vector in coupling fluid	5
k_x	x-component of wave vector	5
k_z	z-component of wave vector	5
L	lens buffer rod length	4
L	longitudinal	2
L	separation of Fabry–Perot interferometer mirrors	7
l	length	2

$\boldsymbol{L}$	scanning length of laser beam	8
$\mathrm{L}_1(k_x)$, $\mathrm{L}_2(k_x)$	characteristic functions of acoustic lens	5
LEED	low-energy electron diffraction	5
LFAM	line-focus acoustic microscopy (*alias* LFB)	5
LFB	line-focus-beam acoustic microscopy (*alias* LFAM)	1, 4, 7
LGM	longitudinal guided mode	7
LM	light microscopy	2
$\mathcal{LPF}\{\ \}$	low-pass filtering operation	4
LPSAW	leaky pseudosurface acoustic waves	8
l_q	roots of Christoffel equation	5
LR	longitudinal resonance (*alias* HFPSW)	7
LSAW	leaky surface acoustic waves	8
LW	longitudinal wave	1
m	slope of line FR	2
MCA	multichannel analyzer	7
MCM	multichip module	1
MD	crack mouth signal	2
MIL-STD	military qualification standard	1
n	optical refractive index	7
$n(t)$	white gaussion noise signal	4
N_0	output noise power	4
N_{AS}	anti-Stokes line channel in MCA	7
N_{cal}	number of channels used in MCA calibration	7
N_{ch}	number of channels used in MCA data acquisition	7
NDE	nondestructive evaluation	8
n_p	average concentration of precipitates	7
N_S	Stokes line channel in MCA	7
OLB	outer lead bonds	1
$\mathbf{P}$	polarization vector	7
P	pupil function	4
$P(t)$	pulse emitted by lens	2
p-p	polarization with incident and scattered light polarized in scattering plane	7
PCB	printed circuit board	1
PCM	polarity comparison methods	1
PFB	point focus beam	1
PF1	peak frequency image	4
PFP	plane parallel Fabry–Perot interferometer	7
PGA	pin grid array	1

PMMA	perspex (*alias* plexiglas)	2
PMT	photomultiplier tube	7
PQAP	plastic quad flat pack	1
PSAW	pseudosurface acoustic wave	7
PVS	phase velocity scanning	8
PW	pulse width	2
Q	parallel transferred momentum	7
Q	phonon normal coordinate	7
$\mathbf{Q}$	transferred wave vector	7
q	focal length of lens	2
$\mathbf{q}$	wave vector	7
q_x	surface acoustic wave vector, wave vector of surface acoustic phonon	7
q_x^{-1}	penetration depth of surface acoustic wave displacement field	7
$\mathcal{R}$	reflection coefficient (at liquid/object interface)	4
R	reflectivity of interferometer mirror	7
R	stress ratio	2
$\boldsymbol{R}$	tip radius	8
r	radial coordinate	4
r	radius of curvature of lens	5
$\mathrm{R}(k_x)$	reflection coefficient	5
rf, r.f., RF	radio frequency	4, 5
RD	crack tip diffracted signal	2
$\mathcal{R}_e$	average reflection coefficient of anisotropic surface	4
Ref.	reference	3
ρ^0	density in reference configuration	6
RR	surface wave reflection	2
$\mathcal{R}_s$	surface reflection coefficient	4
RT	longitudinal lateral wave transmission	2
$\mathcal{R}_T$	reflection coefficient between liquid and topmost layer	4
RT	surface wave transmission	2
$\mathcal{R}_u$	subsurface reflection coefficient	4
RW	Rayleigh wave	7
ρw	density of coupling fluid	5
S	short	2
s	slit aperture width	4
$\mathbf{s}$	tip sample contact stiffness	8
$S(f)$	average spectrum of a full time line	2
$S(f)$	Fourier transform of a line in an $s(t,y)$ image	2

$S(t,y)$	signal (time, position)	2
s-s	polarization with incident and scattered light polarized normal to scattering plane	7
S_0	lowest symmetric mode	8
$S_0(f)$	average spectrum of selected cable echo	2
SAM	scanning acoustic microscope	1, 2, 3, 7, 8
SAW	surface acoustic wave	4, 5, 7, 8
$S_{B(\omega)}$	total Brillouin-scattering spectrum	7
SBS	surface Brillouin scattering	7
$s_e(t)$	echo of time signal	2
SEM	scanning electromicroscopy	2
$S_E^S(\omega)$	power spectrum of scattered light	7
SFM	scanning force microscopy	8
SH	shear horizontal	7
SH	shear wave	1
SIF	scanning interference fringe	8
s_{ij}	compliance constants	5
SIMOX	separation of implantation oxygen	7
SLAM	scanning laser acoustic microscopy	1, 3
SMS	surface wave reflection plus month-diffracted signal	2
SNR	signal-to-noise ratio	4
SOIC	small outline integrated circuit	1
SPDT	single-pole double-throw	2
SR	specular reflection	2
$s_s(t)$	actual time signal	2
SSB	scanning single beam	8
STM	scanning tunneling microscopy	8
SubSAM	subtraction scanning acoustic microscopy	3
SWI	Siemens Nixdorf Information System	1
Synch	amplitude and phase-measuring synchronous detection system	4
T	temperature	7
T	transverse	2
t	time	3
$\boldsymbol{T}$	laser pulse width	8
$T(L/\lambda)$	Airy function	7
T_0	maximum transmission of interferometer mirror	7
TAD	tape automatic bonding	1
TAM	tunneling acoustic microscopy	8
T_c	critical temperature (superconductor)	5

Pij	Pockel coefficients	7
TD_L	longitudinal tip-diffracted signal	2
TD_S	shear wave tip-diffracted signal	2
TEM	transmission electron microscopy	7
T_{fs}	transmission coefficient from coupling fluid to lens	5
t_i	components of traction vector	5
TOFD	time-of-flight diffraction	2
TRAM	time-resolved acoustic microscopy	2
T_{sf}	transmission coefficient from lens to coupling fluid	5
U	volume fraction of the fibular component	3
$\mathbf{u}$	spatial part of the fluctuating displacement vector field	7
$u_1{}^+$	acoustic field at back focal plane of lens	4
UFM	ultrasonic force microscopy	8
UFV	ultrasonic frequency vibration	8
u_i	components of particle displacement	5
u_i	infinitessimal displacement	6
u_{ij}	strain tensor	7
$u_0{}^+$	incident wave field	5
$u_0{}^-$	reflected wave field	5
$u^s{}_i$	finite static displacement	6
u_z	normal component of phonon displacement field	7
V	scattering volume	7
$\boldsymbol{V}$	peak-to-peak cantilever vibration amplitude	8
$\boldsymbol{V}$	scanning velocity of laser beam	8
v	speed of phonon	7
v	velocity	5
v	wave speed in the material	2
$V(f)$	acoustic microscope signal as a function of frequency	4
$V(z)$	acoustic microscope signal as a function of defocus	3–7
V_0	surface wave velocity in [100] direction on (001) plane of cubic crystal	5
v_0	wave speed in water	2
V_{45}	surface wave velocity in [110] direction on (001) plane of cubic crystal	5
V_b	bias voltage	6
v_c	sound velocity in a cell, cytoplasm	3

v_{f}	acoustic velocity in fibrils	3
$\boldsymbol{v}_f$	scanning velocity of laser interference fringes	8
$\boldsymbol{v}_G$	group velocity	8
$V_{\mathrm{G}}(z)$	geometric component of $V(z)$	4
v_{l}	longitudinal velocity	1, 2, 5
VLSI	very large-scale integration	1
$\mathrm{V}_{\mathrm{max}}$	maximum frequency	7
$\boldsymbol{v}_P$	phase velocity	8
v_{R}	Rayleigh velocity	2
$\boldsymbol{v}_R$	Rayleigh wave velocity	8
$V_{\mathrm{R}}(z)$	leaky wave component of $V(z)$	4
$V_{\mathrm{ref}}(z)$	reference $V(z)$	4
v_{s}	acoustic velocity in saline	3
v_{S}	shear velocity	1
v_s	shear wave velocity	2
$V_{\mathrm{s}}(f)$	Fourier transform of actual signal	2
V_{SAW}	velocity of surface acoustic wave	7
v_{SAW}	SAW phase velocity	7
VSOP	very small outline package	1
V_{t}	transverse wave velocity in [110] direction of cubic crystal	5
v_t	transverse wave speed	5
$v_{\mathrm{t}}^{\mathrm{f}}$	shear wave velocity of film	7
$v_{\mathrm{t}}^{\mathrm{s}}$	shear wave velocity of substrate	7
v_w	wave speed in coupling fluid	5
$\boldsymbol{w}$	adhesion energy	8
W	noise power	4
$W(\eta)$	strain energy density	6
x_1', x_2', x_3'	crystalline axes	5
X_i	coordinate in initial configuration	6
x_i	coordinate in reference configuration	6
XTH-2	Xenopus tadpole heart-endothelial cell line	3
Y	Young modulus	3
y	distance	2
yl	ens position	2
z	defocus of microscope, distance between focal plane and object surface	2–6
z_0	total cantilever deflection due to staticforce and ultrasonic frequency vibration	8
z_a	additional cantilever deflection due to ultrasonic frequency vibration	8
z_c	cantilever deflection due to a static force	8

z_{FMM}	cantilever deflection in the force modulation mode	8
z_S	sample-stage displacement	8
α_k	decay constant	7
α_w	coefficient of attentuation in coupling fluid	5
β	ratio of surface wave velocity to transverse velocity in [100] direction	5
Γ	half-width of Lorentzian distribution	7
γ	calibration constant of MCA	7
γ	difference of surface wave velocity in [100] and [110] directions	5
γ_{ij}	Eulerian strain	6
Δy	shift of the vertex of tip-diffracted signal	2
Δ	cubic dilatation	6
δ	incidence angle	2
δt	duration of the waveform	8
$\delta\chi_{ij}$	fluctuating part of susceptibility tensor	7
$\delta\varepsilon_{ij}$	fluctuating part of dielectric constant	7
δ_{ij}	Kroenecker delta function	7
$\delta\nu$	resolving power of Fabry–Perot interferometer	7
Δz	defocus	3
ε	static dielectric constant	7
η	anistropy factor	5
η_{ij}	Lagrangian strain	6
θ	angle of incidence	2, 5
θ	incident of angle of laser beam	8
θ	tilt sample surface or cantilever torsion angle	8
θ_c	critical angle	4
θ_{cr}	critical angle of total reflection	5
θ_i	angle of incidence	7
θ_m	half-aperture angle of lens	5
Λ	period in superlattice	5
Λ	wavelength of surface corrugation	7
λ	wavelength	3, 7
λ_s	wavelength of transverse wave	5
ν	Poisson ratio	5
ν_1, ν_2	Poisson ratio of Tip (1) and Sample (2)	8
ρ	density	1, 3, 5, 7
σ	Poisson ratio	1, 3, 7
σ_{ij}	components of stress tensor	5
τ	length of heat pulse	8
ϕ	azimuthal angle	5

ϕ	scattering angle	7
$\tilde{\nu}^q$	eigenvector	5
χ_{ij}	time-independent isotropic susceptibility tensor	7
Ψ	Bragg angle	7
Ψ	phase shift of carrier	4
Ω	phonon frequency	7
ω	angular frequency	5
ω	carrier frequency	4
ω	frequency of generated surface acoustic wave	8
ω_a	frequency difference of two laser beams	8
$\omega(\mathbf{q})$	frequency as a function of wave vector	7

1

Acoustic Microscopy Analysis of Microelectronic Interconnection and Packaging Technologies

Gabriel M. Crean, Colm M. Flannery, and Séan Cian Ó Mathúna

1.1. Introduction

Several advanced packaging and interconnection technologies are currently under development to meet the requirements of complex, large and/or high-speed microelectronic integrated circuits (IC) and systems. These include the design of multilayer interconnection substrates (see Fig. 1.1) for Multichip Module (MCM) applications[1] and the development of multilayer ceramic pin grid array packages (see Fig. 1.2) for large-area die size, high pin count ICs.[2] A development road map[3] for single-chip packaging technologies is presented in Fig. 1.3.

However, the use of both novel materials and an increasing number of processing steps in such technologies has posed serious yield and reliability issues. The availability of characterization tools for on-line process monitoring and failure analysis of packaging technologies are therefore becoming increas-

Gabriel M. Crean and Colm M. Flannery • National Microelectronics Research Centre, University College, Lee Maltings, Cork, Ireland. Séan Cian Ó Mathúna • Power Electronics Ireland, National Microelectronics Research Centre, UCC, Ireland

Advances in Acoustic Microscopy, Volume 1, edited by Andrew Briggs.
Plenum Press, New York, 1995.

Figure 1.1. Multichip module developed by Siemens Nixdorf Informationsysteme within ESPRIT Project No. 2075 APACHIP—Advanced PACkaging for HIgh Performance. A section of the cold plate has been removed, detailing the populated high-density interconnection substrate.

Figure 1.2. Single-chip cofired ceramic PGA package developed by Hoechst Ceramtec within ESPRIT Project No. 2075 APACHIP—Advanced PACkaging for HIgh Performance. The hermetic lid has been removed to highlight the multilayered structure and connected die.

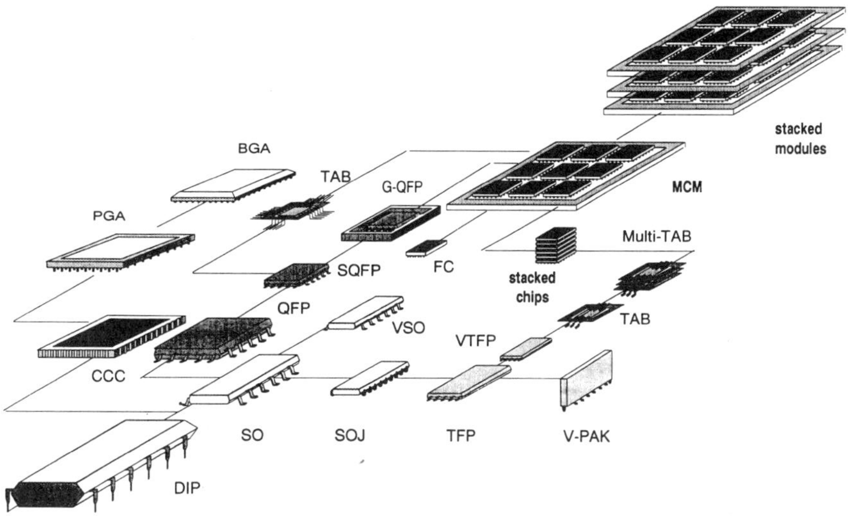

Figure 1.3. Technology road map for single-chip-packaging technologies. [3]

ingly important. Existing commercial techniques for inspecting IC interconnection and packaging technologies include optical systems, X-ray-imaging techniques, dye-penetrant analysis, strength testing, and microsection analysis. These techniques are either destructive or nondepth-specific. Scanning acoustic microscopy (SAM) is finding increasing application in this field due to its unique ability to access subsurface detail, provide depth-specific information, and act as a localized probe of mechanical properties.(4,5) The technique is now used in microelectronics materials research, product development, production process control, quality assurance, and failure analysis.

Chapter 1 reviews the applicability of acoustic microscopy for analyzing microelectronic packaging materials, processes, and technologies. When appropriate, results from acoustic imaging are compared and correlated with other diagnostic techniques, such as X-ray and microsectioning analysis. The major part of the work documented in this chapter has been undertaken at National Microelectronics Research Centre (NMRC Ireland) in a number of European collaborative research projects funded by the Commission of the European Communities under the European Strategie Programme for Research in Information Technology (ESPRIT) and Basic Research in Technology in Europe–European Research in Advanced Materials (BRITE-EURAM) programs and by the European Space Agency.(2,6,7) Chapter 1 therefore presents a European perspective of the current state of the art in this field.

Section 1.2 discusses the application of acoustic microscopy to the analysis of ceramic packaging. Hermetic ceramic packages are used in high-performance, high-reliability applications where performance issues dominate cost considerations. Using acoustic microscopy to analyze the integrity of both solder and glass hermetic seals is presented for side-brazed dual-in-line (DIP) and ceramic quad flat pack (CERQUAD) packages, respectively. Detecting delaminations, which may occur within the complex multilayer structure of cofired ceramic pin grid array (PGAs) packages employed for high pin count packaging is also demonstrated.

Plastic IC packaging accounts for almost 80% of the worldwide packaging market. Improvements in materials, in terms of encapsulants with low-ionic content and low stress, have significantly enhanced the environmental and mechanical reliability of plastic packages. Figure 1.4 is an optical photograph of several single-chip plastic packages. The driving force behind these developments has been the increasing need for low-cost high-density single-chip packages with large-area die (greater than 1 cm^2), reduced package thickness (less than 1mm), and fine-pitch, surface-mount assembly (less than 0.5-mm pitch).[8] There has also been considerable development in plastic power packages for very large scale integration (VLSI) applications where metal heat sinks are encapsulated within the plastic package.[9] Section 1.3 gives examples of the acoustic imaging of several high pin count plastic quad flat pack (PQFP) packages.

Die attach is required in all IC packages to attach the silicon chip to the package.[10] Die attach may also provide an electrical connection to the silicon bulk, and in VLSI applications, it provides the main heat flow path from the silicon die to an external heat sink or ambient. The die-attach layer must also minimize the temperature coefficient of expansion mismatch between the silicon and the package. As the area and power dissipation of silicon chips continues to increase, the integrity of this layer becomes increasingly critical. Although

Figure 1.4. Single-chip plastic packages; *from left to right:* PQFP, a very small outline package, and a small outline IC.

military qualification standard (MIL _ STD _) 883 has provided a basic reference for using acoustic microscopy to assess die-attach integrity,[11] no study to date has reported a detailed interpretation of the acoustic imaging of die-attach materials. Section 1.4 summarizes the work performed at NMRC in this area.

Tape automated bonding (TAB) is an IC-packaging technology that offers advantages in high-circuit density, a low profile, and a facility to test devices immediately prior to assembly.[2] Although flip-chip attachment[12] is now regarded as a viable alternative, because of its attributes, TAB is finding increasing application in low-profile consumer products, smart cards, liquid crystal displays, and in assembling high pin count ICs for use in high-performance computer products. Section 1.5 presents the first evaluation of acoustic microscopy for assessing TAB tape integrity.

With the continuing requirement for high-speed electronic systems, MCM technologies are finding increasing application.[2] These technologies use high-density interconnect substrates, and in most cases, they are interconnected to bare silicon die to obtain maximum density and performance. Three-dimensional MCMs are also under development.[13] The interconnection for MCM technologies is based on high-density organic, printed circuit boards (PCBs) or thick or thin film technology on multilayer cofired ceramic or silicon substrates. Section 1.6 describes acoustic microscopy analysis of high-density interconnection integrity on thin film and high-density organic substrates.

Because of the high costs and short time frames involved in developing new packaging and interconnection materials, technologies and systems, access to predictive design tools is important.[14,15] These design tools are required to undertake system design for all aspects of the system performance, including electrical, thermal, mechanical, and environmental specifications and package or system reliability. However design tools are only as accurate as the available input parameters, and a major limitation in the area of thermomechanical modeling has in fact been the availability of accurate mechanical properties for packaging materials. Section 1.7 presents work performed with a conventional point focus acoustic microscope for measuring the mechanical properties of several packaging materials.

The microelectronics industry has a continuing need for rapid turnaround, user friendly, nondestructive analysis capabilities to assess the integrity and performance of packaging and interconnection technologies. Section 1.8 presents the authors' views on future requirements and potential developments in acoustic microscopy for analyzing microelectronic-packaging technologies.

1.2. Ceramic Packages

In hermetic ceramic packages, lid sealing can significantly influence IC reliability. Failure in a hermetic seal can result in the ingress of moisture and ionic contaminants that may cause corrosion or electrical parameter drift. Recently these problems have become more acute due to the trend toward large-area die. Figure 1.5 shows an optical photograph of several ceramic single-chip DIPs. Existing techniques for inspecting IC package hermeticity include the dye-penetrant test for moisture ingression[16] and the fluorocarbon or helium pressurization leak test[17] for hermeticity failure. However both these techniques are destructive and time-consuming. The first objective of Section 1.2 is to demonstrate the application of acoustic microscopy for lid seal qualification.[18,19] The second objective is to discuss multilayer ceramic package imaging by using as an example a four-layer cofired PGA package. A novel high-resolution shear-wave-imaging technique for deep subsurface imaging of such ceramic-packaging materials is described.

This section discusses the acoustic microscopy analysis of three distinct ceramic-packaging technologies—a side-brazed DIP, a CERQUAD, and a four-layer PGA. A schematic cross section of the side-brazed package is presented in Fig. 1.6. In this IC package, the hermetic seal is provided by reflow soldering. The integrity of the lid seal is critical for high-reliability applications. A schematic cross section of the glass-sealed package is presented in Fig. 1.7. A similar schematic of the four-layer cofired ceramic PGA is shown in Fig. 1.8. This PGA is formed by laminating layers of unfired ceramic green sheet, depositing metal tracks on each layer and interconnecting the layers with laser-drilled or punched vias. The complete package is subsequently fired at 1500°C to sinter the structure together. Tungsten, nickel, and gold layers are then deposited on the package

Figure 1.5. Single-chip DIP; *from left to right:* a glass-sealed ceramic dual-in-line package; a solder-sealed, side-brazed DIP; and a plastic DIP.

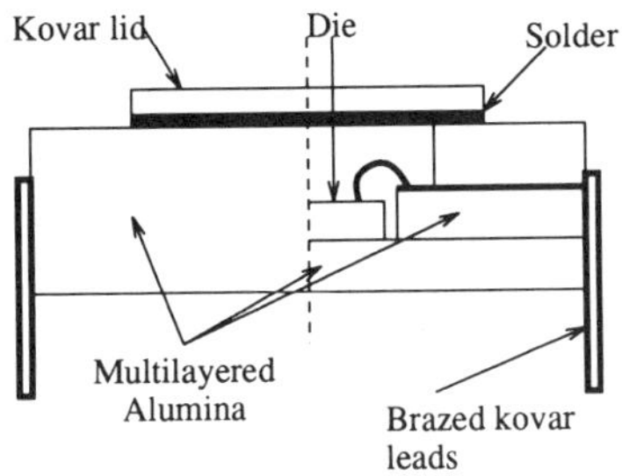

Figure 1.6. A cross-sectional schematic of a side-brazed package with hermetic solder seal.

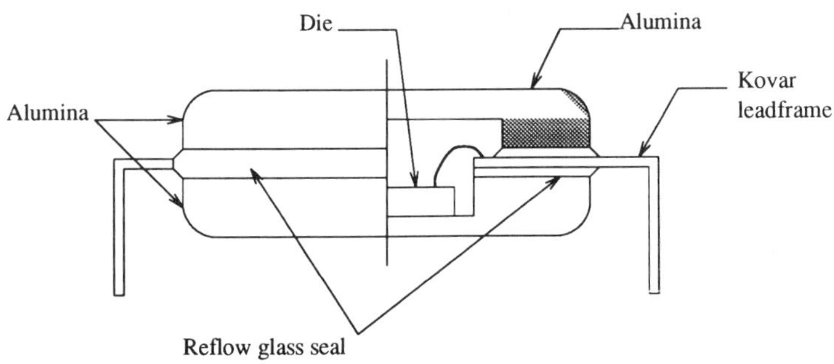

Figure 1.7. A schematic cross section of a hermetic, glass-sealed package.

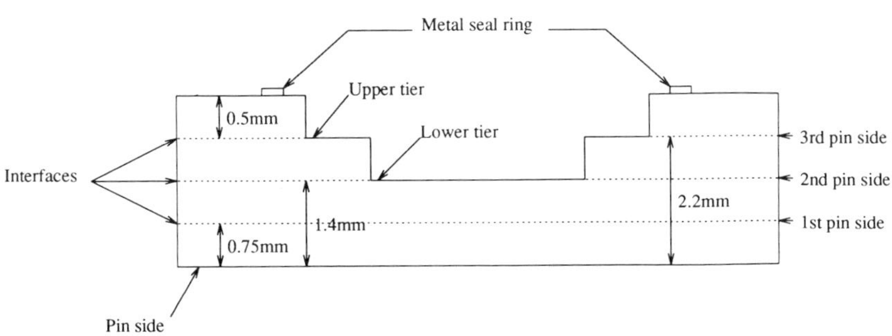

Figure 1.8. A schematic cross section of a four-layer cofired ceramic PGA package.

exterior, to which electrical contacts may be brazed or soldered as needed. The silicon die is then attached to the cavity, and after wire bonding, the package is sealed, typically using a metal lid and solder preform. The four-layer PGA used in this work is specially engineered to contain interlayer delaminations.

A commercial 50-MHz SAM was used in this study. The half-aperture of the acoustic lens was 14°, with a working distance of 12 mm. The X-ray analysis was performed using both Nicolet and Mobilix X-ray-imaging systems. Microsectioning was performed using a Metals Research Macrotome 1 and a Struers DAP-V polisher. Unless otherwise stated, these analysis tools were used for all subsequent work presented in Chapter 1.

An X-ray shadow micrograph of the side-brazed package is shown in Fig. 1.9. The silicon die, package leads, and lid seal are all visible. Voids (marked with arrows) are present around both the edges of the lid seal and more extensively in the upper right-hand corner of the package. An acoustic micrograph of the same package, focused at the lid seal interface is presented in Fig. 1.10. The white regions are indicative of voiding at this interface, and they are consistent with the X-ray shadowgraph image in Fig. 1.9. Grey areas represent regions of good bonding between the lid and package. Black areas in the micrograph result

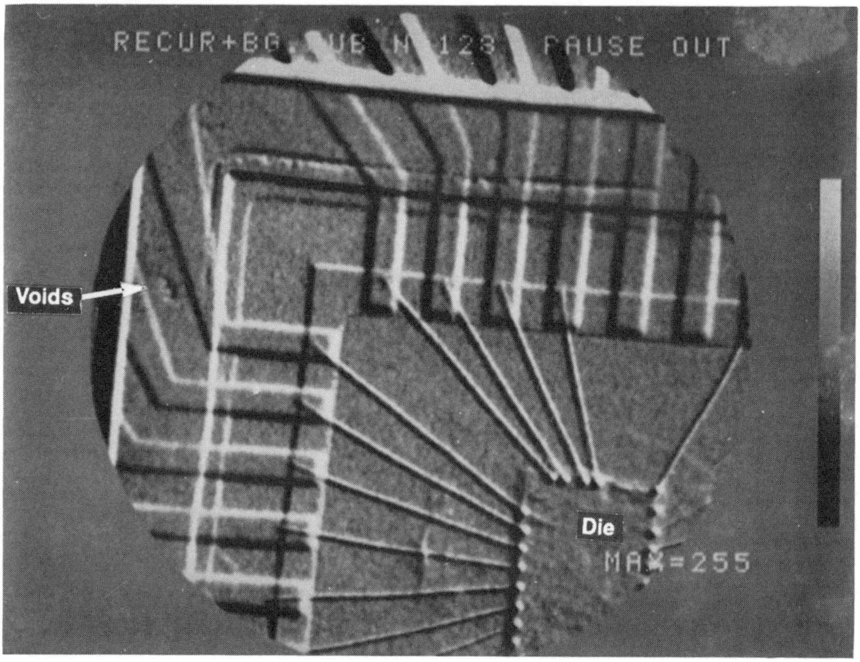

Figure 1.9. An X-ray shadow micrograph of the solder lid seal of the side-brazed package.

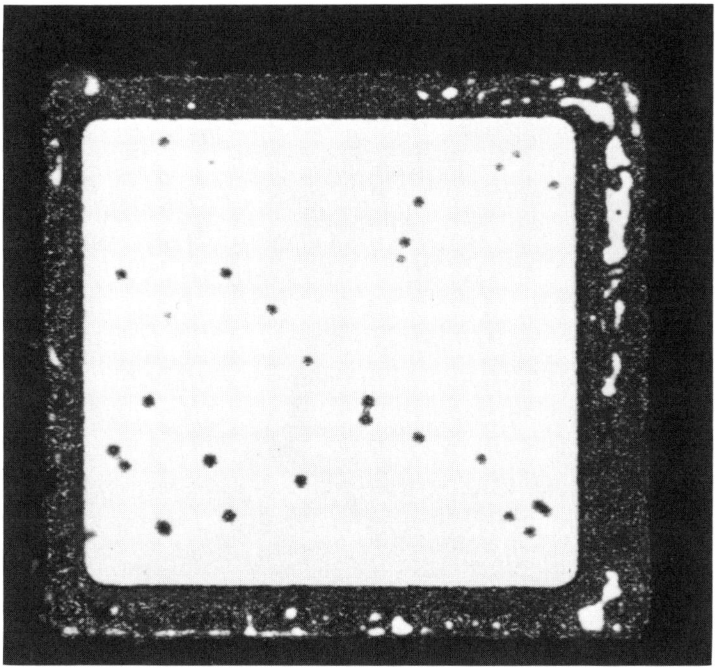

Figure 1.10. Acoustic micrograph of side-brazed package focused at the lid seal.

from nonspecular acoustic-scattering phenomena within the lid or sealing material.

An acoustic micrograph of the glass-sealed CERQUAD package is presented in Fig. 1.11. Bright areas in the four corners of the package indicate voids in the glass seal. This was subsequently confirmed by both X-ray analysis (see Fig. 1.12) and optical microscopy of a microsection of the package, as shown in Fig. 1.13. Region *A* is a void in the glass seal, and Region *B* is the alumina package material (cf. Fig. 1.7).

The preceding results demonstrate the ability of acoustic microscopy to provide depth-specific information, in this case about the lid seal integrity of both solder and glass-sealed packages.[18,19] This depth-specific capability is also important in examining multilayer ceramic packaging technologies, such as the four-layer PGA package. Figures 14(a)–(c) show acoustic micrographs taken at the first, second, and third pinside ceramic interfaces (cf. Fig. 1.8), respectively, of the PGA package. The power and ground planes are detailed in all three micrographs, however extensive delaminations are evident in the acoustic micrographs (white regions) of both the second- and third-layer pin interfaces with this kind of defect.

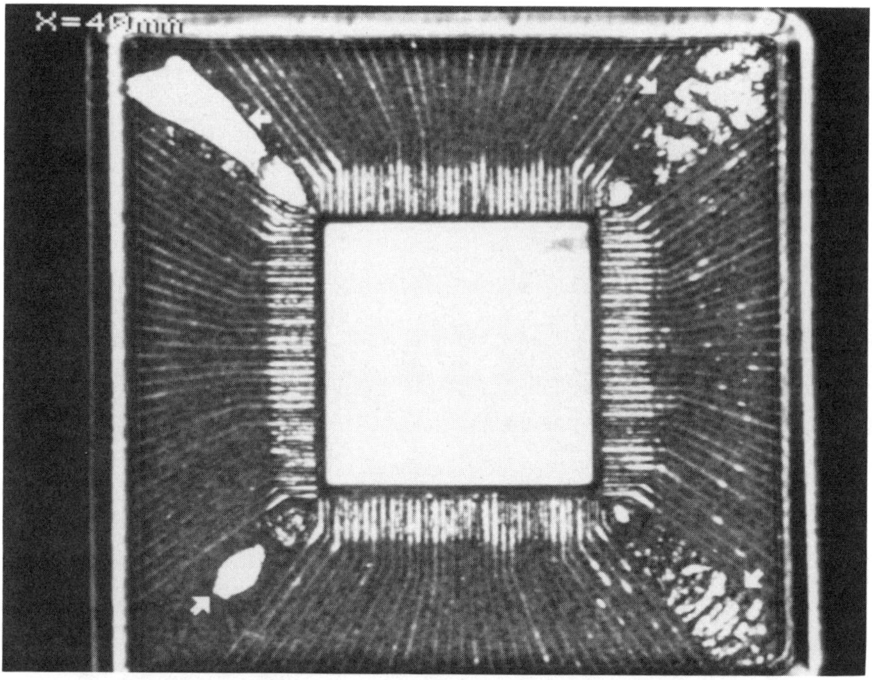

Figure 1.11. Acoustic micrograph of glass-sealed CERQUAD package.

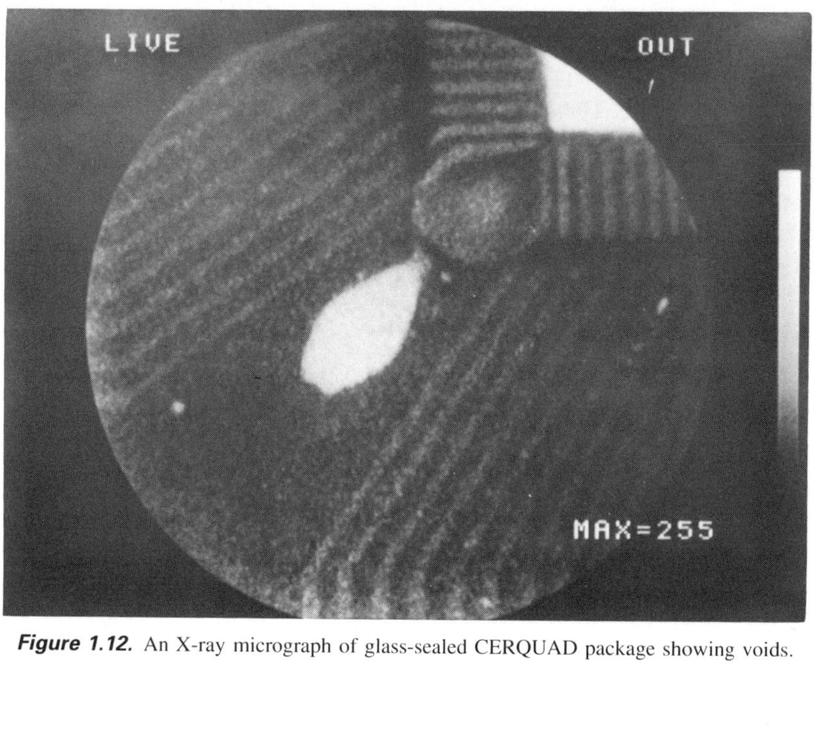

Figure 1.12. An X-ray micrograph of glass-sealed CERQUAD package showing voids.

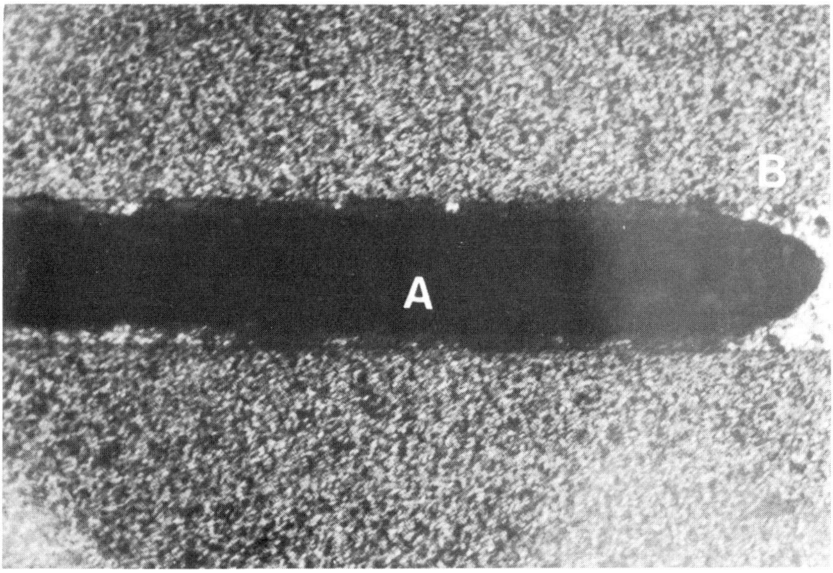

Figure 1.13. Microsection of a glass seal in CERQUAD package. Region *A* is a void in the glass seal and Region *B* is the alumina package material (cf. Fig. 1.7).

An optical micrograph of the PGA package after cross sectioning is shown in Fig. 1.15. Substantial delaminations are visible at both the second- and third-layer interfaces consistent with the interpretation of the acoustic information in Figs. 1.14(b) and (c). The X-ray shadowgraph analysis was unable to differentiate between good or bad PGA packages.

In the preceding work, acoustic microscopy at 50 MHz was employed for nondestructive evaluation of side-brazed DIP, CERQUAD, and PGA ceramic packages. An analysis of the acoustic micrographs of the lid seal of the solder seal package revealed voiding at the package lid interface. This was subsequently confirmed by X-ray shadowgraph analysis. A similar result was observed for the glass seal packages examined. The ability of the SAM to provide depth-specific and localized characterization of the PGA interlayer bonding after cofiring was also demonstrated.

However it was observed that acoustic microscopy was penetration depth-limited for certain ceramic packages (when the thickness of the package base was greater than 1 mm) by the concomitant degradation in lateral resolution that occurred. A requirement for a deep (>1 mm) subsurface high-resolution imaging capability was identified for ceramic-packaging inspection.

There are two problems associated with deep subsurface imaging using longitudinal wave mode acoustic microscopy. These concern the effective working distance and lateral resolution, respectively, in the material system under

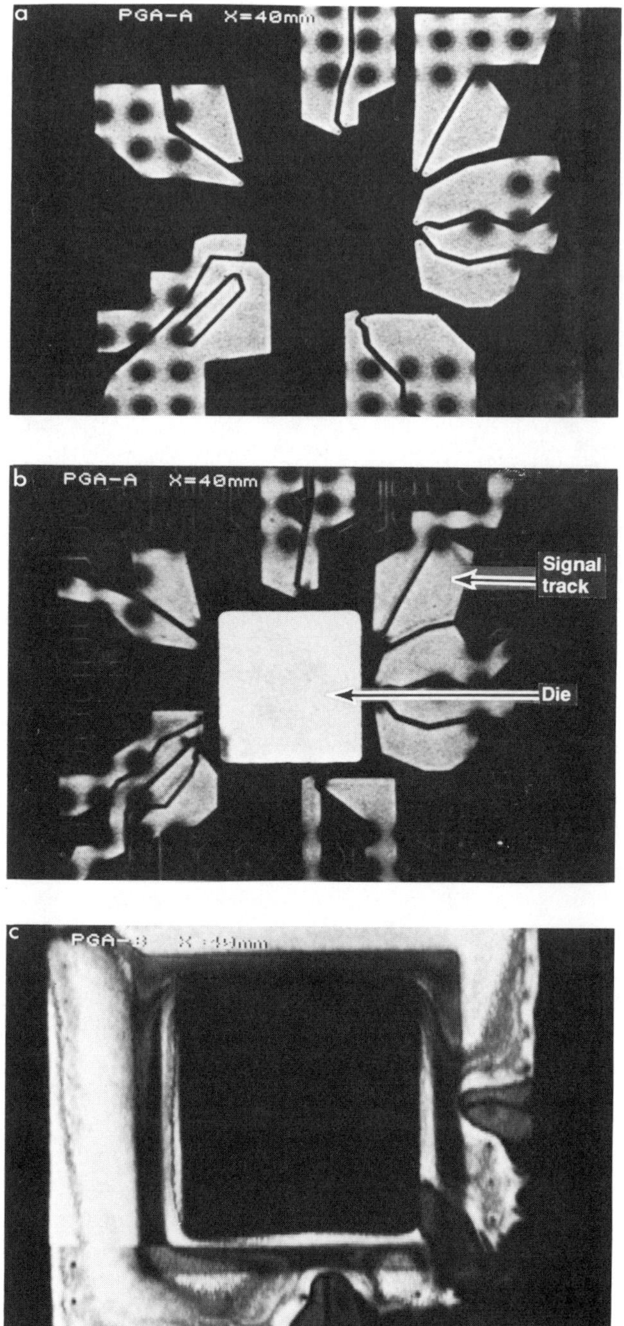

Figure 1.14. Acoustic micrograph of the four-layer, cofired ceramic PGA; (a) focused at the first-pin interface; (b) second-pin interface, and (c) third-pin interface.

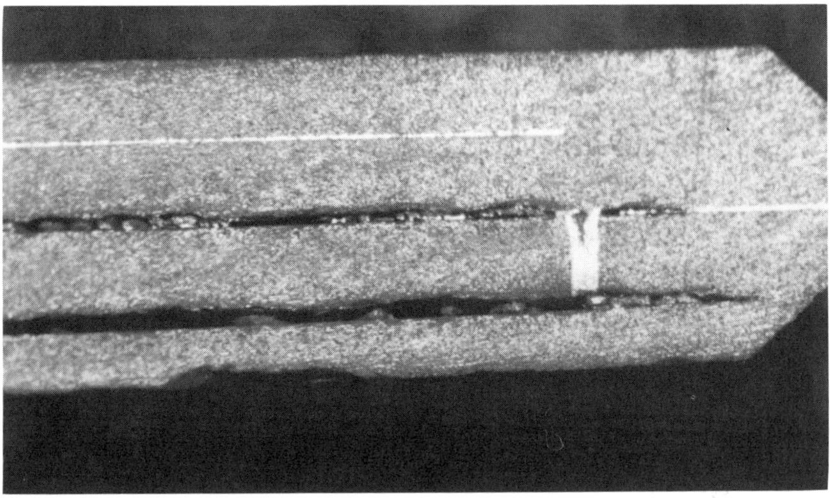

Figure 1.15. Optical cross-sectional image of cofired PGA package detailing delaminations at second- and third-layer pin interfaces.

characterization. For high-acoustic velocity materials, such as the ceramic materials discussed in Section 1.2, these problems become particularly acute. For alumina (Al_2O_3), where the longitudinal wave velocity is approximately 9500 ms^{-1}, the maximum working distance in this material D_m, referred to that of water D_w, is only 15.7% of D_w. In addition the longitudinal wave critical angle[4] is small, thereby limiting the usable aperture of the acoustic lens.

To date several groups have addressed these issues, in particular through the development of optimized acoustic lenses for deep subsurface imaging.[20,21] The novel approach developed at NMRC Ireland[22–25] uses a shear-wave-imaging technique with a conventional (commercially available) acoustic microscope and lens.

The deep-subsurface high-resolution-imaging technique is based on mode conversion phenomena that occur when a focused acoustic beam is incident from water at a material surface. We can predict by calculating the longitudinal and shear wave focal plane distributions in a ceramic substrate that the shear wave will have a lateral resolution significantly smaller than that of a longitudinal wave due to the shear quasiannular source (see Fig. 1.16) formed in the ceramic.[22,24] At an acoustic frequency of 50 MHz in Al_2O_3 the shear wave lateral resolution is 70 μm, compared to 130 μm for longitudinal waves.

This is demonstrated in Figs. 1.17 and 1.18, which present 50-MHz longitudinal and shear wave mode acoustic micrographs, respectively, of the bottom surface of a 1.4-mm-thick ceramic (Al_2O_3) test substrate with 100-μm width metal tracks on its top surface and 350-μm vias drilled through the ceramic. The shear mode has significantly enhanced lateral resolution. The apparent increase in the width of the signal track in Fig. 1.17 is due to the fact that the width of

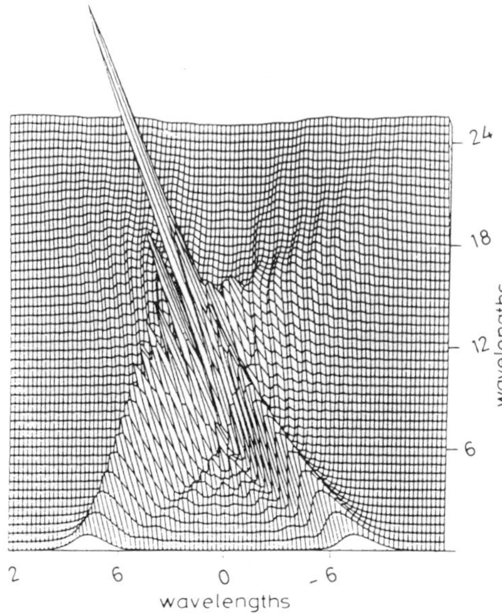

Figure 1.16. Isometric projection of the SH wave amplitude field within an alumina ceramic substrate assuming a 50-MHz acoustic lens with a 28° aperture and 11.5-mm working distance in water.

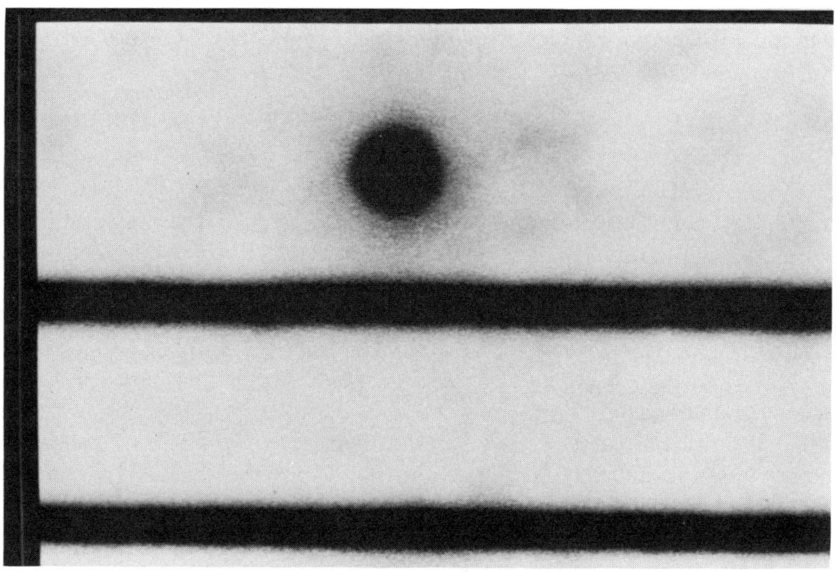

Figure 1.17. Longitudinal mode 50-MHz acoustic micrograph of patterned alumina test substrate (1.4-mm thickness) as viewed from the bottom surface of the substrate. The defocus distance was 10.9 mm.

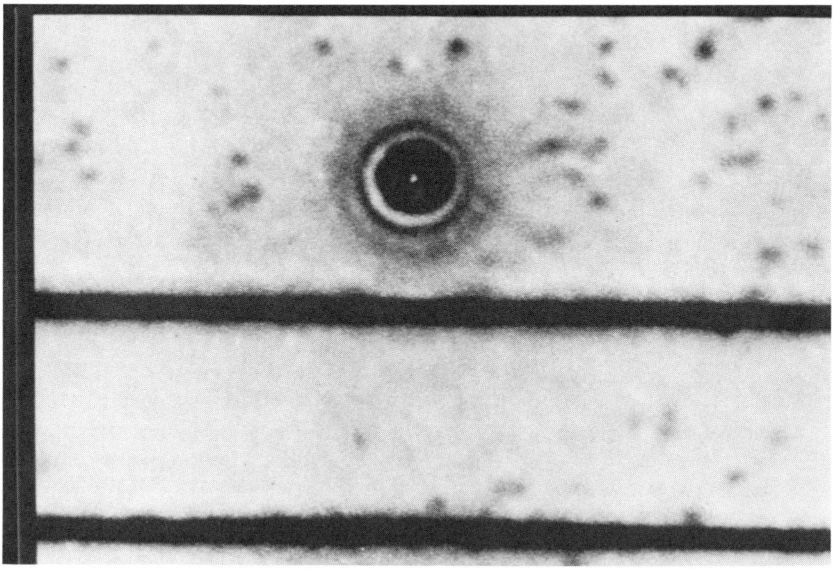

Figure 1.18. Shear mode 50-MHz acoustic micrograph of patterned alumina test substrate (1.4-mm thickness) as viewed from the bottom surface of the substrate. The defocus distance was 7.3 mm.

this track is below the longitudinal wave lateral resolution at this frequency. Further improvements in deep subsurface imaging of ceramic packages were achieved by using an annular source lens.[25]

1.3. Plastic IC Packages

Mechanical failures within plastic IC packages are primarily due to a temperature coefficient of expansion mismatch between various materials within the plastic package, a problem that may be further exacerbated during the surface-mount reflow process (i.e., the popcorn effect, where absorbed moisture in the package resin is vaporized during reflow soldering, typically at the resin–die pad interface. Vapor pressure may then deform the resin and cause cracking). Failures that may occur and result in reliability problems include package cracks, die cracks, delaminations, wire bond failures, die surface damage, enhanced metal corrosion, and electrical parameter shifts. As a result in recent years, considerable research has been undertaken with a view to reducing the impact of the popcorn effect through improved package designs and component-handling procedures.[26–28] There has also been significant research into the development of SAM, both in terms of instrumentation and data analysis, to provide a rapid, user friendly, nondestructive technique for assessing internal cracking and delaminations within plastic IC packages.[29–33]

The following paragraphs present work undertaken at NMRC in the area of plastic package qualification. A range of plastic packages including 132 and 160 lead PQFPs were analyzed using acoustic microscopy in conjunction with other analytical techniques, such as X-ray and microsectional analysis. Finite-element thermomechanical modeling was used to elucidate the failure mechanisms observed.

The acoustic image presented in Fig. 1.19 was obtained at the lead frame–plastic interface of a 132-lead PQFP. The image indicates a deformed lead and a partial delamination between the plastic and the central die paddle. The delamination is represented as the white region in the picture, whereas the central grey area represents good contact between the plastic and the metal paddle. The X-ray image in Fig. 1.20 shows deformed leads with a higher resolution than the corresponding acoustic micrograph. However the X-ray is insensitive to the delamination that is present. The inability of the X-ray system to detect this delamination is due to the fact that the thickness of the delamination (approximately 5–10 μm) does not result in a sufficient density change at the plastic–lead frame interface to make it detectable. This highlights the complementary nature

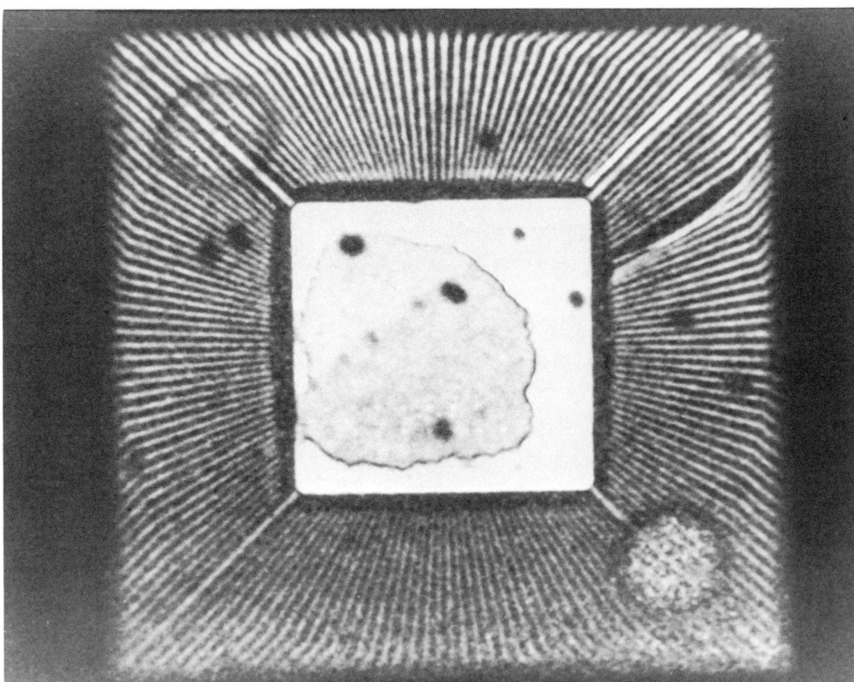

Figure 1.19. Acoustic micrograph of a 132-pin PQFP showing delamination at the die pad–encapsulant interface and the internal deformation of lead.

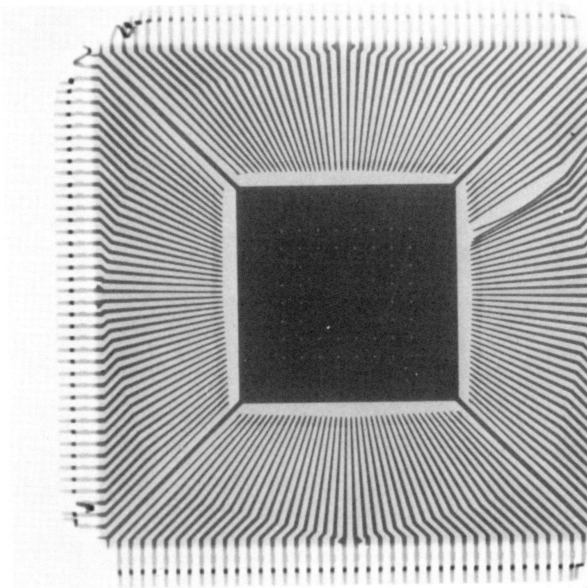

Figure 1.20. An X-ray micrograph of a 132-lead PQFP showing lead deformation but not the delamination at the die pad.

of acoustic microscopy and X-ray microradiography in the nondestructive analysis of IC packages. While X-ray analysis may provide higher resolution imaging, acoustic microscopy can provide depth-specific information relating to mechanical integrity.

Figure 1.21 shows an optical photograph of a 160-lead PQFP package. A number of these PQFPs were analyzed using acoustic microscopy after being subjected to excessive reflow-soldering temperatures in a production environment. An acoustic micrograph obtained at 50 MHz of one of these PQFPs is shown in Fig. 1.22. This image details the lead frame, an insulating tie bar, the silicon die, and the die paddle. The insulating tie bar, which holds the high-density lead frame during the manufacturing process, was only 50 μm thick, and it did not register on the X-ray image because of insufficient variation in density. The die area of the X-ray image is shown in Fig. 1.23. The white region of this image indicates a delamination, while the grey region at the center represents good adhesion between the plastic and the silicon. Dark areas around the periphery of the silicon indicate the location of the wire bonds. The reflected acoustic pulse train detailed in Fig. 1.24 shows a 180° phase shift between reflected acoustic signals from the adhered and delaminated areas consistent with wave propagation theory.[33]

A microsection of the PQFP-160 is shown in Fig. 1.25. This confirmed the presence of the delamination at the die surface with a thickness of approximately

Figure 1.21. A 160-lead PQFP for surface-mount application.

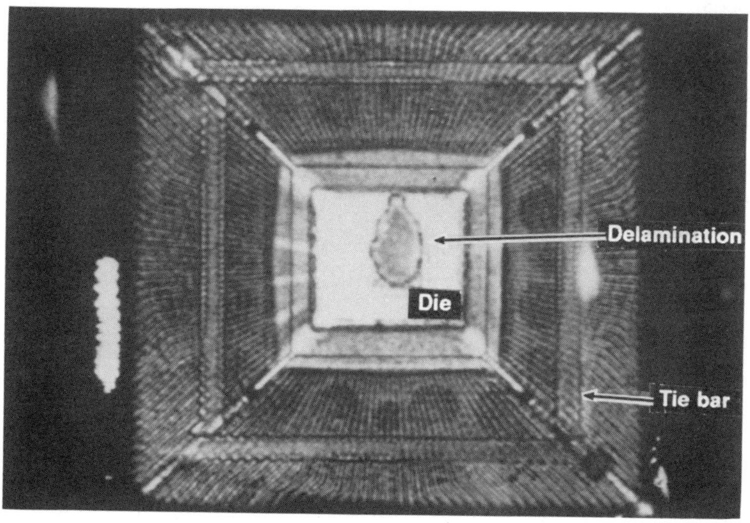

Figure 1.22. Acoustic micrograph of a 160-lead PQFP imaged through the top of the package and detailing the lead frame, tie bar, IC, and die pad.

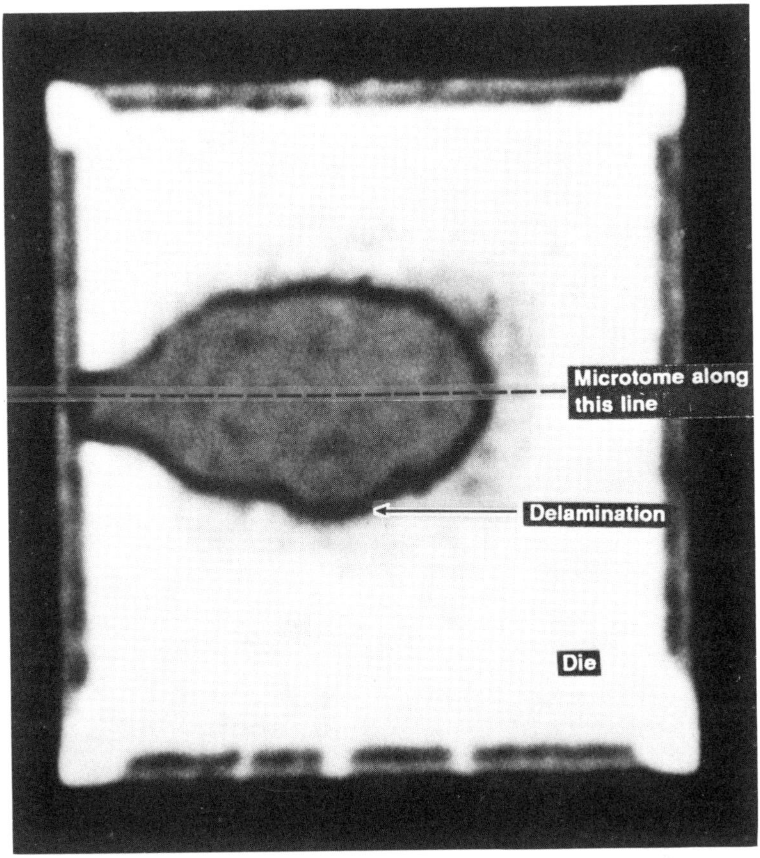

Figure 1.23. Acoustic micrograph of the die–plastic interface of the 160-lead PQFP. The white region corresponds to delaminated areas.

4 μm. The pattern of delamination on the chip surface was correlated with results from finite-element thermomechanical analysis of the package structure.[34] From this modeling, delamination at the material interfaces was found to increase significantly the level of thermomechanical stress. Once delamination commenced, the line of stress concentration followed the adhesion edge on the chip surface, since this was the first point of constraint encountered by the moulding compound. Delaminations may thus propagate and spread throughout the whole package, resulting in the delamination pattern observed in Fig. 1.23.

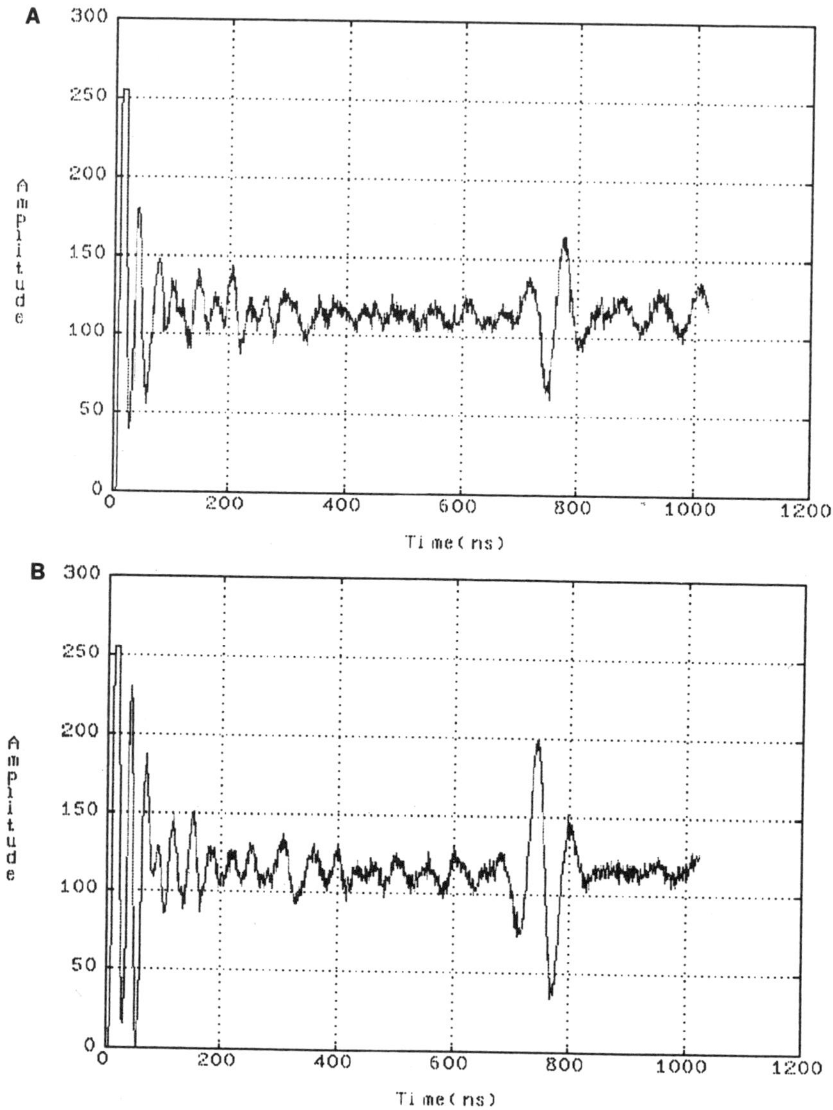

Figure 1.24. Acoustic pulse reflection from the 160-lead PQFP sample. A 180° phase change is evident between the bonded and delaminated regions. (A) Signal from the adhered region of the die surface; (B), Signal from the delaminated region of the die surface.

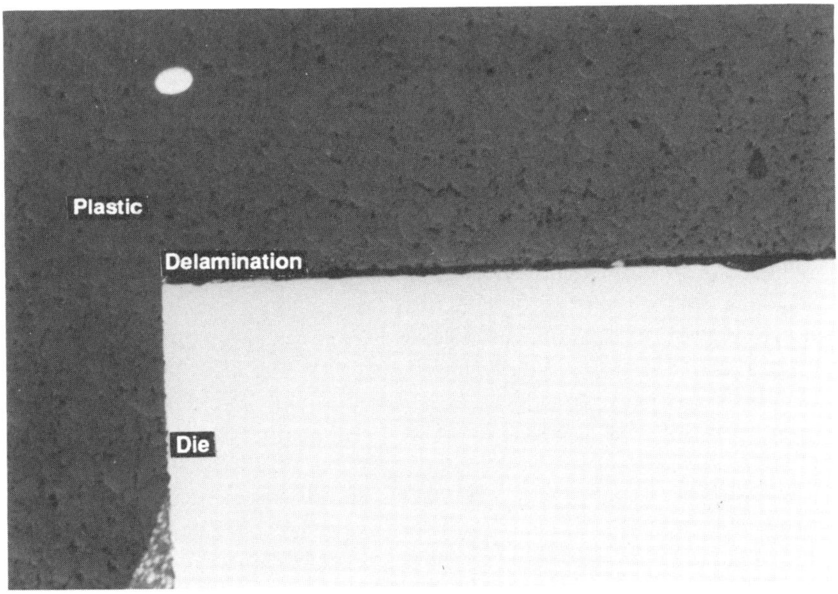

Figure 1.25. Optical micrograph of the 160-lead PQFP confirming delamination at the die surface–plastic interface.

1.4. Die Attach

Die attach is the process of bonding a semiconductor die to a substrate or lead frame. Attachment is achieved with adhesive or solder. The most common attachment materials are epoxy, polyimide, or eutectic adhesives.[10,35] The function of die attach is to form a bond between the chip and substrate providing adequate thermal, mechanical, and electrical contact between the two surfaces. Poor adhesion of the die attach or the presence of voids within the die attach may lead to hot spots in the semiconductor die. The thermal specification of the package is related to the ability of the die attach to transmit heat. Failure due to thermomechanical stresses in the die attach may result in disbonding of the die and crack propagation, which could affect both the quality and long-term reliability of the package. With the trend toward ever larger die sizes, faster curing times, and copper lead frames the characterization of die-attach integrity has become increasingly critical.

Both scanning laser acoustic microscopy (SLAM) and X-ray imaging have been applied to the characterization of die attach.[36] Since each technique has significant disadvantages in terms of not providing depth-specific information, a limited amount of work on applying acoustic microscopy to die attach has been carried out. A comparison between SAM, SLAM, X-ray and destructive physical analysis (DPA) by pull testing silver/glass die attach has been performed.[36] In addition, ceramic dual-in-line package (CERDIP), PQFP, PGA, and MCMs have been examined.[18,37] It has also been demonstrated that waveform analysis by polarity comparison methods (PCM) can be used to detect delaminations.[38]

Interpreting acoustic micrographs of die attach is generally regarded as something of a black art. The need for an in-depth study of die attach with varying cure and substrate parameters and comparison of resultant acoustic micrographs with corresponding X-ray and DPA data was clear to the authors of Chapter 1. Such a study was thought to facilitate interpreting die-attach acoustic micrographs and extracting meaningful quantitative results from them. The project goal was to assess the current ultrasonic inspection standard MIL _ STD _ 883,[11] which was considered inadequate, and develop a consistent methodology to characterize die-attach integrity. A summary of the principal results of this study is presented in this section.

To reflect the trend toward larger die sizes, a silicon die 0.5 mm in thickness with a die size of 12 × 12 mm was chosen. Three substrate types were selected, ceramic (alumina), due to its application as a packaging substrate; metal (brass), which is used as a heat sink material; and glass (Pyrex) to facilitate optical inspection. The thicknesses of the ceramic, metal, and glass substrates were 0.63 mm, 0.9 mm, and 5.0 mm, respectively. The silver expoxy die-attach thickness was 50 μm for all samples except those on brass substrates, which were both 50 μm and 100 μm. A range of samples were manufactured with varying cure parameters to ensure significant variations in die-attach integrity. In addition voids were introduced manually on a selection of these samples. All samples were examined in a prepackaged state (before encapsulation), consisting of a silicon die attached to a substrate by a layer of die attach, as illustrated in Fig. 1.26. Both X-ray transmission shadowgraph and pseudo-three-dimensional images were obtained from each sample.

Acoustic analysis was performed at 50 MHz. The calculated lateral resolution of the system was 68 μm in water and just over 90 μm in silicon. Acoustic micrographs were obtained both through the die (with the lens focused on the die–die-attach interface) and substrate (focusing on the substrate–die-attach interface).

Die shear tests succeeded in removing both the silicon die and die attach from the metal substrates, but the maximum shearing force (20 kg) did not remove the die from the ceramic substrates. Ceramic samples were then set in epoxy, and the die was fully ground away to reveal the die attach underneath. Optical photographs were then obtained.

Figure 1.27 presents a pseudo-three-dimensional X-ray micrograph of a ceramic substrate sample. Voids are observed at both left and right edges as well as a number of horizontal, linearly parallel, evenly spaced internal voids, probably

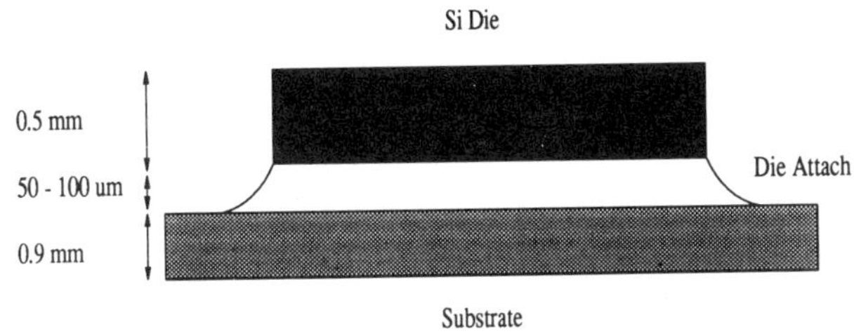

Figure 1.26. Schematic cross section of the die–die-attach–substrate structure.

due to the die-attach-dispensing process. Small line patterns are noticeable throughout the die attach. The acoustic micrograph in Fig. 1.28 through the silicon die shows the edge and linear voids as bright areas; there are also a number of dark features, caused by the die attach scattering the acoustic signal. There is no one-to-one correspondence between the white regions in the acoustic micrograph and the presence of voids in the X-ray shadowgraph. Figure 1.29 presents the corresponding acoustic micrograph of the same sample taken through the ceramic substrate. The linear voids do not show up as strongly, suggesting that these voids do not extend fully through the thickness of the die attach. Distinct dark spots are also visible due to signal scattering by the irregularly

Figure 1.27. A pseudo-three-dimensional X-ray micrograph of the ceramic substrate sample detailing voiding in the silver epoxy die attach.

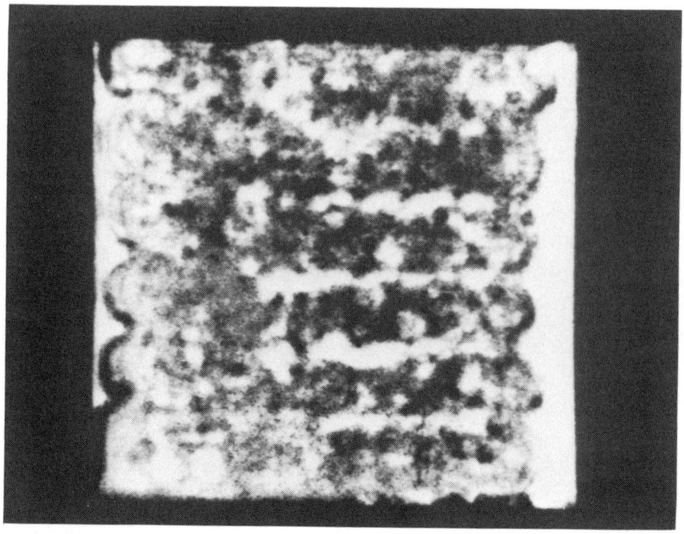

Figure 1.28. An acoustic micrograph of the ceramic substrate sample detailing voiding in the silver epoxy die attach as viewed through the die.

Figure 1.29. An acoustic micrograph of the ceramic substrate sample detailing voiding in the silver epoxy die attach as viewed through the ceramic.

shaped voids. At the bottom of the micrograph, shapes of three voids are visible, again black due to scattering.

An optical micrograph of this sample is presented in Fig. 1.30. The linear, horizontal voids are quite distinct; die-attach patterns and the edge profiles of the die attach are visible. The analysis of the acoustic micrographs in Figs. 1.28 and 1.29 are consistent with these optical data. It is clear that the mottled die attach is present to a larger extent between parallel linear voids, explaining the presence of additional dark spots in the acoustic micrograph in Fig. 1.28. From a close examination of the linear voids, we observe a small amount of die attach on the substrate surface, explaining why these linear voids do not show up quite so clearly in Fig. 1.29.

An X-ray image of a metal substrate sample is presented in Fig. 1.31. A number of large, irregularly shaped voids are visible. The acoustic image through the die as shown in Fig. 1.32 again reveals the presence of these voids. However they do not cover so large an area, and a number of smaller, thinner voids in the optical micrograph show up as black in the acoustic image, probably due to

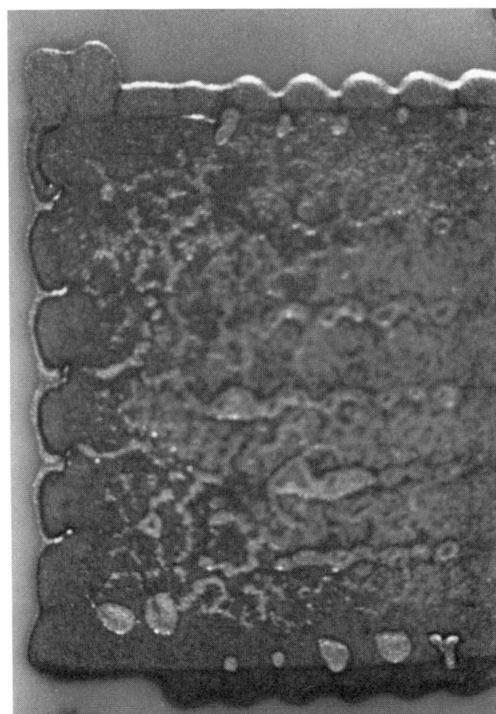

Figure 1.30. An optical micrograph of the die attach after removing the die via grinding.

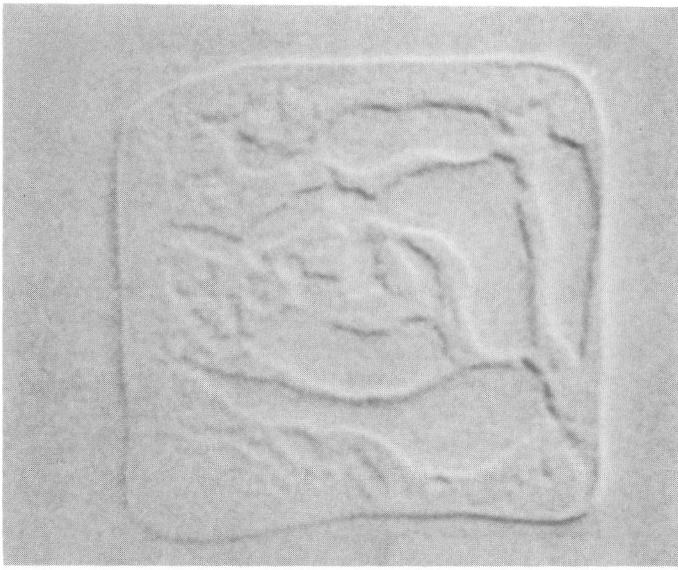

Figure 1.31. A pseudo-three-dimensional X-ray micrograph of the metallic substrate sample detailing voiding in the silver epoxy die attach

both scattering of the acoustic signal at void edges and to voids not extending fully through the thickness of the die attach.

Since the die was removed from the metal substrate with the die attach still adhering, an optical micrograph of the die attach on the silicon die is presented in Fig. 1.33. In regions where no wetting occurred, the dark metal substrate is visible; these correspond to voids extending throughout the thickness of the die attach. Adjacent lighter regions exist where wetting of the metal substrate occurred without adhering to the die. Shapes of the voids as shown in Figs. 1.31 and 1.32 correlate very closely with voids visible in Fig. 1.33. At the edges of the large through-voids, thin layers of the die attach are observable, and the thinner stream like voids do not appear to extend fully through the thickness of the die attach. Enhanced optical microscopy revealed that the die attach at void edges and smaller voids is a thin, textured, 10–15 μm thick layer attached to the back surface of the silicon die. Figure 1.32 shows that all the dark areas in the acoustic micrograph correspond to areas where there are thin, rough die-attach layers, that is, at edges of large voids and at the thinner nonthrough-voids, where the acoustic signal is scattered at the rough, nonplanar void edges.

For several of the metal substrate samples, extensive delaminations were visible in acoustic micrographs at the die-attach–substrate interface, but they were not detected at the die–die-attach interface. This indicated poor adhesion of the metal substrates. Subsequently die shear tests removed the die and die

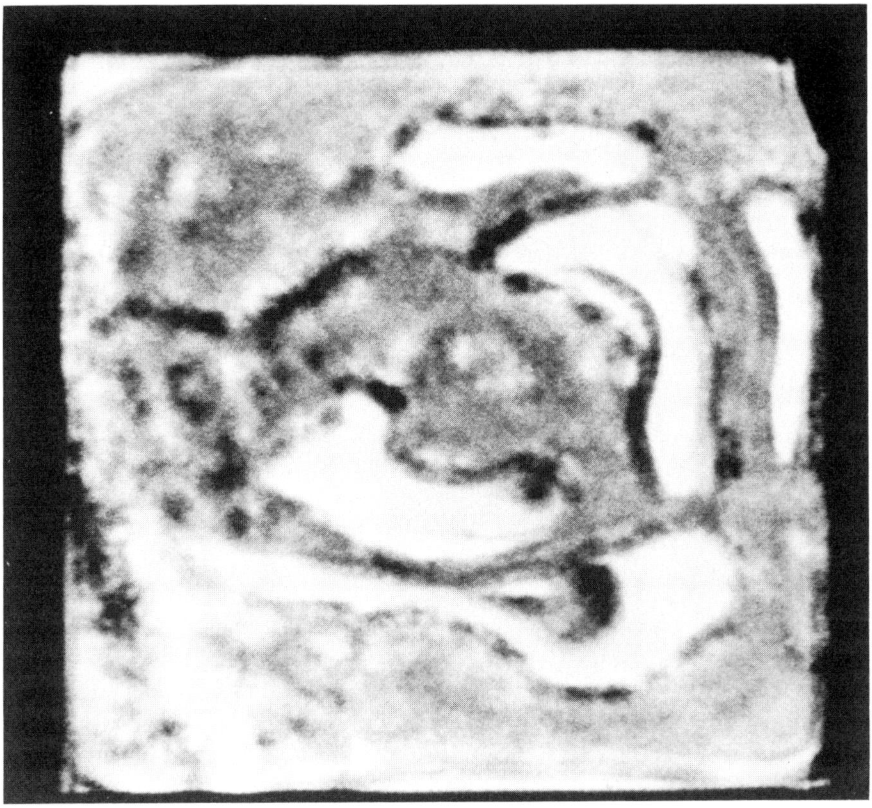

Figure 1.32. An acoustic micrograph of the metallic substrate sample detailing voiding in the silver epoxy die attach as viewed through the die.

attach from the metal substrates with little effort, indicating a wettability-processing issue. This is consistent with the absence of a gold layer to assist adhesion.

It is evident from the preceding results that interpreting acoustic micrographs of die attach is complex. Both bright and dark areas can indicate the presence of voids in die attach depending on their size, thickness, and shape. In addition the sides of large voids may have sloping edges of die attach that scatter the acoustic signal, resulting in an underestimation of the actual void area. The dominant physical mechanism resulting in dark areas in the SAM images appears to be scattering from nonplanar features.

This study has investigated only one particular type of die attach on a number of substrates. For a thorough inspection of die-attach quality, reflection acoustic imaging from both sides and acoustic transmission imaging are desirable;

Figure 1.33. An optical micrograph of the die attach on the silicon die after removing the substrate.

however they may not be possible for some packages, such as those containing a cavity where only reflection imaging from one side is practicable. Further work on the influence of different die-attach materials, curing processes, and the influence of lead frames should be performed. However for the silver epoxy die-attach material investigated, it is possible to elaborate a methodology to determine the presence of voids and to identify adhesion integrity issues using acoustic microscopy.

1.5. Multilayer Interconnect

The requirement to interconnect complex and high-speed ICs has resulted in the design of several multilayer interconnection technologies and the development of multichip modules, such as those in Fig. 1.1.[1,2] The most innovative approaches have been based on thin-film deposition technologies. However novel materials and an increasing number of processing steps in such substrates poses significant reliability concerns. In Section 1.5 we discuss the suitability of existing acoustic microscopes for evaluating interconnection integrity.

Two interconnection substrates were analyzed in this work[39]; an adhesive-polyimide multilayer on a copperclad expoxy glass substrate and a thin-film copper-polyimide multilayer on a cofired alumina ceramic substrate. A schematic cross section of each substrate is detailed in Figs. 1.34 and 1.35, respectively. Commercial reflection acoustic microscopes operating at 200 MHz (Olympus UH3) and 50 MHz (Olympus UH3 and Ultrasonic Sciences Ltd. (USL)) were employed.

Typical Olympus UH3, 50-MHz, reflected acoustic pulse trains from the two copper signal layers of the epoxy glass substrate are shown in Figs. 1.36(a) and (b), respectively. The reflection from the upper conductor track lags that of the surface by approximately 35 ns, while that from the lower layer lags the surface acoustic signal by approximately 80 ns. The corresponding acoustic micrograph of the lower conductor track obtained at a defocus of -200 μm is given in Fig. 1.37. The white speckle apparent in the image is noise due to the jitter in the electronic gate of the Olympus UH3 system whose gate stability is

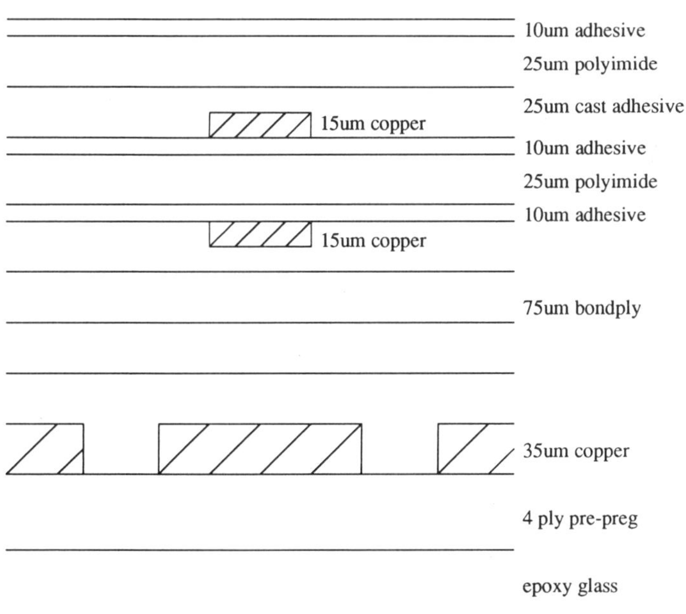

Figure 1.34. A schematic cross section of the adhesive polyimide multilayer on copper-clad epoxy glass substrate. Provided by GEC Marconi, U.K.

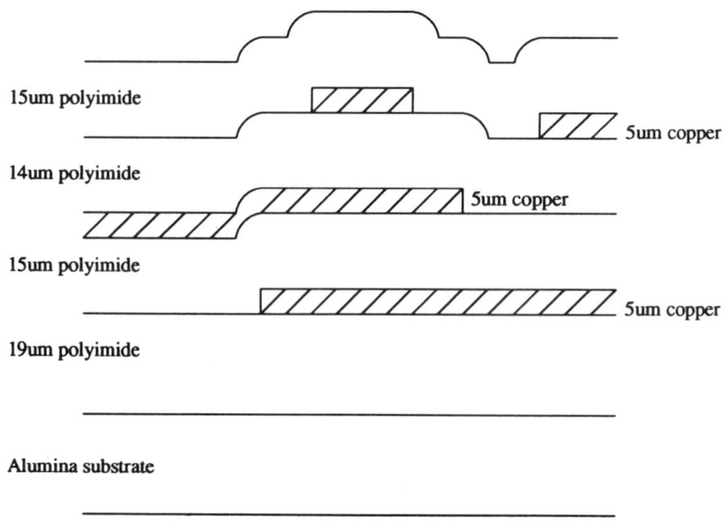

Figure 1.35. A schematic cross section of the thin-film copper polyimide multilayer on a cofired alumina ceramic substrate. Provided by BULL, France.

quoted at ±50 ns. The contrast from the upper conductor track is also present. A pseudo-three-dimensional representation of an acoustic image of the two conductor layers obtained using the USL microscope is presented in Fig. 1.38. This composite image was acquired by moving the position of the electronic gate from the reflection from the upper layer to the reflection from the lower layer in three steps. The significant enhancement of the image vis-a-vis that of Fig. 1.37 is due to the USL gate stability specification of ±1 ns.

Figure 1.39 presents an acoustic micrograph of the upper and lower conductor layers and the ground metallization plane of the ceramic substrate obtained using the Olympus UH3 system at 200 MHz. At this frequency, the microscope has a gate stability of ±10 ns. The micrograph was obtained at a defocus depth of −51 μm. The individual conductor layers are well-resolved; however, it is not possible to image either layer individually due to the 150-ns width of the acoustic pulse. An additional complication in imaging this substrate is the significant surface topography present due to the absence of an interface trigger facility on the Olympus UH3 acoustic microscope.

The preceding results demonstrate that it is possible to obtain acoustic micrographs of conductor tracks from beneath several layers of polyimide and adhesive. However, to resolve the individual conductor layers of the interconnect substrates in this study, would ideally require an acoustic system with a pulse width of 10 ns, gate width of less than 15 ns, and temporal stability better than ±1 ns.

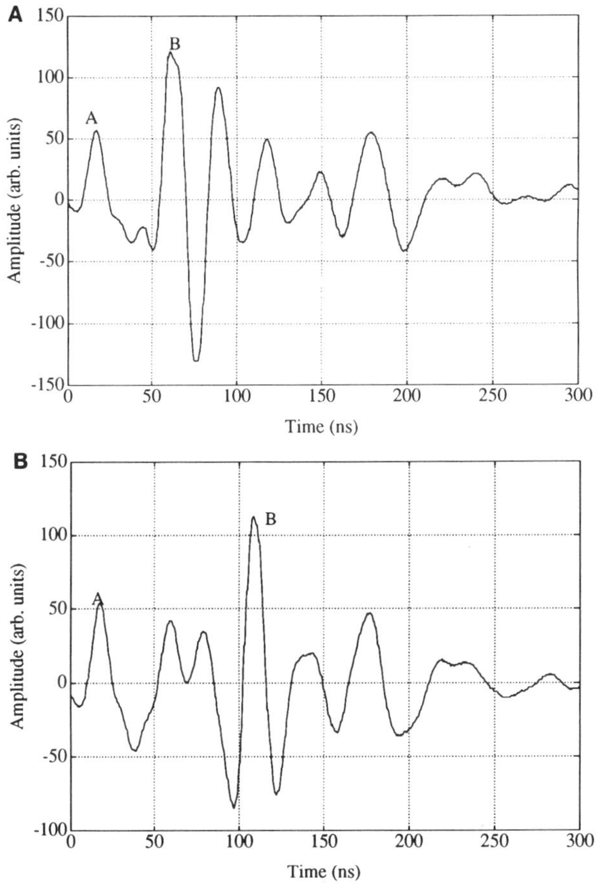

Figure 1.36. Typical Olympus UH3 50-MHz reflected acoustic pulse trains from (a) the upper and (b) lower copper signal layers of the epoxy glass substrate.

1.6. Tape Automated Bonding

Tape automated bonding is an IC-packaging technology that offers advantages in high-circuit density, low profile, and a facility for testing devices immediately before assembly. Because of these attributes, it has increasing application in low-profile consumer products, smart cards, liquid crystal displays, and the assembly of high pin count ICs for use in high-performance computer products.

The TAB tape consists of a layer of copper glued to a polyimide film that is typically in the form of 35-mm or 70-mm photographic film. A plan-view optical micrograph of a populated 70-mm TAB tape is shown in Fig. 1.40.[2] The

Figure 1.37. Acoustic micrograph obtained at 50 MHz of the lower conductor track of the epoxy glass substrate.

film sprocket holes are used for individual TAB tapes or handling purposes in a production enviornment. By using high-resolution photolithography, the copper is patterned to provide leads for attachment to the input/output pads on an IC (Inner Lead Bonds[ILBs]) and for attachment to the circuit board (Outer Lead Bonds [OLBs]).[5,16]

While much research has been devoted to characterizing ILBs,[40,41] OLBs[42,43] and the thermal stress analysis of TAB leads,[44] little work has focused on character-

Figure 1.38. A pseudo-three-dimensional representation of an acoustic image of the two conductor layers obtained using a USL microscope at 50 MHz.

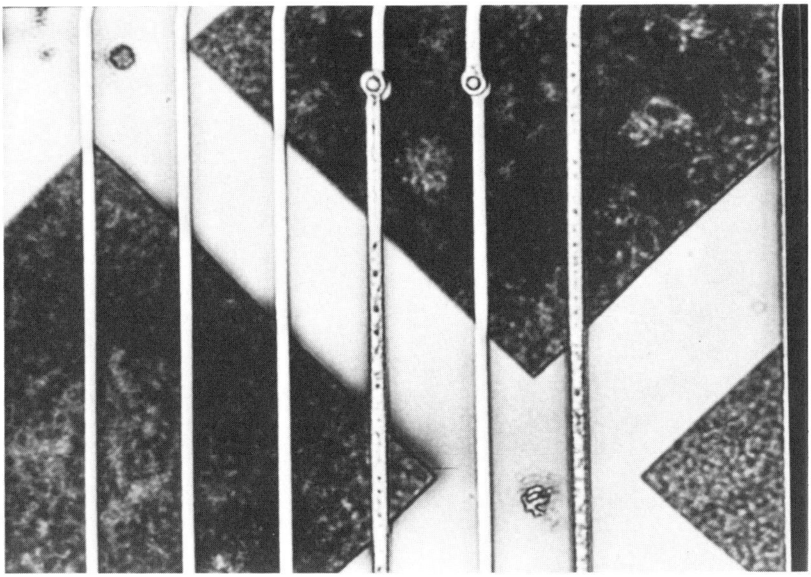

Figure 1.39. An acoustic micrograph of the upper and lower conductor layers and the ground metallization plane of the ceramic substrate obtained using the Olympus UH3 system at 200 MHz.

izing the adhesion of the TAB tape polyimide–glue–copper interfaces. An integrity fault at one of these interfaces could result in delamination and subsequent tape failure. The objective of Section 1.6 is to evaluate the applicability of acoustic microscopy for characterizing TAB tape material systems.[45]

Two commercial TAB tapes were investigated in this study.[45] Both consisted of a three-layer material system comprising polyimide, glue, and copper thin films. A microsection of an unpatterned tape illustrating this three-layer TAB system is presented in Fig. 1.41. The thicknesses of the layers as determined from optical measurements were 77 μm, 21 μm, and 33 μm for the Upilex-S™ polyimide, adhesive, and copper layers, repsectively; and 132 μm, 12 μm, and 39 μm for the Kapton-V™ polyimide tape system.

Acoustic micrographs were obtained using an Olympus UH3 SAM operating in pulse mode at a frequency of 100 MHz. The lateral resolution of the lens was approximately 38 μm in the materials examined. During SAM analysis, samples were mounted in a 35-mm slide holder under tension to ensure tape flatness. The X-ray transmission imaging analysis was performed using a LIXI MOBILIX X-ray system operating at an energy of 70 KeV and a spot size of 10 μm. The tapes were temperature/humidity stressed in a Weiss Technik temperature/humidity chamber. A Struers DAP-V grinding and polishing machine was used for microsectioning.

Figure 1.40. A plan-view optical micrograph of a populated 70-mm TAB tape developed within ESPRIT Project No. 2075 APACHIP—Advanced PACkaging for HIgh Performance.

Figure 1.41. Optical microsection of a TAB tape showing a three-layer TAB tape system.

A reflected acoustic pulse echo trace from the Upilex-S tape is shown in Fig. 1.42. Two distinct peaks are observed; Peak *A* from the polyimide surface of the tape and Peak *B* from the polyimide–glue–copper interface. Both peaks are well-separated in time (0.074 μs), and it is therefore possible to monitor the reflected acoustic signal from the tape interface.

Using a pulse echo analysis technique,[46] the acoustic longitudinal wave (LW) velocity of each polyimide film was determined from the measured time delay between peaks and the film thickness. The LW velocity of Upilex-S was found to be 2700 ms^{-1} and that of Kapton, 2200 ms^{-1}. Errors in these measurements are approximately 5%, mainly due to the uncertainty in film thickness.

Acoustic imaging was performed, focusing on the polyimide–glue–copper interfaces of the sample. A large-area acoustic micrograph of Sample 1 is shown in Fig. 1.43. There was an area on the interface (as indicated by the arrow) where dark patches and bubblelike features were evident. These features are consistent with the presence of voids. A magnified view of this area is presented in Fig. 1.44.

Optical micrographs, such as the one in Fig. 1.45 of a microsection through the sample in Fig. 1.44, revealed voids (indicated by arrows) that varied in diameter from 65–10 μm, with a typical diameter of 25 μm. We would expect such voids to appear as bright areas in an acoustic image due to the high acoustic impedance mismatch between polyimide and air. However the voids observed in the optical micrographs of the microsectioned samples had a diameter less than the lateral resolution of the 100-MHz acoustic lens. These voids therefore act as localized scattering centers, and hence they result in a loss of reflected acoustic signal intensity, with a consequent dark appearance in the acoustic image.

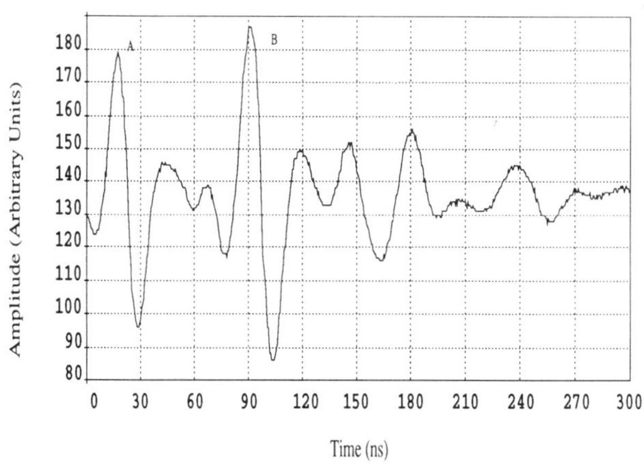

Figure 1.42. Reflected acoustic pulse echo train from a TAB tape through a polyimide surface; Peak *A* is from the tape surface and Peak *B* from the polyimide–glue–copper interface. The acoustic frequence was 100 MHz.

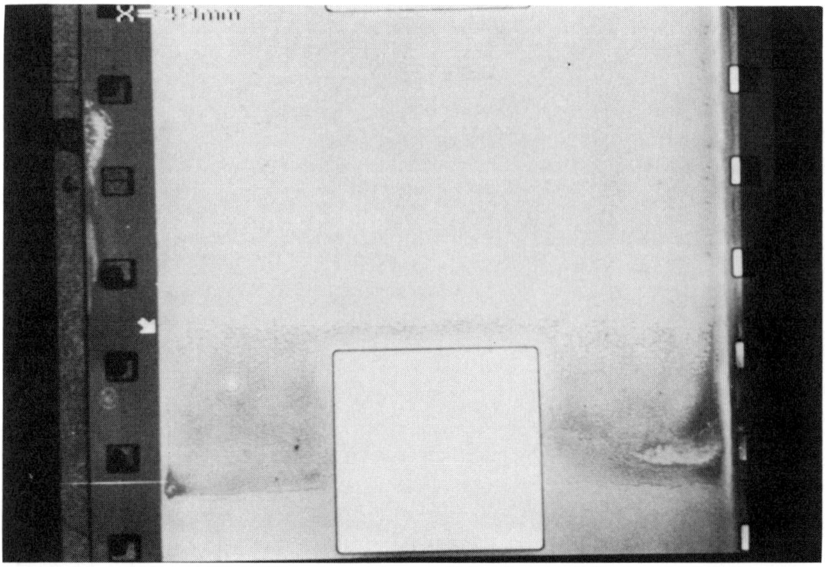

Figure 1.43. An acoustic micrograph of 35-mm TAB tape focused at a polyimide–glue–copper interface. A 100-MHz acoustic lens was employed. The white arrow indicates a large number of voids.

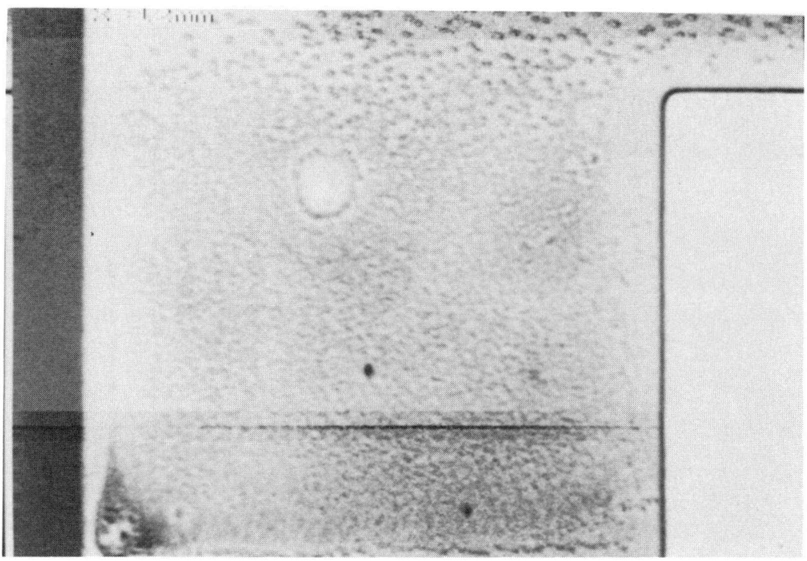

Figure 1.44. A 12 × 9 mm close up acoustic micrograph of 35-mm TAB tape detailing large-scale voiding.

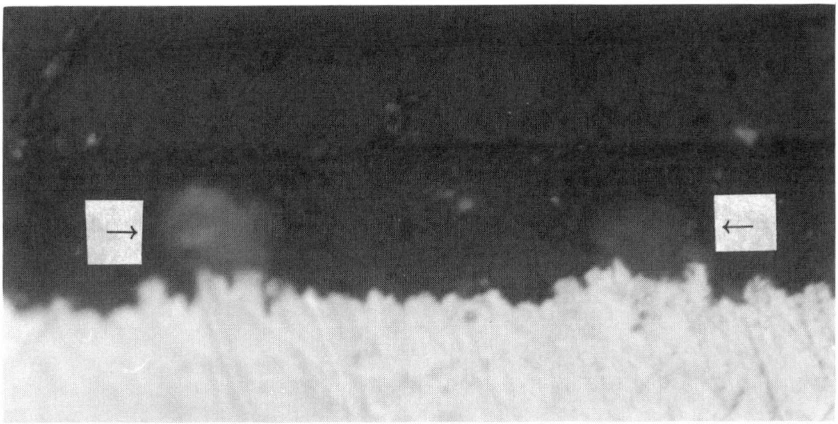

Figure 1.45. An optical microsection of TAB tape showing typical voids of 25-um diameter (indicated by arrows).

It is observed from Fig. 1.46 that no voids are evident in the corresponding X-ray micrograph. This is due to the negligible attenuation experienced by X-rays in traversing the polyimide and glue layers, and this is consistent with the observation that it is not possible to image the sprocket holes at the edge of the TAB tape.

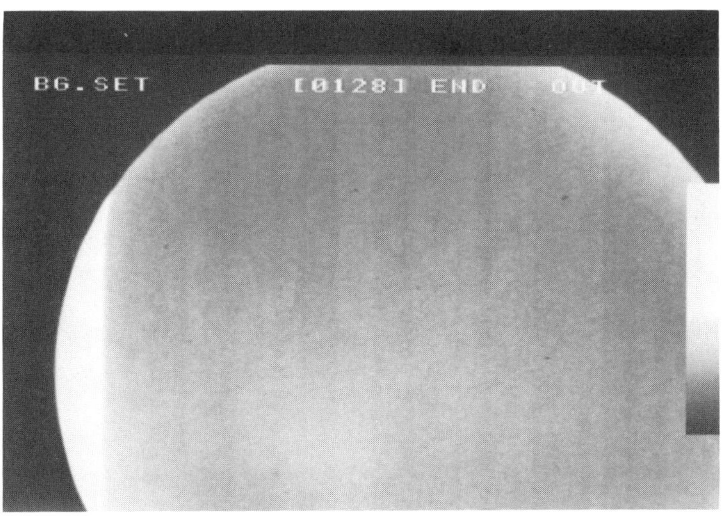

Figure 1.46. The X-ray image to corresponding Fig. 1.42. No voiding is visible.

Temperature and humidity stressing at 85 °C and 85% humidity was performed for a total of 16 days. After one, three, five, and nine days, samples were removed, inspected optically and by acoustic imaging, and returned to the stressing chamber. At the end of this stressing period, acoustic imaging revealed no change in the state of the samples.

In summary it is demonstrated that SAM can detect voids at polyimide–glue–copper TAB tape interfaces. Optical microsectioning was undertaken to confirm the presence of these voids and to measure their dimensions. The X-ray imaging yielded no information on interface quality. No change in sample state was observed after thermal stressing.

At present a 40 mm × 30 mm acoustic image takes several minutes to obtain when using a mechanical-scanning system, such as that on the Olympus UH3 SAM employed in this study. This is not acceptable for on-line process control of TAB tape in a manufacturing facility. A discussion of potential solutions for implementing a real-time incoming inspection system for on-line TAB tape process control is presented in Section 1.8.

1.7. Mechanical Properties

A wide range of material systems are currently under investigation for high-performance IC packaging applications. The incorporation of novel ceramic substrate materials, such as aluminum nitride or diamondlike carbon (DLC), within a specific packaging technology is determined by a number of competing requirements relating to thermal, electrical, and mechanical integrity criteria. In addition the trend toward multilayer interconnect structures and large-area die attach poses significant mechanical integrity problems, and it has led to the rapid use of finite-element computer-aided stress-modeling codes for optimum package reliability.(14,15) However the accuracy of such simulations depends critically on accurately determining the material mechanical properties of the constituent material systems used as input in the code.(34)

At present several techniques exist for determining bulk- and thin-film material mechanical properties, including X-ray diffraction,(47) mechanical indentation and flexural techniques,(48) optical techniques,(49) and ultrasonic techniques.(50–52)

A goniometric method(53) and both SLAM(54) and line focus beam (LFB)(55) acoustic microscopy have been used to determine elastic constants of materials. A conventional point focus beam (PFB) acoustic microscope can also be used to determine mechanical properties of materials nondestructively.(46) The application of the PFB technique for characterizing packaging ceramic material systems is presented in Section 1.7. Experimental results obtained using a 50-MHz acoustic lens are detailed for a range of ceramic materials, and these are compared with published data.

The operation principle of the PFB technique is based on using mode-conversion phenomena,[46] which occur when an acoustic wave impinges on a water–solid or solid–water interface. Both LW and shear (SH) acoustic waves can be transmitted to a layer when a focused acoustic beam is incident at a water–solid interface from water and subsequently reflected from the bottom of this layer. Figure 1.47 shows a typical reflected acoustic pulse train received at the acoustic lens transducer from a silicon nitride substrate. Pulse *A* is the reference signal reflected by the upper surface of the solid layer. Pulses *B* and *D* are the LW and SH wave mode pulse signals reflected from the bottom of the layer; Pulse *C* is the mixed mode signal, in which the LW wave (or SH) is incident at the lower surface of the layer. Mode conversion causes an SH (or LW) wave to be reflected from this surface. The acoustic velocities of this layer can now be obtained by analyzing the LW and SH wave paths via the time difference between the arrival of Pulses *A* and *B* (or *C, D*) at the lens transducer. In addition a normally incident LW transmitted to a ceramic substrate can undergo multiple reflections between its front and rear surfaces. A typical LW wave pulse train in silicon nitride is shown in Fig. 1.48. The increased time base vis-a-vis Fig. 1.47 facilitates a more accurate measurement of the LW velocity of the

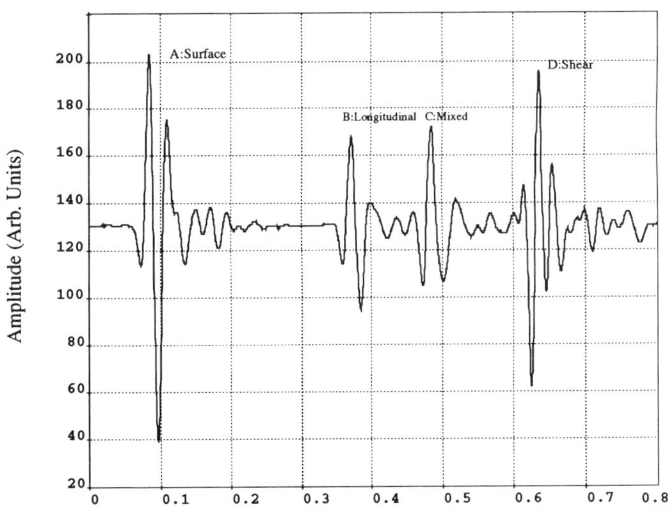

Figure 1.47. A typical reflected acoustic pulse train received at the acoustic lens transducer from a silicon nitride substrate. Pulse *A* is the reference signal reflected by the upper surface of the solid layer. Pulses *B* and *D* are the LW and SH wave mode pulse signals reflected from the bottom of the layer. Pulse *C* is the mixed mode signal, in which the LW wave (or SH wave) is incident at the lower surface of the layer. Mode conversion causes an SH (or LW) wave to be reflected from this surface.

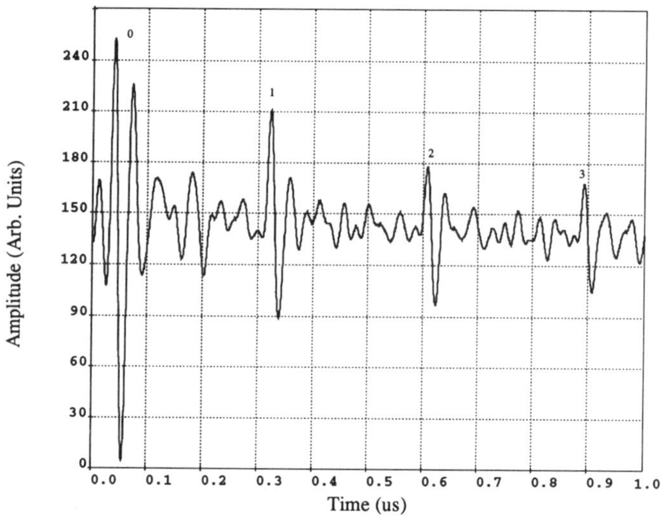

Figure 1.48. A typical reflected acoustic pulse echo train received at the acoustic lens transducer from a silicon nitride substrate detailing LW wave multireflections.

ceramic. From a measurement of both acoustic velocities, the mechanical properties of the ceramic can be calculated.[33,4]

Acoustic measurements were obtained over a number of points on each sample surface. Longitudinal wave multireflections were obtained first, then the lens was defocused to obtain a strong SH wave signal, and SH wave echoes were obtained over the same points. An average value of acoustic velocity measurements at different points was calculated to give the overall v_l and v_s velocities for a given sample.

Fused silica (SiO_2) and E6 borosilicate glass were chosen as reference samples to validate the accuracy of the technique. The measured acoustic velocities are shown in Table 1.1 with previously published data for comparison.[56]

Values of v_l for fused silica and E6 glass are within 0.2% and 0.1%, respectively, of published values. Values of v_s are within 1.3% and 0.7% of reference values. These deviations are well within calculated error margins. The larger error in v_s is primarily due to the uncertainty of the phase shift of the reflected SH wave pulse and interference between the second LW pulse and the SH wave pulse.

Several ceramic material systems were subsequently characterized using the PFB acoustic velocity measurement technique. These included alumina (Al_2O_3), silicon nitride (Si_3N_4), zirconia (ZrO_2), titanium diboride (TiB_2), beryllium oxide (BeO), aluminium nitride (AlN), and silicon carbide (SiC).[46,52,57] For a number of these materials, published acoustic velocity data were unavailable.

Table 1.1. Measured Acoustic Velocities in Fused Silica and E6 Glasses

Sample	Density ρ (Mg m^{-3})	Longitudinal velocity v_l (m s^{-1})	Shear velocity v_s (m s^{-1})
Fused silica			
Quoted values[4]	2.202	5968	3764
Measured (this work)	—	5954	3814
E6 glass	—	—	—
Quoted values[56]	2.18	5403	3304
Measured (this work)		5408	3328

A preliminary report on the acoustic, optical and X-ray diffraction phase properties of silicon nitride substrates processed using a number of different methods, including sintering, hot pressing, and hot isostatic pressing (HIP), has been published.[52,58] Significant variations in acoustic velocity were observed with processing route and microstructure. Table 1.2 details acoustic wave velocity measurements for alumina, beryllia, aluminium nitride, and silicon carbide.[57]

The measured acoustic velocities of the three alumina substrates are consistent with published data. The calculated Young's modulus values are in good agreement with manufacturer's data. A monotonic increase in both v_l and v_s values with increasing sample purity was recorded. The two beryllia samples studied using this technique were also characterized by using a goniometric acoustic velocity measurement method.[53] The measured v_l values are in good agreement, but there is a difference in measured SH wave velocities that may be attributed to the errors previously discussed. Young's modulus values are within 7% of manufacturer's data. Similarly the aluminium nitride and silicon carbide acoustic velocity measurements are in good agreement with published work.

The measurement technique just described has been demonstrated to provide consistent and accurate LW and SH acoustic wave velocity measurements, complementary to the line focus *V(z)* technique, which primarily measures a surface acoustic wave velocity. The PFB technique provides a rapid, bulk, nondestructive, localized testing method for monitoring materials by using a conventional acoustic microscope and lens. Further work is in progress to improve the accuracy of the measurement technique and in particular to reduce the error in the SH wave velocity measurement.

1.8. Future Perspectives

The work undertaken at NMRC Ireland that is presented in Chapter 1 has demonstrated the application of acoustic microscopy to the analysis of microelec-

Table 1.2. Mechanical Properties of Electronic Ceramics

Sample type	Density (mgm^{-3}) ρ	v_l ($m\ s^{-1}$)	v_s ($m\ s^{-1}$)	Poisson's ratio σ	Young's modulus E (1)[a]	(GPa) (2)[b]
Alumina	—	—	—	—	—	—
Ref. 4	3.97	10822	6163	0.26	—	—
95%	3.8	9950	5630	0.264	300	305
96%	3.8	10140	5680	0.271	300	311
99.9%	3.96	10810	6110	0.265	385	375
Beryllia	—	—	—	—	—	—
99.5% (Ref. 53)	2.86	11636	7059	—	—	—
99.5%	2.86	11720	6750	0.252	350	325
97.9% (Ref. 53)	2.98	11953	7175			
97.9% (2%Z_rO_2)	2.98	11960	6990	0.241	361	375
Aluminum nitride	—	—	—			
Ref. 46	—	10800	6360			
A1N (5% yttrium)	3.25	10690	6320	0.232	300	319
Silicon nitride	—	—	—			
Ref. 4	3.185	10607	6204	0.24		
Si_3N_4 (HIP)	3.17	11290	6000	0.303	320	298
Silicon carbide	—	—	—			
Ref. 4	3.21	12099	7485	0.19		
SiC	3.1	12200	7640	0.178	410	426

[a]Manufacturer's data
[b]Measured in this work

tronics packaging materials (measuring mechanical properties), product development (cofired ceramic PGA manufacture), production process control (hermetic seal integrity, die attach, and TAB tape manufacture), quality assurance (plastic packages) as well as failure analysis and analysis of field returns.

To date the industrial implementation of SAM for analyzing microelectronic interconnection and packaging technologies has been limited by system costs, image acquisition time, system complexity issues, and the need for skilled personnel to interpret images. Because of these limitations, acoustic microscope systems are being used mainly in research and development (R&D), production process control, quality assurance, and failure analysis, all of which are essentially off-line. On-line use of acoustic microscopy in high-volume production environments is a requirement, particularly in the plastic IC packaging industry. However while major improvements have been made to remove the preceding limitations and provide first-generation on-line systems, further developments are required.

In the area of system cost, the main reason for high-cost acoustic microscopes in the past has been the development of hardware-based systems. This issue and a reduction in system complexity are addressed by implementing hardware functions in software. In addition image acquisition speed has been improved by developing low-mass acoustic lenses for low-frequency operation. Further improvements in this area can be achieved by using an acoustic transducer array, as shown in Fig. 1.49,[45,59] which depicts the potential design and implementation of such a system for inspecting TAB tape. Such a system uses electronic scanning to provide real-time image acquisition. Finally it is probable that the expert systems currently under development for analyzing acoustic images in medical

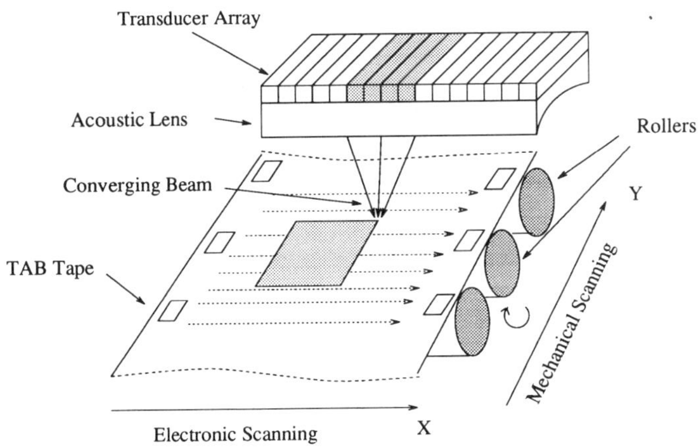

Figure 1.49. A schematic diagram of a potential acoustic transducer array system for rapid-scanning on-line inspection of a TAB tape.

and other advanced non-destructive test (NDT) areas will also be implemented in the new generation of acoustic process control systems for electronic-packaging technologies.

While R&D and failure analysis facilities will continue to require general-purpose acoustic microscope systems, there is a continuing need for low-cost acoustic microscopes for use in on-line production control in an IC package-manufacturing environment. These systems are likely to be dedicated systems built in an application-specific configuration providing real-time output in the form of go/no-go decisions regarding the production process.

1.9. Conclusions

Chapter 1 reviews the applicability of acoustic microscopy for analyzing microelectronic-packaging materials, processes, and technologies. The technique has been successfully applied at NMRC Ireland to a wide range of advanced packaging technologies, including TAB tape, hermetic lid seals, multilayer PGAs, high-density multilayer interconnect, single-chip plastic packages, and large-area die attach. This work, as documented in Chapter 1, represents the first comprehensive study of this nature, and it presents a European perspective of the current state of the art in this field.

Acknowledgments

The authors wish to acknowledge the significant contributions of Professor De Zhang (on leave from the Institute of Acoustics, Nanjing University, China), J. Flannery, and W. Lawton to the work presented in Chapter 1. Professor De Zhang was instrumental in establishing the deep-subsurface SH-wave-imaging technique at NMRC Ireland. The authors would like to thank the Interconnection and Packaging group at NMRC Ireland for assistance in microsectioning, X-ray analysis, and environmental stressing. The authors would like to acknowledge the assistance of Mr. C. Gartside for the USL acoustic microscopy images.

One of the authors (C. F.) would like to acknowledge the award of a postgraduate research studentship from NMRC Ireland.

This work has been partially funded by the Commission of the European Community's ESPRIT Programme within Project no. 2075; APACHIP—Advanced Packaging for High Performance, and we gratefully acknowledge the interaction with our industrial partners.

The TAB tape was supplied by MCTS, France; the multilayer interconnect substrates were provided by GEC Marconi, U.K., and by BULL, France; and the single-chip cofired ceramic pin grid array packages were supplied by Hoechst Ceramtec, Germany.

The work on mechanical property determination is partially funded by the Commission of the European Community BRITE-EURAM Project no. 4398; Ceramic evaluation using Micro-Acoustic Waves (C-MAW).

The authors would like to thank our colleagues in this project, in particular R. Cundill and B. Bles of SKF, The Netherlands, for providing advanced silicon nitride substrates. Finally the expertise of Dr. P. Kelly, T. Flaherty and S. Lynch of the NMRC Materials Analysis Group in thermal, mechanical, and optical characterization of these substrates is much appreciated.

References

1. Reche, J. J. (1990). Fabrication of high-density multichip modules. *IEEE Trans. Components Hybrids Manuf. Technol.* **13**(3)565–69.
2. Chantraine, P., Chandler, N., Brandenburger, J., Chilo, J., de Maquillé, Y., Ó Mathúna, S. C., Koschnick, W., Bargain, R., Dümcke, R. (1992). In: *Proceedings Annual ESPRIT Conference,* Commission of the European Communities, Brussels, Belgium, pp. 101–111.
3. Reichl, H. Packaging aspects of single- and multichip modules. Proceedings 1st European Conference on Electronic Packaging Technology (EuPac 1994), Essen, German Welding Society, Dusseldorf, Germany, Feb. 1994, pp. 6–9.
4. Briggs, G. A. D. (1992). *Acoustic Microscopy,* Oxford University Press, Oxford, UK.
5. Quate, C. F., Atalar, A., Wickramasinghe, H. L. (1979). Acoustic microscopy with mechanical scanning—a review. *Proc.. IEEE* **67**, 1092–1114.
6. BRITE-EURAM Project no. 4398 (0000). Ceramic evaluation using microacoustic waves (C-MAW).
7. European Space Agency Contract MTSL 018 (Acoustic Microscopy)
8. Buda, L. R., Gedney, R. W., Kelley, T. F. (1992). High-density packaging: future outlook. *Proceedings 42d Electronic Components and Technology Conference,* San Diego, IEEE, New York, pp. 36–41.
9. Kelly, G., Lyden, C., Ó Mathúna, S. C., Barrett, J., Exposito, J., Pape, H. (1993). Development and use of numerical modelling tools for advanced packaging technologies—a European perspective. Presented at *Annual Winter Meeting of ASME,* New Orleans.
10. White, G. L. (1985). An introduction to die attach. *Welding Institute Research Bulletin,* pp. 268–271.
11. Mil-Std-883C Method-2030 (1987). *Ultrasonic Inspection of Die Attach.* Notice 5, US Dept. of Defense, Washington, DC.
12. Heinen, K. G., Schroen, W. H., Edwards, D. R., Wilson, A. M., Stierman, R. J., Lamson, M. A. (1989). Multichip assembly with flipped integrated curcuits. *Proceedings Electronic Components Conference,* IEEE, New York, pp. 672–80.
13. (1993). Packaging takes on a new dimension. *European Semiconductor Production,* 21–22.
14. Groothius, S., Schroen, W., Murtuza, M. (1985). Computer-aided stress modelling for optimising plastic package reliability. *23d Annual Proc. IEEE International Reliability Physics Symposium,* New York, pp. 184–91.
15. Ó Mathúna, S. C., Fromont, T., Koschnick, W., O'Connor, L. (1993). Test Chips, Test Systems and Thermal Test Data for Multichip Modules in the ESPRIT–APACHIP Project, *IEEE Transactions on Components, Packaging and Manufacturing Technology, Part A,* **17**(3). IEEE, New York, pp. 117–126.

16. Tummala, R. R., and Rymaszewski, E. J. (1989). *Microelectronics Packaging Handbook,* Van Nostrand Reinhold, New York.
17. Mil-Std-883C Method 1014.5 (1983). Seal. U.S. Dept. of Defense, Washington, DC.
18. Flannery, J., Crean, G. M., O'Mathuna, S. C. (1992). In: *Acoustical Imaging,* vol. 19, Plenum Press, New York, pp. 717–21.
19. Crean, G. M., Zhang, D., Flannery, J., and O'Mathuna, S. C. (1992). In: *Proceedings 14th International Conference on Acoustics* (Li Peizi, ed.), Beijing, pp. L5–1.
20. Nongaillard, B., Rouvaen, J. M., Logette, P., Bridoux, E., Torquet, R. (1986). Zooming lens for acoustic microscopy applications. *J. Appl. Phys.* **59**, 3028–32.
21. Atalar, A. and Koymen, H. (1989). A high-efficiency Lamb wave lens for subsurface imaging. *Proceedings IEEE Ultrasonics Symposium,* IEEE, New York, pp. 813–16.
22. Zhang, D. and Crean, G. M. (1991). Shear wave imaging for deep nondestructive evaluation. *Electr. Lett.* **27**(24), 2248–49.
23. Zhang, D. and Crean, G. M. (1992). Deep subsurface imaging using shear wave mode acoustic microscopy. *Proceedings Institute of Physics (IOP) Conference on Developments in Acoustics and Ultrasonics* (M. Povey and D. J. Clements, eds.), IOP Publishing, Bristol, UK, pp. 215–18.
24. Zhang, D. and Crean, G. M. (1992). Calculation of the focussed acoustic shear wave field in a ceramic substrate. *Proceedings of the 14th International Conference on Acoustics* (Li Peizi, ed.), Beijing, pp. L5–3.
25. Zhang, D. and Crean, G. M. (1993). Shear wave imaging using an annular lens. *Appl. Phys. Lett.* **62**(3), 318–20.
26. IPC-SM-786 Standard (1991). Impact of moisture on plastic IC package cracking; and IPC-Test Method 650-2.6.20 (1991). Plastic surface mount component cracking. Institute for Interconnecting and Packaging Electronic Circuits (IPC), Lincolnwood, IL.
27. Moore, T. M. and Kelsall, S. J. (1992). The impact of delamination on stress-induced and contamination-related failure in surface mount ICs. *Proceeding IEEE International Reliability Physics Symposium,* IEEE, New York, pp. 169–76.
28. Van Doorselaer, K., Hente, A., Saboui, A., Moscicki, J. P. (1993). Thin type packaging: an effective way to improve the popcorn resistance of plastic-packaged ICs. *Proceedings IEEE International Reliability Physics Symposium,* IEEE, New York, pp. 244–49.
29. Moore, T. M. (1992). C-Mode acoustic microscopy applied to integrated-circuit-package inspection. *Solid-State Electron.* **35**(3), 411–21.
30. Van Doorselaer, K., Moore, T. M., Tiziani, R., Baelde, W. (1992). Evaluation of methods for delamination detection by acoustic microscopy in plastic-packaged integrated circuits. *Proceedings 18th International Symposium for Testing and Failure Analysis,* ASM International, Ohio, pp. 425–31.
31. Shook, R. L. and Conrad, T. R. (1993). Accelerated life performance of moisture-damaged plastic surface mount devices. *Proceedings IEEE International Reliability Physics Symposium,* IEEE, New York, pp. 227–35.
32. van Gestel, R., de Zeeuw, K., van Gemert, L., Bagerman, E. (1992). Comparison of delamination effects between temperature-cycling test and highly accelerated stress test in plastic-packaged devices. *Proceedings IEEE International Reliability Physics Symposium,* IEEE, New York, pp. 177–81.
33. Brekhovskikh, L. M. (1980). *Waves in Layered Media,* 2d ed., Academic Press, New York.
34. Kelly, G., Lyden, C., O'Mathúna, S. C., Campbell, J. S. (1992). Investigation of thermomechanically induced stress in a PQFP 160 using finite element techniques. In: *Proceedings 42d Electronic Components and Technology Conference (ECTC),* IEEE, New York, pp. 467–72.
35. Mirasole, M. J. (1988). In: *Proceedings of the International Symposium for Testing and Failure Analysis (ISTFA),* ASM International, Materials Park, Ohio, pp. 77–88.
36. Kessler, L. W., Semmens, J. E., Agramonte, F. (1985). Nondestructive die attach bond evaluation

comparing scanning laser acoustic microscopy (SLAM) and X-radiography. In: *Proceedings of the Electronics Component Conference.* IEEE, New York, pp. 250–58.

37. Semmens, J. E. and Kessler, L. W. (1991). Application of acoustic microimaging to the evaluation of multichip modules (MCM). In: *Proceedings of the Technical Program of the Surface Mount International Conference,* Surface Mount International, Edina, Minnesota, pp. 44–52.
38. Siettmann, J., Dias, R., Fiebelkorn, K. (1992). In: *Proceedings of the International Reliability Physics Symposium,* IEEE New York, pp. 309–14.
39. Flannery, J., and Crean, G. M. (1990). Scanning acoustic microscopy of multilayer interconnect structures. In: *Proceedings Symposium International Society for Hybrid Microelecronics, ISHM,* Reston, Virginia, pp. 358–63.
40. Zakel, E., Azdasht, G., Reichl, H. (1991). Investigations of laser-soldered TAB inner lead contacts. *IEEE Trans. on Components, Hybrids, and Manufacturing Technology,* **14**(4), 672–79.
41. Teoh, H., McGeary, M., Ling, J. (1990). Laser/Infrared evaluation of TAB inner lead bond integrity. *Proceedings Electronic Components and Technology Conference,* IEEE, New York, pp. 442–49.
42. Semmens, J. E. and Kessler, L. W. (1990). Nondestructive evaluation of TAB outer lead bonding using acoustic microscopy. *Proceedings International Society for Hybrid Microelectronics Symposium,* ISHM, Reston, Virginia, pp. 644–51.
43. Scharr, T. A. and Nagarkar, M. D. (1989). Outer lead and die bond reliability in high-density TAB. *Proceedings Electronics Components Conference,* IEEE, New York, pp. 177–83.
44. Lau, J. H., Rice, D. W., Harkins, C. G. (1989). Thermal stress analysis of tape-automated bonding packages and interconnections. *Proceedings Electronics Components Conference,* IEEE New York, pp. 456–63.
45. Flannery, C. M., Crean, G. M., O'Mathuna, S. C. (1994). In: *Proceedings of 1st European Conference on Electronic Packaging Technology (EuPac '94),* Essen, Germany.
46. Zhang, D. and Crean, G. M. (1991). In: *Proceedings Materials Research Society Symposium,* vol. 239, Materials Research Society, Pittsburgh, pp. 281–86.
47. Segmulleller, A., and Murakami, M. (1988). In: *Treatise on Materials Science and Engineering,* vol. 27 (H. Herman, ed.), pp. 143–200. Academic Press, New York.
48. Pethica, J. B. and Oliver, W. C. (1990). In: *Proceedings of the Materials Research Society Symposium,* vol. 130, Materials Research Society, Pittsburgh, pp. 13–17.
49. Flinn, P. A. (1990). In: *Proceedings of the Materials Research Society Symposium,* vol. 130, Materials Research Society, Pittsburgh, pp. 41–51.
50. Crean, G. M., Golanski, A., Oberlin, J. C. (1989). Effective elastic constants of thin-film tungsten silicide from surface acoustic wave analysis. *Appl. Phys. Lett.* **50**(2), 74–76.
51. Crean, G. M. (1990). In: *Proceedings 14th Conference on Scanning Microscopy in Materials Testing,* pp. 55–67. DVM, Berlin.
52. Flannery, C. M., Flaherty, T., Zhang, D., Cundill, R. T., Crean, G. M. (1993). In: *Proceedings of the International Conference on Advanced Materials (ICAM), Symposium on Nondestructive Evaluation of Materials,* Tokyo, Elsevier, Amsterdam.
53. Alper, T., Challis, R. E., Crean, G. M., Zhang, D. (1992). Longitudinal and shear ultrasonic velocities in AlN, Al_2O_3, and SiC ceramics. *Proceedings Institute of Physics (IOP) Conference on Developments in Acoustics and Ultrasonics,* pp. 209–214, (M. Povey and D. J. Clements, eds.) IOP Publishing, Bristol, UK.
54. Oishi, M., Noguchi, K., Murayama, K. (1990). Measurement of elastic constants of ceramics by a scanning laser acoustic microscope. *JSME International Journal, Series 1* **33**(1), 96–100.
55. Mihara, T., and Obata, M. (1992). Elastic constant measurement using line-focus-beam acoustic microscope. *Exper. Mech.* **32**(1), 30–33.
56. Rowe, J. M. (1987). Quantitative acoustic microscopy of surfaces. Ph.D. diss., Oxford University.
57. Flannery, C. M., Flaherty, T., Zhang, D., Cundill, R. T., Crean, G. M. (1994). In: *Proceedings*

of 1st European Conference on Electronic Packaging Technology (EuPac '94), Essen, German Welding Society, Dusseldorf, Germany, pp. 160–164.

58. Flaherty, T., Lynch, S., Kelly, P. V., O'Keeffe, M., Cundill, R. T., Crean, G. M. (1993). In: *Proceedings of the International Conference on Advanced Materials, Symposium on Fabrication of Silicon-Based Ceramics,* Tokyo, Elsevier, Amsterdam.
59. Kubota, J., Okada, H., Musha, Y., Takishita, Y., Iwasaki, A., Saski, S. (1988). Electronic scanning of 25-MHz ultrasound for imaging IC packages. IEEE Ultrasonics Symposium, IEEE, New York, pp. 767–70.

2

Measuring Short Cracks by Time-Resolved Acoustic Microscopy

D. Knauss, T. Zhai, G. A. D. Briggs, and J. W. Martin

2.1. Introduction

Detecting defects, for example cracks,[1] is important in predicting the lifetime of a material. The growth behavior of short cracks[2] plays an essential role in the lifetime of a component, since the lifetime is mainly controlled by the time required for a crack to grow from a certain initial size to about 1 millimeter. Cracks are defined as short when for example the crack length is small compared with the microstructure of the specimen or when the crack is simply shorter than ≈0.5 mm.[3] The growth of short surface breaking cracks can be measured by light microscopy (LM) or scanning electron microscopy (SEM). A common method of studying short cracks is a replica technique based on taking several plastic replicas at various stages of crack growth and subsequently examining these replicas with LM or SEM.[4,5] The disadvantage of these techniques however, is that they give information about crack development only on the surface of the specimen, so that the depth of the crack has to be determined indirectly by assuming the shape of the crack. For long cracks this may be appropriate because a local change in propagation direction does not alter the overall crack geometry

D. Knauss, T. Zhai, G. A. D. Briggs, and J. W. Martin • Department of Materials, University of Oxford, Oxford, England.

Advances in Acoustic Microscopy, Volume 1, edited by Andrew Briggs.
Plenum Press, New York, 1995.

on which the driving force of the crack depends.[6–8] However for short cracks, a change in propagation direction can alter the crack geometry significantly and thus change the driving force for the crack propagation. If its size is comparable with the microstructure of the material, a deflection of the crack at a grain boundary can alter the overall crack geometry. To understand the behavior of short cracks, it is therefore necessary to measure their three-dimensional growth. This can be achieved by using acoustic waves, which can penetrate into the material. In this way the crack depth can be measured directly.

Before crack depth can be measured, it must be detected. Both crack detection[9–11] and depth measurement can be achieved by scanning acoustic microscopy (SAM) using a combination of continuous wave imaging and the short-pulse (time-resolved) technique.[12] Chapter 2 briefly describes the short-pulse system before discussing the measurement technique in some detail. We present experimental measurement examples of this technique for determining the depth of short cracks in transparent plastic materials and an Al-alloy.

2.2. Short-Pulse system for the SAM

The short-pulse system follows the same principles as other pulsed ultrasonic systems used for nondestructive testing. The main difference between the two systems is pulse length. To detect small defects of about 100 μm, the acoustic pulse must be very short (≈8 ns). A schematic diagram of the short-pulse system is shown in Fig. 2.1.[10–13] The short pulse is generated by a step recovery diode and the pulse length matching the center frequency of the lens to transmit the maximal energy. The maximal voltage of the pulse applied to the transducer of the lens is limited to the breakdown voltage of the transducer. The pulse $p(t)$ emitted by the lens can be approximately described for example by a raised cosine

$$p(t) = \left\{1 - \frac{\cos(2\pi ft)}{p_w}\right\}\cos(2\pi ft) \qquad 0 \leq t \leq \frac{p_w}{f}$$

and zero otherwise, where f is the center frequency of the lens and p_w the pulse length.[14] The pulse generated by the step recovery diode passes through the single-pole double-throw (SPDT) switch (in position T) to the lens and the specimen. Switching SPDT to position R, the lens receives signals from the specimen, which are then passed to the receiving amplifier. After amplification the signal is fed into a sampling oscilloscope to digitize the received pulse. Sampling and digitizing the signal are achieved by moving the time position at which the signal is sampled step-by-step from the beginning to the end of the selected time window of the sampling oscilloscope, thereby allowing sufficient time for each point to be converted from an analog into a digital value. The digitizing process steps

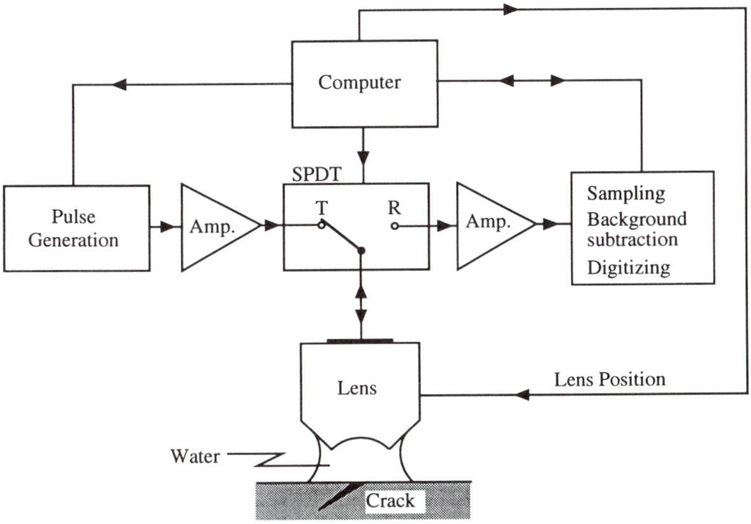

Figure 2.1. Schematic diagram of the short-pulse system.

are controlled by a computer. To remove spurious signals due to switching transients, a so-called background subtraction is employed to measure the signal without acoustic frequency excitation. The digitized signal is stored in the computer for further processing. Signal averaging can be used to enhance the signal-to-noise ratio.

2.3. *Time-Resolved Measurement Technique*

The principle on which time-resolved acoustic microscopy is based is the time-of-flight diffraction (TOFD) technique,[15,16] in which the transit time of a short acoustic pulse from a transducer to the crack tip and back is measured. The TOFD was developed for example to inspect nuclear vessels for cracks millimeters in size. In the conventional TOFD technique, the pulse-emitting transducer and the receiving transducer are separate, whereas in the SAM the point focus lens of the microscope acts as both transmitter and receiver for the pulses.

To measure the depth of a short crack, the SAM can be operated in two different modes: a gated and continuous wave excitation mode to image the specimen surface and locate the crack and an impulse excitation mode for time-resolved measurements to determine the crack depth. Once the crack on the surface is located, the lens is moved in equally spaced steps across the crack in the y direction (see Fig. 2.2) with the SAM operating in the pulse mode. At each

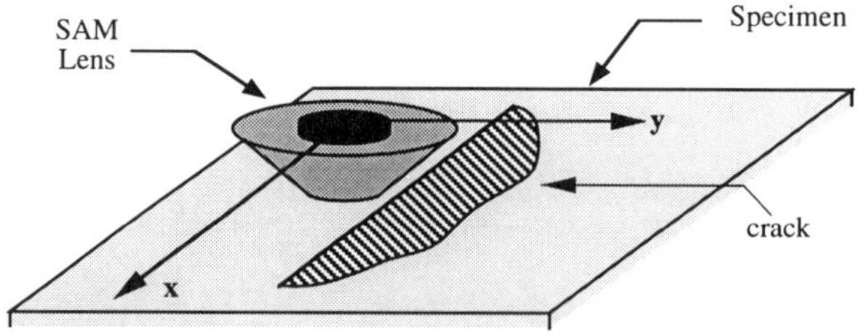

Figure 2.2. Generating an s(t,y)-image. The lens of the acoustic microscope is moved step-by-step across the crack in the *y* direction.

step the lens emits a short pulse (≈8 ns long) and receives back-scattered acoustic signals. These signals are amplified and passed to a sampling oscilloscope, sampled at 512 equally spaced time points, digitized, and finally stored in a computer. After the scan across the crack is completed, the stored information is converted into a grey scale and displayed on a frame store (see Fig. 2.3). In the following discussion, this display is referred to as an s(t,y) image (signal, time, *y* position), and it consists of 512 × 512 image points, with the horizontal axis representing time and the vertical axis representing the lens position. Figure 2.3 shows a schematic s(t,y)-image when no crack is present and the sample is at focus. The only signal occurring in this s(t,y)-image is the reflected pulse from the specimen surface. In the presence of a crack and appropriate defocus, various signals can occur in an s(t,y)-image. These signals are represented in Fig. 2.4 by ray diagrams; Fig. 2.5 shows traces of signals in a schematic s(t,y)-image. The signals in Figs. 2.4 and 2.5 follow.

Specular reflection (SR): Part of the incident pulse is reflected directly by the specimen surface. The transit time of the SR is independent of the lens position if the lens moves parallel to the surface, and therefore its time position

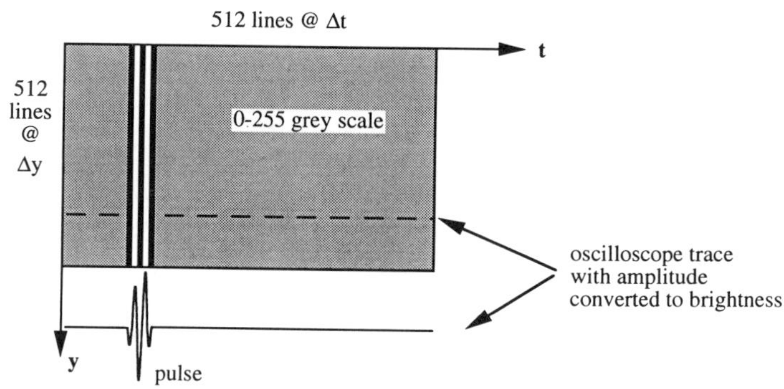

Figure 2.3. Grey scale s(t,y)-image with the reflection from the surface of the specimen.

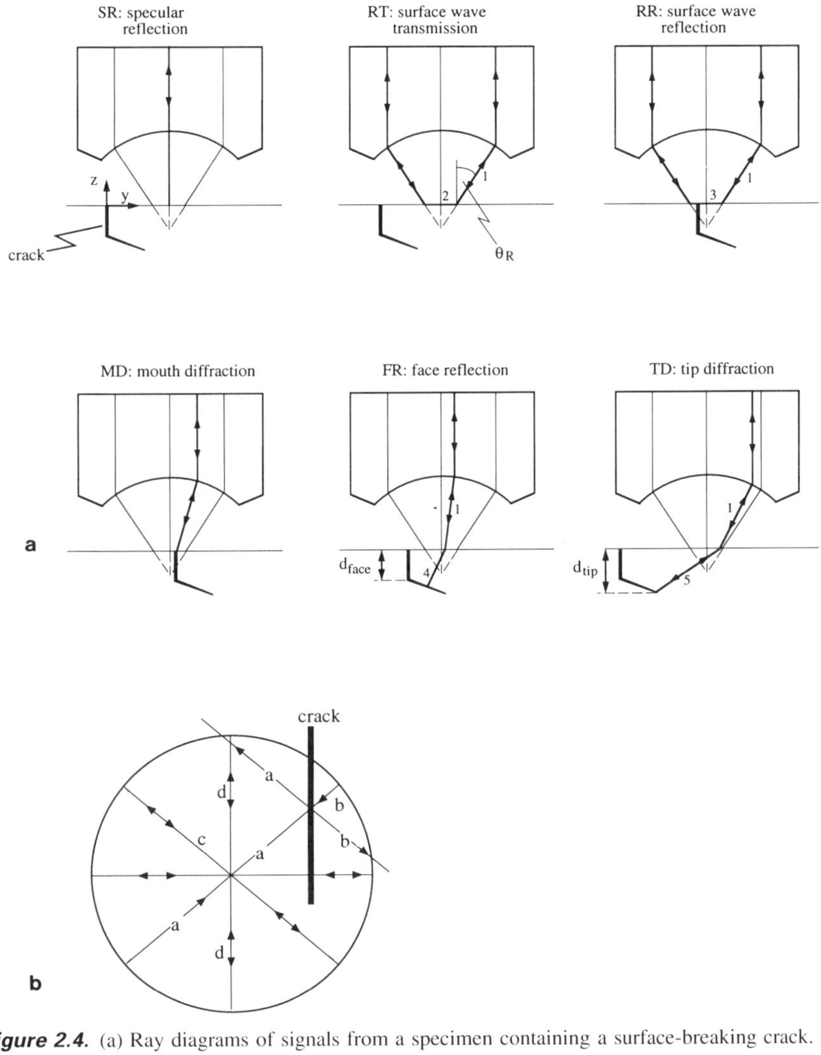

Figure 2.4. (a) Ray diagrams of signals from a specimen containing a surface-breaking crack. (b) Ray diagram in top view.

in an s(t,y)-image is constant. The transit time of the specular reflection from the lens surface to the specimen surface and back is given by

$$t_{SR} = \frac{2}{v_o}(q + z) \tag{1}$$

where v_o is the wave speed in water, q the focal length of the lens, and z the defocus of the microscope (usually $z < 0$).

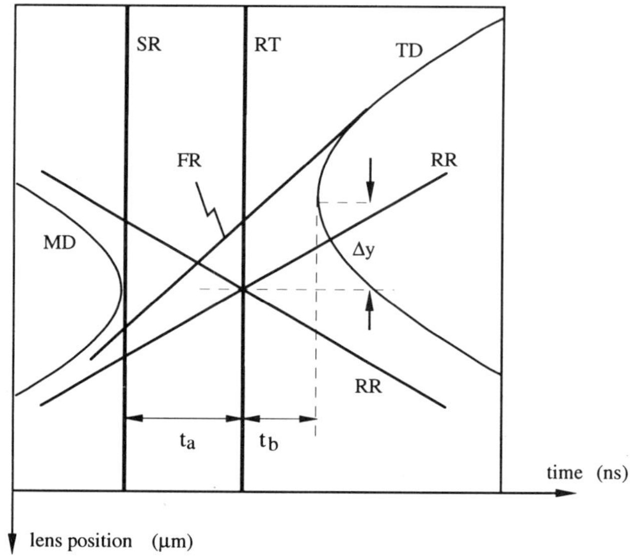

Figure 2.5. Schematic s(t,y) image with the signals indicated in Fig. 2.4. The depth d of the crack can be determined from $\Delta t = t_a + t_b$ and v, the sound velocity in the material as $d = 1/2(v \cdot \Delta t)$.

Surface wave transmission (RT): A surface wave, generated and received by a ray at the critical angle, is transmitted along the surface. The length of the path along the surface, and therefore the transit time, depends on the defocus of the lens, but it is not affected by the crack. Thus its transit time is independent of the lens y position, so the time position of RT in an s(t,y)-image is constant. The RT appears throughout the scan, since surface waves can always be transmitted inside the insonified area without being reflected by the crack [e.g., Rays c and d in Fig. 2.4(b)].

In relatively low-stiffness materials, such as polystyrene or perspex (perspex is also known as PMMA or Plexigas), propagation along the surface is a lateral longitudinal wave (or surface-skimming compression wave).[17] In high-stiffness materials, such as aluminium, the surface wave is a Rayleigh wave.[18] The transit time of RT is given by twice the time t_1 of a longitudinal wave traveling in water from the lens to the specimen plus the time t_2 of a surface wave traveling along the sample surface:

$$t_{\mathrm{RT}} = 2t_1 + t_2 = \frac{2}{v_o}\left(q + \frac{z}{\cos\theta_{\mathrm{R}}}\right) - \frac{2}{v_{\mathrm{R}}} z \tan\theta_{\mathrm{R}} = \frac{2}{v_o}(q + z\cos\theta_{\mathrm{R}}) \tag{2}$$

where θ_{R} is the critical angle of the surface wave and v_o is its velocity.

Surface wave reflection (RR): A surface wave propagates along the surface and is reflected back to the lens by the edge of the crack.[19,20] There are actually two surface wave reflections because a surface wave can be generated and reflected on each side of the crack. (There are also two RT waves, but they are indistinguishable.) Transit times of the two RR waves depend linearly on the distance between the mouth of the crack and the lens position, with one signal arriving earlier and one later than the RT signal.[21] In an s(t,y)-image, the two RR signals form a crossing pattern. At the crossing point of the two signals, their transit time and the transit time of the RT signal are the same.[12] At this position, the lens is directly over the crack mouth, and thus this point can serve as a reference point to determine the scanning position of the lens in regard to the crack trace on the surface. At larger distances, signals vanish due to attenuation. The transit times of the two RR signals are given by:

$$t_{RR} = 2t_1 + 2t_3 = \frac{2}{v_o}\left(q + \frac{z}{\cos\theta_R}\right) - \frac{2}{v_R}(z\tan\theta_R \pm y)$$

$$= \frac{2}{v_o}(q + z\cos\theta_R \pm y\sin\theta_R) \qquad (3)$$

where y is the distance between the axis of the lens and the crack mouth.

Crack mouth signal (MD): Part of the incident pulse is scattered directly by the mouth of the crack back to the lens. The shape of this signal in an s(t,y)-image is hyperbolic, and it occurs before the specular reflection SR. The MD and SR coincide when the lens is directly over the mouth of the crack.[22] The transit time of this signal is given by:

$$t_{MD} = \frac{2}{v_o}[q - (y^2 + z^2)^{1/2}] \qquad (4)$$

Crack face reflection (FR): Part of the incident pulse is transmitted into the material and reflected by the face of an inclined crack. The crack face acts like a mirror for the pulse traveling along the ray path in Fig. 2.4. The path of the ray is determined by applying Snell's law to the reflection at the crack surface. In an s(t,y)-image, the FR signal appears as a straight line whose slope depends on the inclination of the crack. Its length and position depend on the length and inclination angle of the crack segment, the depth, and y coordinates of the segment. The transit time of FR (when both ray paths to and from the crack face coincide) is given by

$$t_{FR} = 2t_1 + 2t_4 = \frac{2}{v_o}\left(q + \frac{z}{\cos\theta_i}\right) + \frac{2}{v}[(-z\tan\theta_i + y)\sin\beta + d_{face}\cos\beta]$$

$$= \frac{2}{v_o}(q + z\cos\theta_i + y\sin\theta_i) + \frac{2}{v}d_{face}\cos\beta \qquad (5)$$

where β is the inclination angle of the crack, d_{face} the depth of the crack segment, and v the wave speed in the material (longitudinal or shear wave). The variable θ_i is given by Snell's law

$$\sin\theta_i = \frac{v_o}{v}\sin\beta$$

(The calculation of the mixed-mode face reflected signal is described in the appendix to Chap. 2.)

Crack-tip diffracted signal (TD): Part of the transmitted pulse is diffracted by the crack tip and returns to the lens. In an s(t,y)-image, the signal appears approximately hyperbolic. The transit time of the signal is at a minimum when the lens is directly over the crack tip. The transit time of the signal is given as

$$t_{TD} = 2t_1 + 2t_5 = \frac{2}{v_o}\left(q + \frac{z}{\cos\theta_i}\right) + \frac{2}{v}\frac{d_{tip}}{\cos\theta_t} \tag{6}$$

The relation between the angle θ_i, θ_t, the lens position y, and the depth of the crack tip d_{tip} is given by the following equation and Snell's law ($y > 0$, $z < 0$):

$$\theta_t = \arctan\left(\frac{y - z\tan\theta_i}{d_{tip}}\right)$$

and

$$\sin\theta_t = \frac{v_o}{v}\sin\theta_i$$

Signals FR and TD contain information about the inclination of the crack and its depth, and these can be exploited to obtain the crack geometry. Referring to Fig. 2.5, for a crack perpendicular to the surface of the specimen, the vertex of the signal TD is at the same position as the crossing point of the two surface-reflected signals. For an inclined crack however, the tip-diffracted signal is shifted in the y direction away from the crossing point, and a reflected signal from the crack face should appear. The inclination angle β of the crack can be determined from either the slope, $m = \delta y/\delta t$, of the line FR and the wave speed v in the material or from the shift Δy of the vertex of the tip-diffracted signal and the depth of the crack tip d_{tip}

$$\beta = \arcsin\frac{1}{2}\frac{v}{m} = \arctan\frac{d_{tip}}{\Delta y} \tag{7}$$

The time interval $\Delta t = t_a + t_b$ between the specular reflection, the minimum time position of the tip-diffracted signal, and knowledge of the elastic wave speed v in the material serve to determine the depth of the crack as

$$d_{\text{tip}} = \frac{1}{2}(v\,\Delta t) \tag{8}$$

However because the contrast of the specular reflection signal is often saturated in an s(t,y)-image, it is difficult to measure the time distance Δt directly. A more precise way of determining Δt is to measure the time distance t_b between the crossing point of the two surface-reflected signals and the vertex of the tip-diffracted signal, because the crossing point is considerably more unambiguously defined in the image. The remaining time distance t_a between the specular reflection and the crossing point can be calculated from

$$t_a = \frac{-2z}{v_o}\left\{1 - \left[1 - \left(\frac{v_o}{v_R}\right)^2\right]^{1/2}\right\} \tag{9}$$

In Eq. 9, v_o and v_R are the wave speed in water and the wave speed of the surface wave, respectively.[17,18]

This two-dimensional ray model serves to identify various signals in an experimental measurement and to determine the geometry of the crack. For calculations of the time-of-flight of the pulses, we assume that the material is

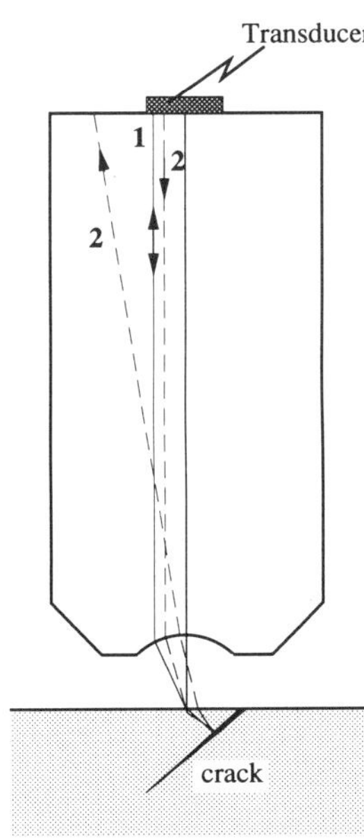

Figure 2.6. Acoustic lens. 1: principle ray, 2: ray not returning to the transducer.

isotropic and the transit time is given by pulses traveling along the principle rays for various signals. The principle rays (see Fig. 2.4) are those rays perpendicular to the transducer. Thus for the shear and longitudinal face reflected signals for example, the paths to and from the specimen are identical. (This is also true for the shear and longitudinal tip diffracted signal.) For mixed-mode signals, the two paths are different, but each one is a principle ray (see the appendix to Chap. 2). Due to the dimension of the lens and the great speed of acoustic waves in the lens (sapphire), only the principle rays and rays close to them can return to the transducer and contribute to the signal (see Fig. 2.6). To validate the technique, cracks in the transparent model materials, polystyrene and perspex, have been used.

2.4. Crack Depth Measurements in Plastic Material

In polystyrene[23] or perspex,[24] most cracks have a simple geometry, and therefore they are ideal for testing time-resolved acoustic microscopy for the depth measurement of cracks. Two other advantages of polystyrene and perspex are first that they are transparent, which allows a comparison between acoustic and direct optical measurements; and second the impedance mismatch between these materials and water (compared for example with the impedance mismatch between metals and water) is small, so that more energy is transmitted to the specimen.[25] This is important because of the small amplitude of the tip-diffracted signal.

The first example gives the depth measurement of a crack in polystyrene. The density and velocities for polystyrene are $\rho = 1.08$ g cm^{-2}, longitudinal wave velocity $v_L = 2.35$ μm ns^{-1}, shear wave velocity $v_S = 1.12$ μm ns^{-1}, and the Rayleigh velocity $v_R = 1.048$ μm ns^{-1}. Since the Rayleigh velocity in polystyrene is slower than the wave speed in water, no Rayleigh waves are generated, and the surface waves in this material are longitudinal lateral waves. Figure 2.7 shows an s(t,y)-image of a crack in polystyrene taken at a defocus of $z = -60$ μm. The first strong set of vertical stripes on the left in Fig. 2.7 is the specular reflection from the specimen surface (labeled SR). The second strongest set of stripes further to the right is an electronic echo (labeled CE) of SR. This echo is caused by impedance mismatches at the ends of the coaxial cable connected to the SPDT.

The weaker signal between the specular reflection and the cable echo is the longitudinal lateral wave transmission (RT). The crossing pattern of the surface wave reflection can be recognized between these two signals. The strong, hyperbola-shaped signal on the left in Fig. 2.7 is the crack mouth diffraction; its echo appears in the center of the image. The other weaker sets of vertical stripes are remnants of spurious signals after the background subtraction.

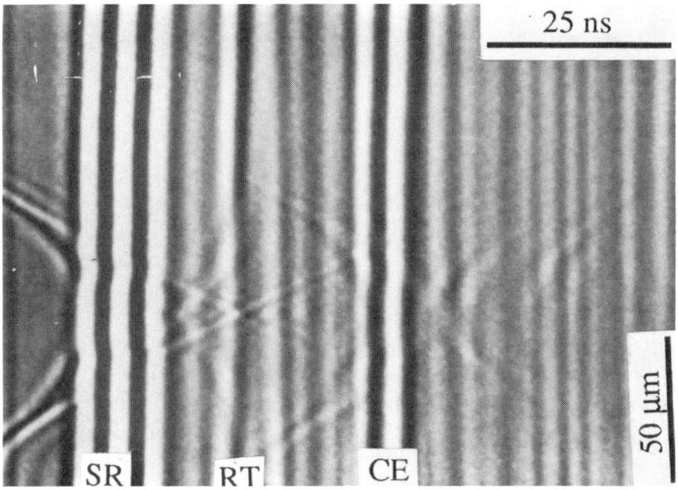

Figure 2.7. An s(t,y)-image at the indicated position; z = −60 μm; SR-specular reflection, RT-surface wave transmission, and CE-cable echo.

Two approximately hyperbola-shaped signals can be seen on the right beyond the cable echo. These are two diffracted signals from a kinked crack. From the time location of the vertices and the longitudinal wave speed in polystyrene, the depth of the kink and the crack tip can be calculated from Eqs. 7 and 8. Depths obtained from the s(t,y)-image are 53 μm for the kink and 76 μm for the tip, respectively. There is also a small shift Δy in the scanning direction between the two diffracted signals of about 5 μm from which an inclination angle β of the second crack segment of β = 77.7° can be calculated. Due to this large angle and the fact that the segment is curved rather than straight (observed with the light microscope), no face reflection can be seen. Figure 2.8 shows a schematic cross section through the crack deduced from acoustic measurements, and Table 2.1 compares acoustic and optical measurements.

Figure 2.9 shows a second example of an acoustic measurement of a crack in a plastic material and an s(t,y)-image of a long oblique crack in perspex after removing the cable echo.[24] The inclination angle of the crack is β = 25°, and

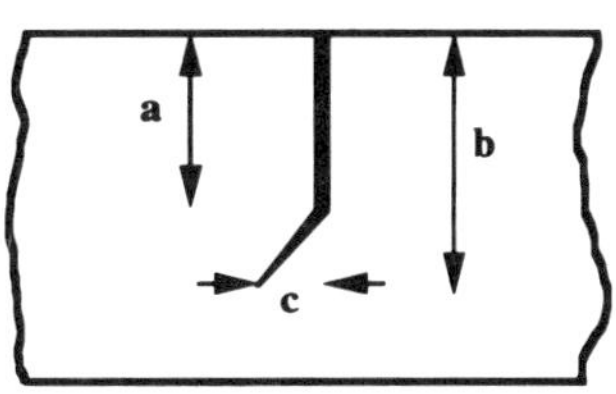

Figure 2.8. Geometry of the crack determined from acoustic measurements in Fig. 2.7.

Table 2.1. Comparison between the Crack Depth Determined from the s(t,y)-Image in Fig. 7 and Direct Optical Measurements with a Light Microscope

	Crack Geometry (μm)		
Method	*a*	*b*	*c*
Optical	54	75	10
Acoustic	53	76	5

the three inclined lines in Fig. 2.9 are the face reflected signals of the longitudinal (a), the mixed mode (c), and the shear wave (b). The mixed-mode face reflection is a combination of longitudinal waves traveling to the crack and shear waves, reflected by the crack, traveling back to the lens or vice versa. The vertical set of stripes on the left side is again the specular reflection. In this case it is easy to identify signals and interpret the s(t,y)-image, since the face-reflected signals are strong. But as we see in the polystyrene example, diffracted signals from the kink and the crack tip are weak compared to the specular reflection. Thus it is necessary to use image processing to remove the cable echo and specular reflection to enhance the contrast of the diffracted signal. Another purpose of this process is to be able to measure the time distance between the crossing point of the two surface wave reflections and the vertex of the diffracted signal more precisely. An example of an s(t,y)-image after removing the crack-independent signals

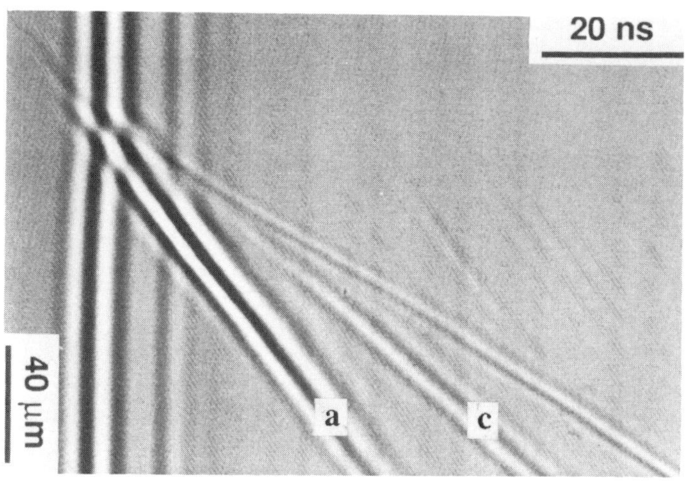

Figure 2.9. Acoustic measurement of a long 25° oblique crack in perspex; (a) longitudinal, (b) shear, and (c) mixed-mode reflection from the crack face.

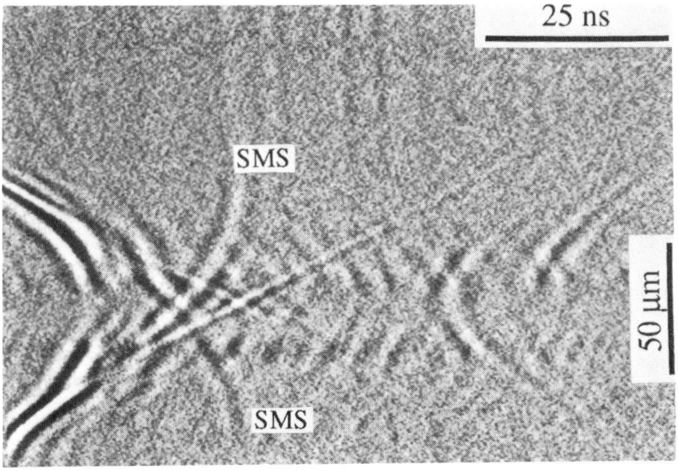

Figure 2.10. The s(t,y)-image in Fig. 2.7 after removing the cable echo, specular reflection, and surface wave transmission signal (SMS; see text).

(specular, surface wave transmission) and the cable echo is given in Fig. 2.10, which shows the s(t,y)-image in Fig. 2.7 after processing. The remaining signals in Fig. 2.10 are the strong mouth-diffracted signal on the left, the crossing pattern of the surface wave reflection, and the two diffracted signals from the kink and the crack tip. Furthermore two additional lines (labeled SMS) appear in Fig. 2.10 after processing. These two signals are surface waves scattered by the crack mouth into rays in the fluid that then travel directly back to the lens. The SMS is a combination of a surface wave reflection and a mouth-diffracted signal. Finally Fig. 2.11 shows an overlap between the s(t,y)-image in Fig. 2.10 and the time-of-flight traces (fine white and black lines) calculated by ray theory using Eqs. 1–7, and based on the optically obtained crack geometry. As we see from Fig. 2.11, agreement between calculated signal traces and the measurement is good, so we can use our ray model for analyzing s(t,y)-images. Figures 2.10 and 2.11 also show the improvement in s(t,y)-images after image processing; therefore we briefly describe how image processing is done.

Processing the original s(t,y)-images is accomplished in two steps: First the cable echo is removed from the original s(t,y)-image, then the specular and surface wave transmission signals are subtracted.[26] To remove the temporally separated cable echo, a time line $s(t)$ of the s(t,y)-image is selected, where only SR, RT, and their cable echo are present. This full time line $s(t)$ is then separated into a part containing only the actual signals $s_s(t)$ and a part containing only their echoes $s_e(t)$. The full time line can therefore be written as

$$s(t) = s_s(t) + s_e(t)$$

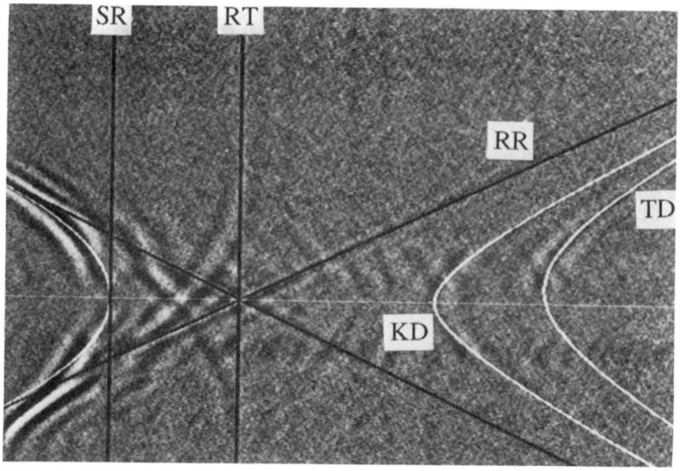

Figure 2.11. Overlap between Fig. 2.10 and ray theory calculations. The KD = diffraction from a crack kink.

Taking the Fourier transforms (FT) of these signals, the FT of the actual signal can be written as

$$S_s(f) = H(f) \cdot S(f)$$

where $S(f)$ is the FT of a line in an s(t,y)-image and $H(f)$ is the filter function given by

$$H(f) = 1 - \frac{\bar{S}_e(f)\bar{S}^*(f)}{|\bar{S}(f)|^2 + Q^2}$$

where $\bar{S}(f)$ is the average spectrum of a full time line $s(t)$ and $\bar{S}_e(f)$ is the average spectrum of the selected cable echo $s_e(t)$. The asterisk denotes complex conjugation, and Q is a real constant. Average spectra are calculated by taking the FT of the average of a number of lines in the original s(t,y)-image. Once we have found the filter function, we can remove the cable echo by multiplying the FT of each line of an s(t,y)-image by $H(f)$.

The second processing step is to remove the specular reflection (SR) and the surface wave transmission (RT). This is achieved by aligning each line $s(t)$ in the s(t,y)-image with a reference line, then subtracting the reference line from the actual line. The reference line is obtained from the same image by taking the average of a number of $s(t)$ lines (e.g., 20) at a position in the image where no signals from the crack are present. Due to time shifts of $s(t)$ lines in the original image, the reference line and the actual line have to be aligned in time. This alignment is accomplished in two steps. The first step in the time domain is to interpolate the maximum in the cross correlation between the reference line and the actual line to find the necessary shift. Because shifts in the time domain cannot be smaller than one digitization interval, i.e., shifts in the time domain are integer numbers, a second shift in the frequency domain is applied, which permits time shifts smaller than one digitization interval.

Since the amplitudes of $s(t)$ can change during the scan (due to surface roughness near the crack), the amplitude of the reference line has to be adjusted. This is accomplished by taking the reference line as a moving average of previously shifted original s(t,y)-lines. This image processing is an essential part in measuring the depth of cracks for more complex cases, e.g., complicated crack shapes, weak diffracted signals.

2.5. Applying TOFD to Metals

Having established the principles of crack depth measurement using time-resolved acoustic microscopy in transparent polymers, we now demonstrate the measurement of a crack in an Al–Li alloy.[23] The low anisotropy[27] of this material allows us to treat it as isotropic, so we can use our ray model to interpret measurements. In addition relatively low acoustic impedance for a metal allows quite good coupling to bulk waves.

An acoustic micrograph of a fatigue crack in Al–Li alloy 8090 is shown in Fig. 2.12, taken at a frequency of 350 MHz and zero defocus. Unlike cracks with a simple geometry that occur in polystyrene and perspex, cracks in aluminium develop with a more irregular shape, both along the surface and into the specimen. The s(t,y)-depth measurement of the crack in Fig. 2.12 is given in Fig. 2.13 after image processing. While acoustic waves in water couple well only with longitudinal waves in plastics, both longitudinal and shear waves are excited by the SAM in aluminium. Thus we can expect to receive longitudinal, shear, and mixed-mode signals diffracted from the crack tip, which complicate evaluating the s(t,y)-image. In Fig. 2.13 parts of the longitudinal (TD_L) and the shear wave (TD_S) tip-diffracted signals are visible. The vertex of the shear wave tip-diffracted signal is identified as marked in Fig. 2.13. The depth of the crack obtained from the time distance measured between the surface wave reflection crossing point and the vertex of the tip-diffracted signal and the calculated time distance between the specular reflection and the surface wave transmission using Eq. 9 ($z = -40$ μm) is 42 μm. The tip-diffracted signal is shifted by about 7 μm in the y direction, which means that the tip of the crack is shifted by 7 μm from directly below the mouth of the crack.

To confirm the acoustic depth determination, the specimen was cut to provide a cross section of the crack. Figure 2.14 shows an acoustic micrograph taken at a frequency of 500 MHz and zero defocus of the crack depth profile. The depth

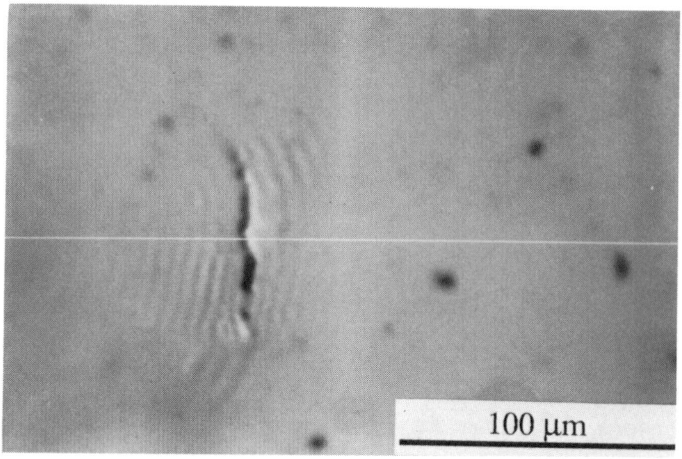

Figure 2.12. Acoustic micrograph of a short crack in Al-alloy; f = 350 MHz, z = 0 μm. The white line marks the position for the depth measurements.

of the crack obtained from this section was 43 μm, and the shift in the scanning direction between the crack mouth and crack tip (indicated by an arrow) was 6.5 μm. The crack kink close to the surface could not be determined from the acoustic measurement in Fig. 2.13 because the kink was not deep enough, and thus signals from it were obscured by strong specular and surface wave signals.

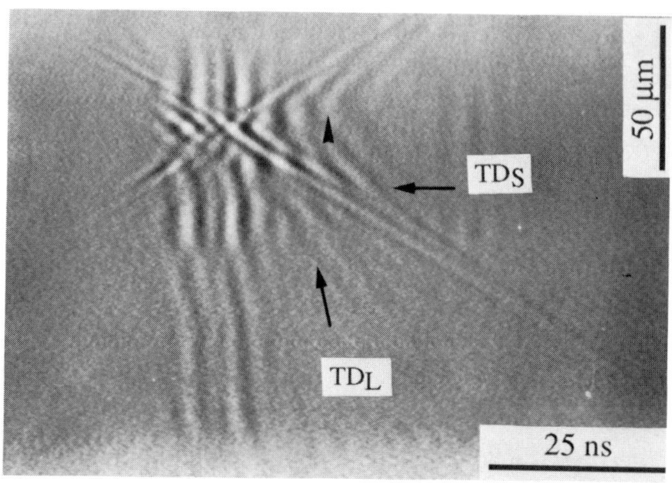

Figure 2.13. An s(t,y)-image along the line indicated in Fig. 2.12 after processing; z = −40 μm. The vertex of the shear wave tip-diffracted signal (TD_S) is indicated by the small arrow (d = 42 μm); TD_L: longitudinal tip-diffracted signal.

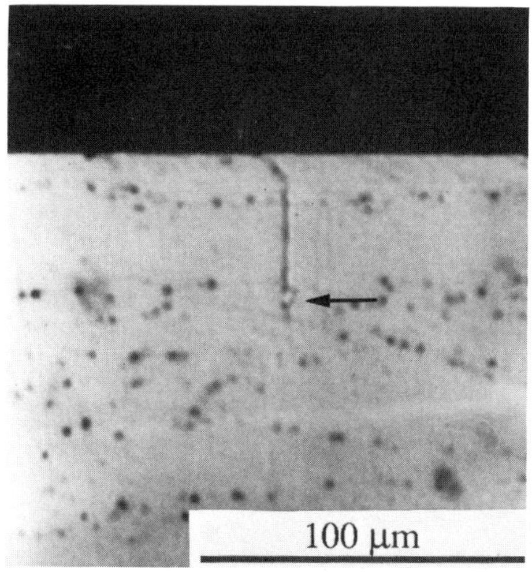

Figure 2.14. Acoustic micrograph showing a cross section of the crack; f = 350 MHz, z= 0 μm.

So far we have seen that time-resolved acoustic microscopy can be used to determine the depth of cracks in polystyrene and Al alloy. The next step is to apply this technique for studying the growth behaviour of short cracks.

2.6. Crack Growth Measurement

The material we used to measure crack growth was Al alloy.[28] Grains in this material had an elongated shape, with a typical size in longitudinal (*L*), transverse (T), and short (S) direction of 200, 170, and 25 μm, respectively. Two sets of bend specimens were machined from this material, so that in one set elongated grains are parallel to the surface of the specimen (crack growth in the S direction), and in the other set, they ran perpendicular (crack growth in the *L* direction) to the surface. After grinding and polishing, specimens were fatigued at a stress ratio of $R \approx 0$ ($R = \sigma_{min}/\sigma_{max}$) and a stress amplitude of ≈70% of the yield strength, and crack growth on the surface and in depth were mesaured.

2.6.1. Crack Propagation in the S Direction

In the first set of specimens, elongated grains ran parallel to the surface. Figure 2.15 shows an acoustic micrograph of a crack (a) before and (b) while loading the specimen elastically. Both pictures were taken at a frequency of 350 MHz and zero defocus, and they show a significant difference in the acoustic contrast due to the closure of the crack.[29,30] In the unloaded case, parts of the

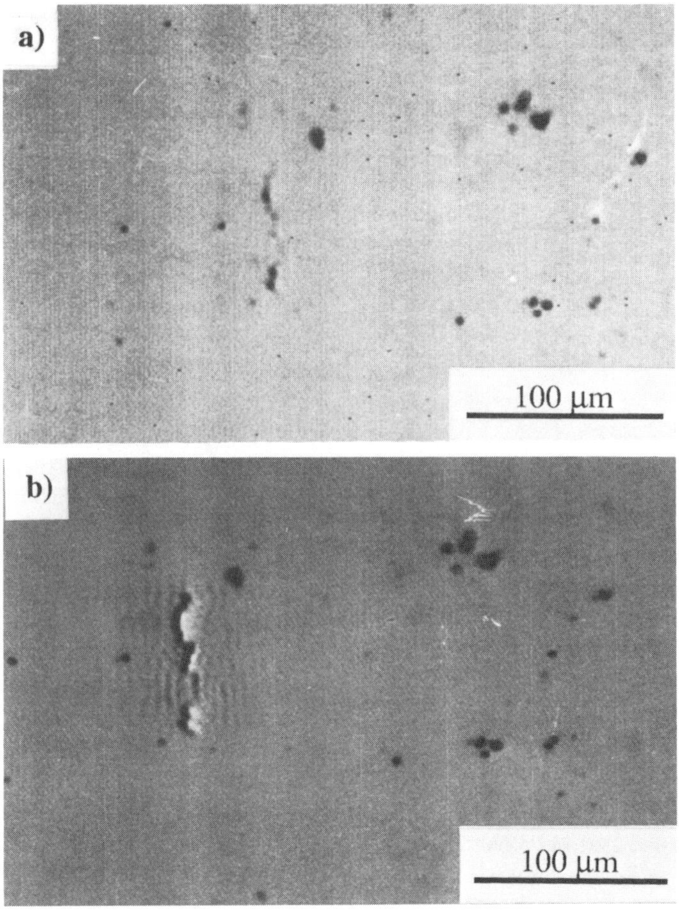

Figure 2.15. (a) Acoustic micrograph of a unbent Al-Li alloy specimen with a 80-μm-long crack. (b) The same crack while bending the specimen elastically; f = 350 MHz, z = 0 μm.

crack are closed along the surface, and contrast on the micrograph vanishes along much of the crack, indicating the amount of closure occurring on the surface. The length and depth of this crack were monitored during the fatigue process, with measurements taken about every 3000 cycles. Development of the crack surface length and crack depth is given in Fig. 2.16 as a graph. On the surface, the crack grew uniformly from an initial length of 80 μm to a final length of 176 μm. However development of the crack depth determined from the time-resolved measurements revealed quite a different behavior. The crack depth remained at about 30 μm for 15,000 cycles, after which it advanced to a depth of about 57 μm and remained at that depth for at least 10,000 cycles before growing to its final depth of 75 μm between 25,000–28,000 cycles. Another

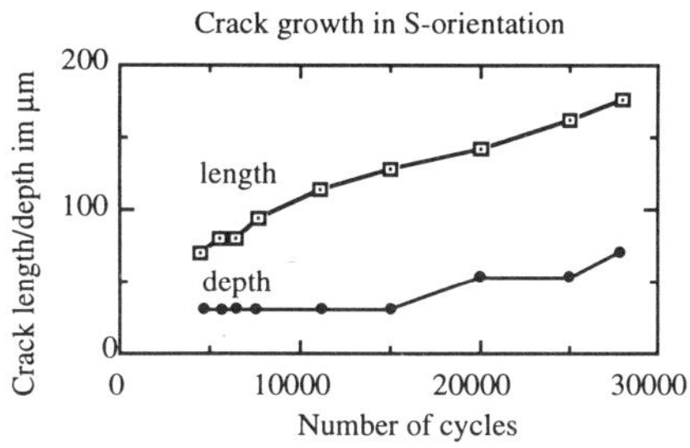

Figure 2.16. Crack growth as a function of the applied number of cycles for a crack growing in direction with a short distance between the grain boundaries (S direction).

interesting aspect of the acoustic depth measurements is the change in crack geometry during the fatigue process. During the last 3000 cycles, the crack changed its geometry completely from a nearly straight crack to a crack with a large kink. Figure 2.17(a) shows the s(t,y)-depth measurement of the crack before the growth of the kink. From the diffracted signal [labeled TD in Fig. 2.17(b)], a crack depth of 57 μm was determined by fitting the calculated ray traces to the s(t,y)-image. An overlap between the traces (fine white lines) and the depth measurement is given in Fig. 2.17(b). From the fit a shift of 16 μm between the vertex of the tip-diffracted signal and the crack mouth was obtained, which corresponds to an inclination angle β of ≈74°.

The depth measurement at the end of the fatigue process after the kink occurred is given in Fig. 2.18(a). It shows a significant difference compared to the measurement in Fig. 2.17(a). The two strong signals, labeled FR_L and FR_S, are the longitudinal and shear wave reflection from the face of the kink. At the upper end of the shear wave signal FR_S, a weaker line with a different slope can be seen, which is part of the shear wave tip-diffracted signal. From slopes of the two lines and the longitudinal (≈6.8 μm ns^{-1}) and shear wave (≈3.6 μm ns^{-1}) velocity, an inclination angle with a kink of 24° was obtained. The length ℓ of the kink given by the length of the two lines FR_L and FR_S and the inclination angle was $\ell = 44$ μm, which with the depth d_1 determined from Fig. 2.17(a), yielded a crack depth of 75 μm

$$(d_{\text{crack}} = d_1 + \ell \cdot \sin 24°)$$

Figure 2.18(b) shows a ray theory calculation based on the geometry obtained from acoustic measurements. Signals in Fig. 2.18(b) are the longitudinal (FR_L), the mixed-mode (FR_M), and the shear wave (FR_S) face reflection from the crack kink; the diffracted longitudinal (TD_L) and shear wave (TD_S) signals from the 75-μm-deep crack tip; and the crossing pattern of the surface wave reflection (Rayleigh wave). The fine, vertical white line on the left indicates the position of the specular reflection.

After the end of fatigue testing, a through-section was made by cutting the specimen along the scanning line of the acoustic depth measurement. An acoustic micrograph of the crack depth profile, taken at a frequency of 350 MHz and zero

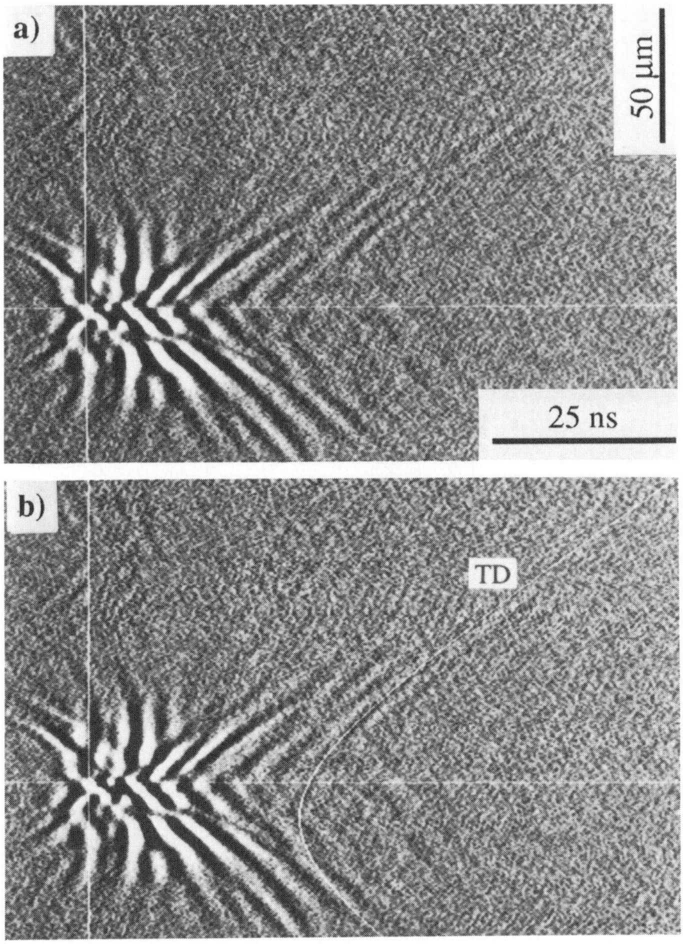

Figure 2.17. (a) Depth measurement of the crack in Fig. 2.15 after 20,000 cycles. Fine white lines indicate the position of the specular reflection (vertical) and the crack mouth (horizontal). (b) Overlap between (a) and ray theory calculations (fine white hyperbola-shaped line). The depth of the crack determined from this measurement was 57 μm (z = −40 μm).

defocus, is given in Fig. 2.19(a). The crack geometry determined from this section and the grain structure (obtained with a light microscope after etching the specimen) are shown schematically in Fig. 2.19(b). The section yielded the following crack geometry: a kink depth of 55 μm, a kink length of 45 μm, an inclination angle of the kink of 24°, and a total crack depth of 73 μm. The shift between the crack mouth and the end of the crack segment running from the surface to a depth of 55 μm is 25 μm, which yielded an inclination angle

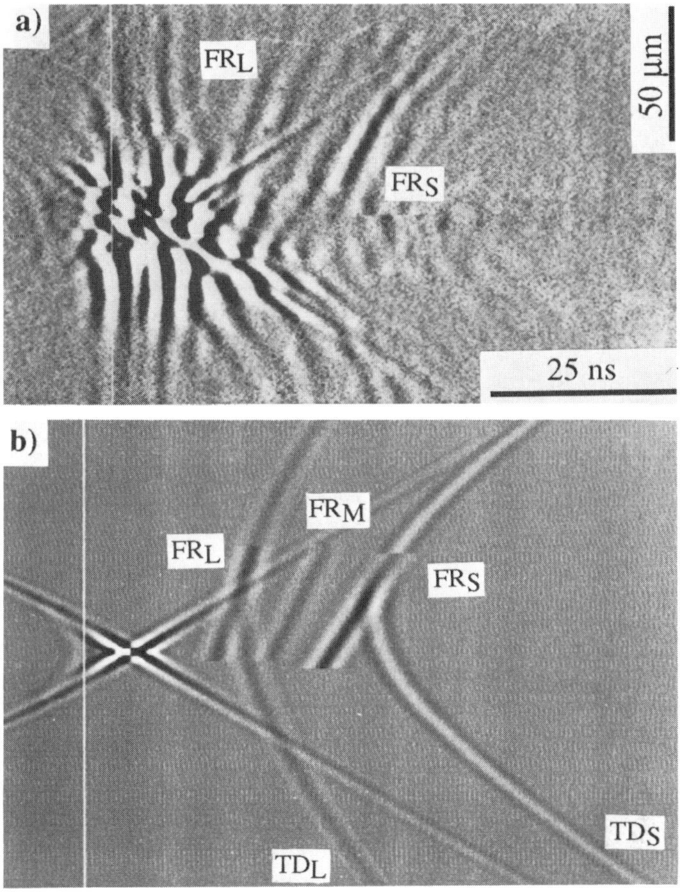

Figure 2.18. (a) An s(t,y)-image of the crack in Fig. 2.15 after 28,000 cycles. The FR_S and FR_L are the longitudinal and shear wave reflection, respectively, from a 44 μm long and 24° inclined kink; z = −40 μm. (b) Ray theory calculations for the crack geometry determined from the through-section [see Fig. 2.19(a)]; FR_M: mixed–mode face reflection.

of 65°. A comparison with the crack geometry determined from the acoustic measurements shows a good agreement. The difference of 9° in the inclination angle of the first segment is mainly due to the difficulty in localizing the exact position of the crack mouth in an s(t,y)-image (given by the crossing point of the two surface wave reflections).

The crack kinking in the section coincided with the grain boundaries.[31,32] The first acoustically obtained crack depth was about 30 μm, which is the depth of the second grain boundary. As previously described, the crack remained at

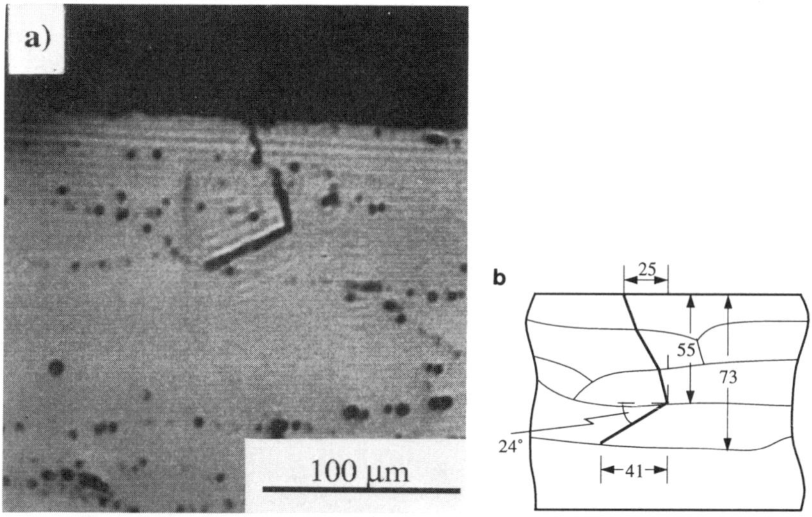

Figure 2.19. (a) Acoustic micrograph of the through-section of the crack measured in Fig. 2.18; f = 350 MHz, z = 0 μm. (b) Crack geometry (in μm) and grain structure obtained by light microscopy after etching the specimen.

this depth for about 10,500 cycles before propagating to a depth of 55 μm, which again coincided with the depth of a grain boundary. The final depth of 75 μm is also the depth of a grain boundary. To investigate further the influence of grain boundaries on crack propagation, crack growth was measured in a second set of specimens whose elongated grains ran perpendicular to the surface and therefore had a much larger distance between grain boundaries.

2.6.2. Crack Propagation in the L Direction

The average distance between the grain boundaries in the L direction was about 200 μm. In this direction, we expect a different crack growth behavior if grain boundaries influence the propagation of cracks. Figure 2.20 shows an acoustic micrograph of a 48-μm-long (*left*) and a 81-μm-long (*right*) crack in such a specimen. The picture was taken at a frequency of 350 MHz and a defocus of $z = -8$ μm. At this defocus, elongated grains are visible in the micrograph. The line in Fig. 2.20(a) indicates the position for the depth measurement. As

Figure 2.20. (a) Acoustic micrograph of crack propagating in the L direction; f = 350 MHz, z = −8 μm. The black line marks the position of the depth measurement. (b) Depth measurement of the crack shown in (a) after 3100 cycles. The depth obtained from this measurement was 62 μm. (c) An overlap between (a) and the ray calculation for a 62-μm-deep crack (TD_1); z = −40 μm.

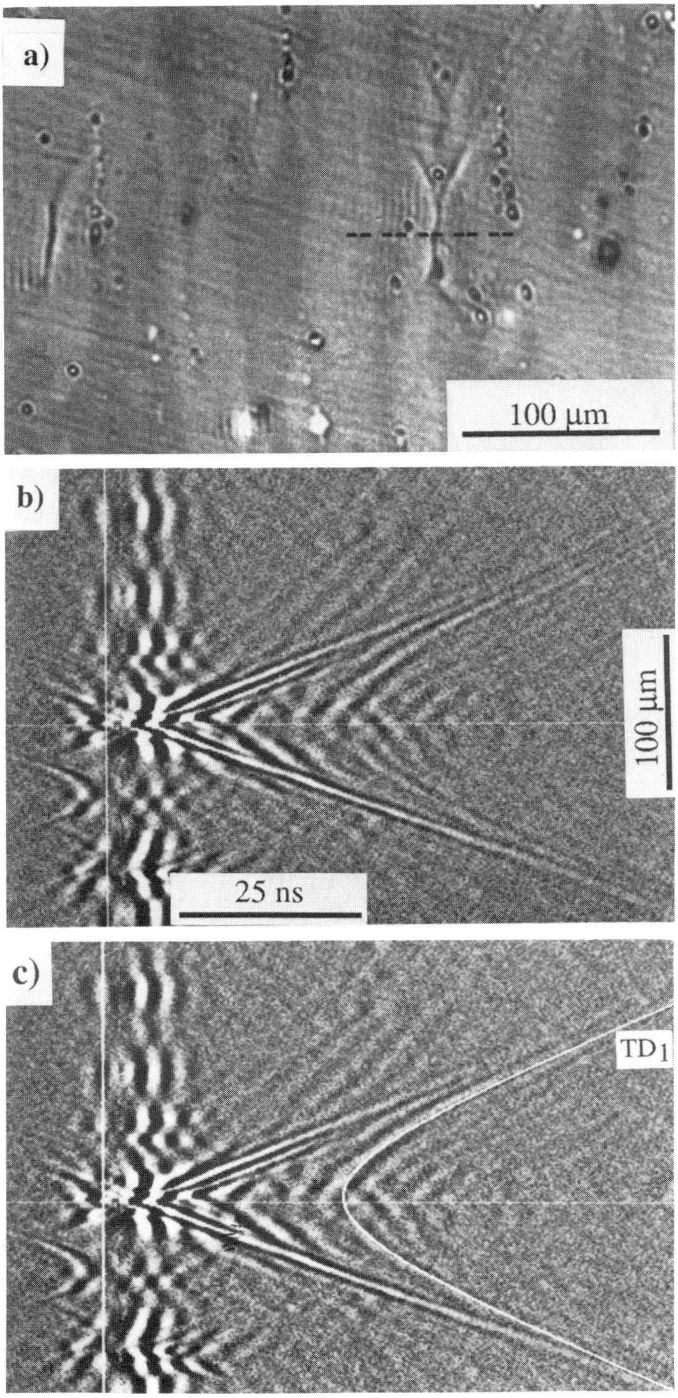
a)
100 μm
b)
100 μm
25 ns
c)
TD1

before the specimen was fatigued, and propagation of the crack along the surface and in the depth direction was monitored with the acoustic microscope. The specimen was fatigued for about 10,000 cycles, and measurements of the crack length and depth were taken approximately every 1500 cycles. The first acoustic measurement yielded a depth of 62 μm at a crack length of 81 μm; it is given in Fig. 2.20(b) after removing crack-independent signals (SR and RT). The signals visible in Fig. 2.20(b) are the crossing pattern of the surface wave transmission, the remaining parts of SR and RT on the left, and a shear wave diffracted signal [labeled TD_1 in Fig. 2.20(c)] from the crack tip. In Fig. 2.20(b) and (c), vertical and horizontal, fine white lines indicate the position of the specular reflection and the position of the crack mouth, respectively. From 3100–6800 cycles, no change in crack depth was detected, but the surface length grew from 81 to 90 μm. During the next 1500 cycles ($\Sigma_{cycles} = 8300$), the crack advanced to a depth of 106 μm and a surface length of 100 μm, and finally after 10,000 cycles, the crack reached a depth of 130 μm, with a slight change in the propagation direction and a surface length of 110 μm. The acoustic depth measurement after 10,000 cycles is shown in Fig. 2.21, with ray traces of diffracted signals. In Fig. 2.21 three diffracted signals were identified: TD_1, the diffracted signal from the first depth measurement after 3100 cycles (d = 62 μm); TD_2, the diffracted signal detected after 8300 cycles (d = 106 μm); and the new tip-diffracted signal (TD_3) after 10,000 cycles (d = 130 μm). A shift among the crack mouth, TD_2, and TD_3 in the scanning direction of 9 and 12 μm, respectively, was measured. In all three depth measurements, only the shear wave diffracted signal was considered to determine the depth of the crack. Figure 2.22 shows a schematic depth profile of the crack obtained from acoustic measurements.

2.7. Discussion

From the preceding examples of time-resolved acoustic crack measurements, we see how irregular cracks behave and how their shape changes during crack growth, which has a significant influence on the crack-driving force and thus on the prediction of the crack growth rate. In particular, short cracks (when, for example, crack dimensions are comparable with the microstructure) show a different behavior depending on their propagation direction in relation to for example grain boundaries. To measure the depth of these short cracks with the acoustic microscope, the pulse length must be on the order of a few nanoseconds and thus the bandwidth in the high MHz range. For lens with a center frequency of 220 MHz, the corresponding pulse length is ≈8 ns. In Al for example with a shear wave velocity of 3.11 μm/ns, the minimum distance between two diffraction points that can be resolved with a pulse length of 8 ns is about 12 μm. To increase the resolution, pulses must be even shorter. At a high

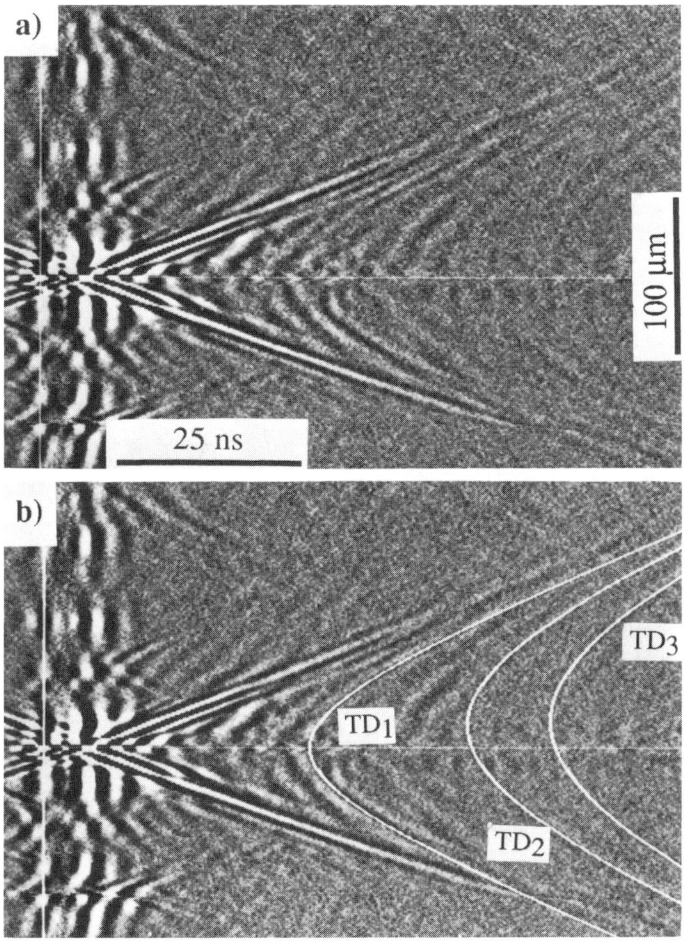

Figure 2.21. (a) Depth measurement of the same crack in Fig. 2.20 after 10,000 cycles. The crack grew to a depth of 130 μm (TD_3). (b) An overlap between (a) and the ray calculation for the shear wave diffracted signals of diffraction locations at the following depths: 62, 106, and 130 μm; z = −40 μm.

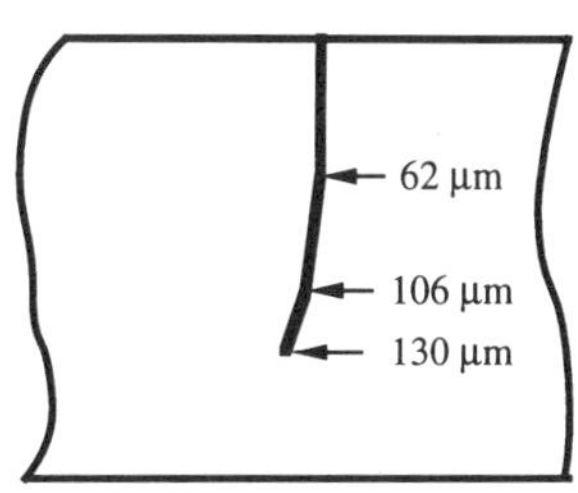

Figure 2.22. A schematic crack geometry determined from the acoustic measurement in Fig. 2.21.

frequency on the other hand, wave attention is stronger, which makes it more difficult to detect scattered signals in the crack and to measure deeper cracks. In the Al-Li alloy for example, the maximum depth we could measure was about 220 μm.

2.8. Summary

Examples of time-resolved acoustic microscopy (TRAM) measurements presented here demonstrate the applicability of this technique to detecting and measuring the depth of short cracks not only in ideal materials, such as polystyrene and perspex, but also in metals. Most importantly TRAM can be used to measure the growth of short fatigue cracks. These measurements show how irregular the three-dimensional growth of short cracks can be and how the geometry can change during the fatigue process.

Acknowledgments

We would like to thank the Science and Engineering Research Council, National Power, and the European Union for funding this project; D. D. Bennink for developing the image-processing programs; and the Run Run Shaw foundation for providing a scholarship to T. Zhai.

References

1. Marsh, K. J., Smith, R. A., Ritchie, R. O. (1991). *Fatigue Crack Measurement: Techniques and Applications,* Engineering Materials Advisory Services Ltd., Warley, West Midlands, UK. Pp. 554.
2. Miller, K. J. and d. l. Rios, E. R. (1986). *The Behaviour of Short Fatigue Cracks,* Mechanical Engineering Publications Ltd., London. Pp. 560.
3. Suresh, S. and Ritchie, R. O. (1984). Propagation of short fatigue cracks. *Int. Met. Rev.* **29**, 444–73.
4. Nicholls, D. J. and Martin, J. W. (1990). The use of ΔK when examining microstructure effects on small crack growth. *Int. J. Fatigue* **12**, 469–73.
5. Nicholls, D. J. and Martin, J. W. (1990). Small crack growth in the aluminium–lithium alloys 8090 and 8091. *Fatigue and fracture of engineering materials structures* **13**, 83–94.
6. Suresh, S. (1991). *Fatigue of Materials,* Cambridge University Press, Cambridge.
7. Ewalds, H. L. and Wanhill, R. J. H. (1984). *Fracture Mechanics,* Edward Arnold, London.
8. Broek, D. (1987). *Elementary Engineering Fracture Mechanics,* Martinus Nijhoff Publishers, Dordrecht, Netherlands.
9. Yamanaka, K. and Enomoto, Y. (1982). Observation of surface cracks with scanning acoustic microscopy. *J. Appl. Phys.* **53,** 846–50.
10. Briggs, G. A. D. (1992). *Acoustic Microscopy,* Oxford University Press, Oxford.

11. Somekh, M. G., Bertoni, H. L., Briggs, G. A. D., Burton, N. J. (1985). A two-dimensional imaging theory of surface discontinuities with the scanning acoustic microscope. *Proc. R. Soc. London* **A 401**, 29–51.
12. Weaver, J. M. R., Daft, C. M. W., Briggs, G. A. D. (1989). A quantitative acoustic microscope with multiple detection modes. *IEEE Transaction on Ultrasonics, Ferroelectrics, and Frequency Control* **36**, 554–60.
13. Sinton, A. M., Briggs, G. A. D., Tsukahara, Y. (1989). Time-resolved acoustic microscopy of polymer coatings. In: *Acoustical-Imaging,* vol. 17 (H. Shimizu, N. Chubachi, J. Kushibiki, eds.), pp. 210–12. Plenum Press, New York.
14. Minachi, A., Mould, J., Thompson, R. B. (1993). Beam propagation through a bimetallic weld-A comparison of predictions of Gauss–Hermite beam model and finite element method. *J. Nondestr. Eval.* **12**, 151–58.
15. Charlesworth, J. P. and Temple, J. A. G. (1989). *Engineering Applications of Ultrasonic Time-of-Flight Diffraction,* Research Studies Press, Somerset, England.
16. Ogilvy, J. A. and Temple, J. A. G. (1983). Diffraction of elastic waves by cracks: application to time-of-flight inspection. *Ultrason.* **21**, 259–69.
17. Chan, K. H. and Bertoni, H. L. (1991). Ray presentation of longitudinal lateral waves in acoustic microscopy. *IEEE Transaction on Ultrasonics, Ferroelectrics, and Frequency Control* **38**, 27–34.
18. Bertoni, H. L. (1984). Ray-optical evaluation of V(z) in the reflection acoustic microscope. *IEEE Trans. Sonics Ultrason.* **SU-31**, 105–16.
19. Achenbach, J. D., Gautesen, A. K., Mendelson, D. A. (1980). Ray analysis of surface wave interaction with an edge crack. *IEEE Trans. Sonics Ultrason.* **SU-27**, 124–29.
20. Kundu, T. and Mal, A. K. (1981). Diffraction of elastic waves by a surface crack on a plate. *J. Appl. Mech.* **48**, 570–76.
21. Briggs, G. A. D., Jenkins, P. J., Hoppe, M. (1990). How fine a surface crack can you see in a scanning acoustic microscope? *J. Microsc.* **159**, 15–32.
22. Tew, R. H., Ockendon, J. R., Briggs, G. A. D. (1988). In: *Recent Developments in Surface Acoustic Waves* (Parker, D. F. and Margin, G. A., eds.), pp. 309–16. Springer, Berlin.
23. Knauss, D., Bennink, D. D., Zhai, T., Briggs, G. A. D., Martin, J. W. (1993). Depth measurement of short cracks with an acoustic microscope. *J. Mat. Sci.* **28**, 4910–17.
24. Zhai, T., Bennink, D. D., Knauss, D., Briggs, G. A. D., Martin, J. W. (1993). Depth measurement of short cracks in perspex with the scanning acoustic microscope. *Materials Characterization* **31**, 115–26.
25. Silk, M. G. (1979). The transfer of ultrasonic energy in the diffraction technique for crack sizing. *Ultrason.* **17**, 113–21.
26. Bennink, D. D., Knauss, D., Zhai, T., Briggs, G. A. D., Martin, J. W. (1992). In: *Acoustical Imaging,* vol. 20 (Y. Wei, ed.), Plenum Press, New York.
27. Müller, W., Bubeck, E., Gerold, V. (1986). In: *Aluminium-Lithium Alloys III,* Proceedings of the 3d International Aluminium-Lithium Conference, London, pp. 435–41.
28. Davidson, D. L. (1988). Small and large fatigue cracks in aluminum alloys. *Acta Metall.* **36**, 2275–82.
29. Thompson, R. B., Skillings, B. J., Zachary, L. W., Schmerr, L. W., Buck, O. (1982). In: *Review of Progress in Quantitative Nondestructive Evaluation,* vol. 2 (D. O. Thompson and D. E. Chimenti, eds.), pp. 325–41, Plenum Press, New York.
30. Thompson, R. B., Fiedler, C. J., Buck, O. (1983). In: *Nondestructive Methods for Material Property Determination* (C. O. Ruud and R. B. Thompson, eds.), pp. 161–70. Plenum Press, New York.
31. Navarro, N. (1988). Short and long crack growth: a unified model. *Phil. Mag.* **A 57**, 15–36.
32. Navarro, A. and d. l. Rios, E. R. (1992). Fatigue crack growth modelling by successive blocking of dislocations. *Proc. R. Soc. London A* **43**, 375–90.

Appendix. Mixed-Mode Face Reflection

The mixed-mode face reflection is a combination of longitudinal waves traveling to the crack face and shear waves, reflected by the crack, traveling back to the lens. For the shear or longitudinal face reflection signals, normal incidence on the crack face was assumed. For the mixed-mode reflection however, incidence and reflection are not normal to the crack face, which makes calculating the transit time slightly more difficult. The first step in calculating the time-of-flight for the mixed-mode face reflection is to set the distance y_1 of Ray 1 and then to calculate its incidence angle θ_1 (see Fig. A.1). By applying Snell's law to Ray 1 at the water–sample interface, the angle θ_2 of the longitudinal Ray 2 in the material can be calculated. This ray is reflected at the crack face as a shear ray, again obeying Snell's law. The incidence angle δ_2 of the longitudinal ray on the crack face is given by the crack angle β and the angle θ_2. The δ_3 is calculated from Snell's law. Finally angles θ_3 and θ_4 can be calculated.

The second step is to calculate the distance y_4 of Ray 4 from the given lens position y, the chosen position y_1, and the calculated angles. The y_4 has to satisfy the condition

$$y_4 + z \cdot \tan\theta_4 = 0 \qquad (z < 0 \qquad 0^\circ \leq \theta, \beta \qquad \delta \leq 90^\circ)$$

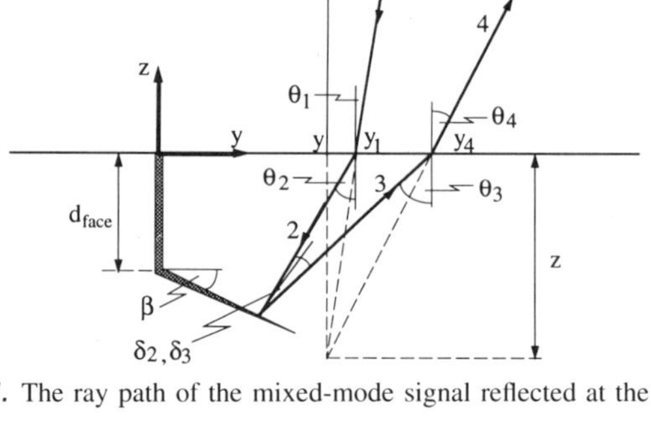

Figure A.1. The ray path of the mixed-mode signal reflected at the crack face.

In practice the calculation follows these steps.

1. Set y_1.
2. Calculate θ_1 and all other angles, including θ_4.
3. Determine y_4.
4. Compare y_4 with $z \cdot \tan\theta_4$.
5. Change y_1 until $y_4 + z \cdot \tan\theta_4 = 0$ $(z < 0)$.

Equations for the angles are

$$\sin\theta_2 = \frac{v_L}{v_o} \sin\theta_1$$

$$\delta_2 = \theta_2 - \beta$$

$$\sin\delta_3 = \frac{v_s}{v_L} \sin\delta_2$$

$$\theta_3 = -\delta_3 + \beta$$

$$\sin\theta_4 = \frac{v_o}{v_s} \sin\theta_3$$

The distance y_4 is given by

$$y_4 = -(y \sin\beta + d_{face} \cos\beta) \frac{\tan\delta_2 + \tan\delta_3}{\sin\beta \tan\delta_3 + \cos\beta} - y_1 \frac{\sin\beta \tan\delta_2 - \cos\beta}{\sin\beta \tan\delta_3 + \cos\beta}$$

Finally the time-of-flight for a pulse traveling along the mixed-mode ray path is given by

$$t_{FR}^{M} = t_1 + t_2 + t_4 + t_3 = \frac{1}{v_o}\left(q + \frac{z}{\cos\theta_1}\right) + \frac{1}{v_L} \frac{[d_{face} \cos\beta + (y + y_1) \sin\beta]}{\cos\delta_2} +$$

$$\frac{1}{v_o}\left(q + \frac{z}{\cos\theta_4}\right) + \frac{1}{v_s} \frac{[d_{face} \cos\beta + (y + y_4)\sin\beta]}{\cos\delta_3}$$

For the shear wave face reflection for example, δ_1 and δ_3 become zero, $y_1 = y_4 = -z \cdot \tan\theta_1$, and the time-of-flight simplifies to

$$t_{FR} = \frac{2}{v_o}\left(q + \frac{z}{\cos\theta_1}\right) + \frac{2}{v_s}[(-z \tan\theta_1 + y)\sin\beta + d_{face} \cos\beta]$$

which is identical to Eq. 5 for the shear wave face reflection.

3

Probing Biological Cells and Tissues with Acoustic Microscopy

Jürgen Bereiter–Hahn

3.1. Introduction

Acoustic microscopy provides unique possibilities for probing mechanical properties of cells and tissues. Two types of microscopes are presently used and commercially manufactured, the scanning laser acoustic microscope (SLAM) operated at about 100 MHz and the scanning acoustic (SAM) microscope operated in reflection mode. In the SAM frequencies up to 2 GHz allow lateral resolutions in the submicrometer range (see Fig. 3.1) and in the direction of the acoustical axis (z-axis) resolution of 30–50 nm is easily achieved. Nevertheless applications of this method in cell biology and histology are still very limited. Very few laboratories are equipped with only an acoustic microscope. The reasons for this situation lie primarily in difficulties of image interpretation and in uncertainties whether the characterization of mechanical properties of cells will be of general significance according to modern cell biology. Shear induced on the specimens by the scanning motion of the acoustical lens and the high sensitivity of the method to variations in surface topography (specimens must be attached to a very flat surface) impose further limitations.

JURGEN BEREITER–HAHN • Cinematic Cell Research Group, Zoological Institute, Johann Wolfgang Goethe University, Frankfurt am Main, Germany

Advances in Acoustic Microscopy, Volume 1, edited by Andrew Briggs.
Plenum Press, New York, 1995.

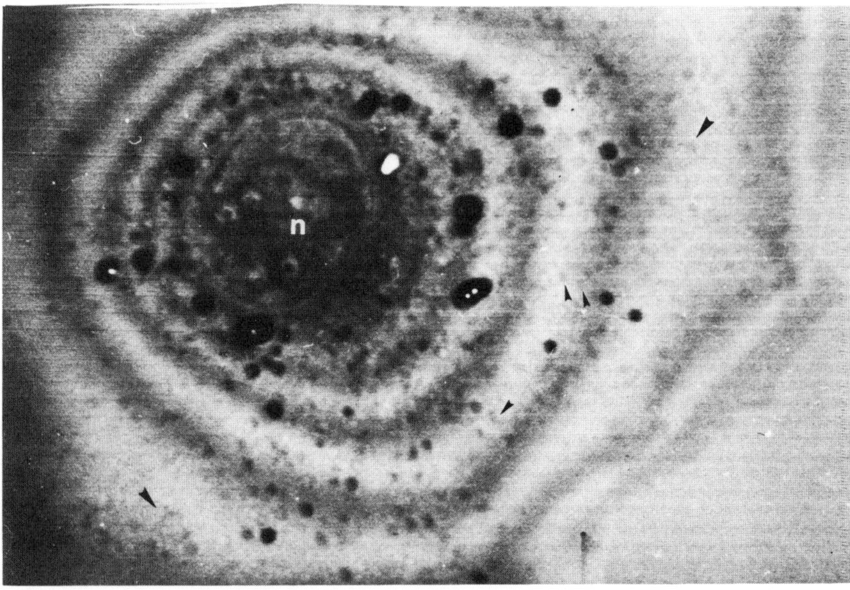

Figure 3.1. A SAM image of a well-spread XTH-2 cell on glass; 1.2 GHz: numerous small granular inclusions (black) and organelles (grey particles, arrowheads) are resolved. The nucleus (n) is clearly delineated, and some nucleoli are visible. Cell thickness is marked by the interference fringes; image width: 100 μm.

The SAM images of cells and tissues require careful interpretation. Therefore several methods have been developed to extract the information contained in SAM pictures by means of mathematical calculations. Because of several unknowns when determining a SAM image of a cell, assumptions have to be made, and hypotheses underlying the equations have to be simplified. Extreme simplifications may lead to procedures that visualize relative changes in the mechanical properties of cells, most often sufficient for physiological interpretations. The development of such semiquantitative methods must precede detailed calculations and mathematical modeling, otherwise the reliability of a simplified approach remains uncertain.

Chapter 3 summarizes the possibilities offered by SAM (and SLAM) to cell biologists and expands the perspectives outlined by Briggs.[1] Methods easily applied with widespread image analysis equipment should stimulate and further encourage acceptance of acoustic microscopy in biology and medicine.

3.2. Probing Cells in Culture

Cells in culture can be grown in suspension or attached to a plain solid, either the surface of glass or the surface of plastic material. *In situ* the frequently used fibroblasts span between a fibrillar matrix, or the leukocytes squeeze through narrow spaces and crawl along the collagen fibrils in the connective tissue. Epithelial cells are the only cell type that *in situ* has a chance to reside on a flat surface (basal lamina). Nevertheless, most of our knowledge of cell behavior, its control and physiology, has been derived from studying the artificial situation of cell cultures. The same holds true for mechanical forces determining cell shape. In sparse cultures, endothelial and epithelial cells tend to spread, forming a basal layer of actin fibers reinforcing the basal plasma membrane adhering to glass or plastic surfaces (see Fig. 3.2). The central cytoplasm enclosing the nucleus and most of the organelles arches up on this basal layer. In some cell types, a typical fried egg shape with a sharply separated cell center and periphery may result, while in others the cell center continuously tapers toward the periphery (see Fig. 3.3). If cells are moving or connected with adjacent cells, an asymmetrical shape arises. These differences have to be considered when comparing reports on the subcellular distribution of longitudinal sound velocity (v_c) and attentuation (see Refs. 2–5). Cell shape always results from interacting mechanical forces, therefore shape-specific force distribution has to be expected. In addition cytoplasm is motile, so cell shape is dynamic and depends on internal and external factors influencing force generation. The rate of changing form and the speed of cell locomotion are restricted by adhesion to the environment and by the viscosity of the cytoplasm.

The SAM measures v_c, attenuation, cell thickness (topography), and reflectivity of sound. The approximate cell shape (topography) is immediately obvious

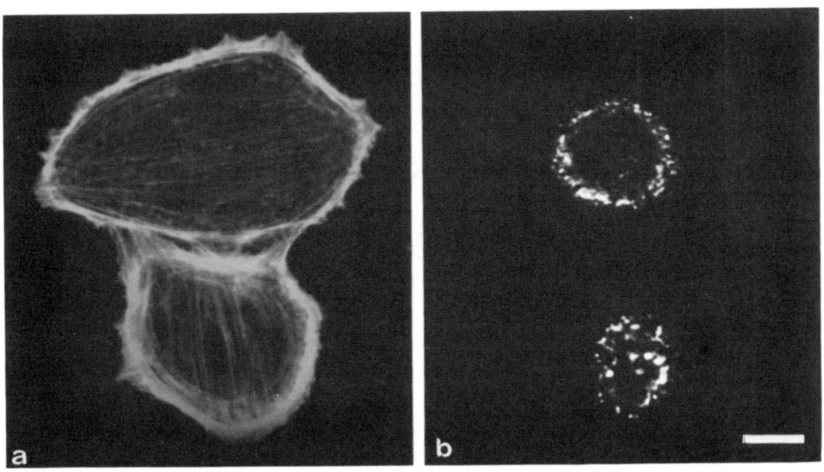

Figure 3.2. An Actin distribution in two XTH-2 cells as revealed by TRITC-phalloidin fluorescence shown with a confocal laser-scanning microscope (CLSM, TCS®, Leica). (a) Basal fibrillar actin layer (close to the supporting glass). (b) Optical section of the same cells through the dome-shaped central cytoplasm; F-actin is present in the cortical zone beneath the membrane only. Bar: 10 μm.

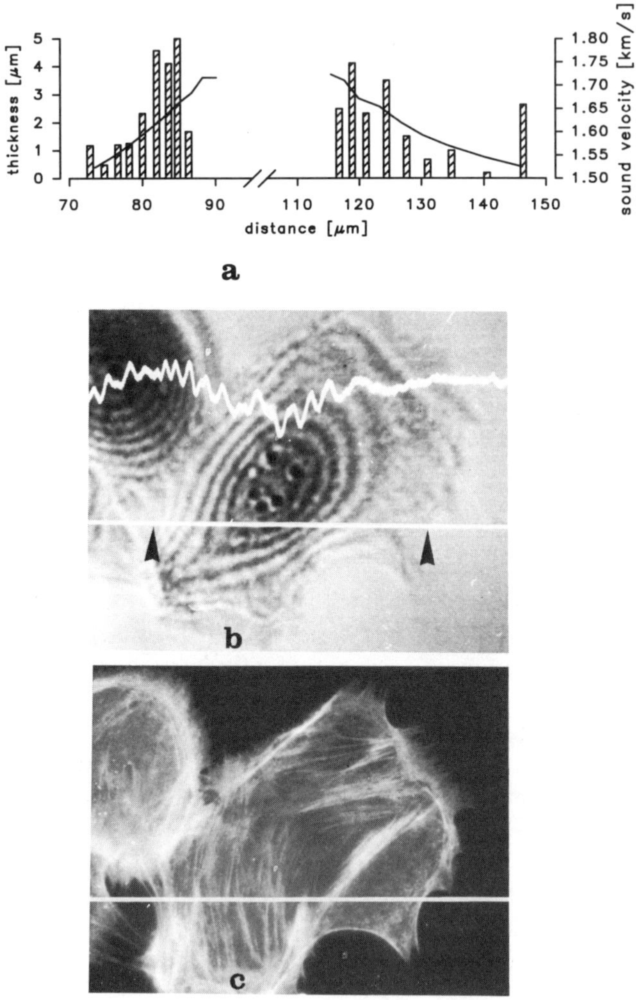
thickness [μm]
0
1
2
3
4
5
sound velocity [km/s]
1.50
1.55
1.60
1.65
1.70
1.75
1.80
70
80
90
110
120
130
140
150
distance [μm]
a
b
c

from the pattern of interference fringes in SAM images. These interference fringes delineate zones of equal acoustical path lengths between reflecting boundaries of the cell/culture medium and cell/substratum. Local variations in v_c shift the position of an interference fringe; therefore their pattern is only a first approximation of surface topography.

Although attenuation is the parameter most intensively studied in tissue sections and macromolecular solutions (e.g., Ref. 6), on the subcellular level, interpretation of attenuation in terms of physicochemical properties of cytoplasm is almost impossible. Viscosity of the cytoplasm is only one factor responsible for the attenuation of sound; scattering on rough surfaces is another. Cross linking cytoplasmic proteins during chemical fixation is revealed by an increase of attenuation, which can be assumed to reflect higher viscosity.[7,8] On the other hand, fixation may induce phase separation between proteins and their hydration shell, thus enhancing scattering.

One of the basic assumptions made when interpreting SAM images of cells is the homogeneity of cytoplasm. In fact this is not valid because of the distribution of cytoskeletal elements, granules, vacuoles, and the endomembrane system. Vacuoles or granules exceeding the limits of the lateral resolution of SAM can be excluded from calculations because they are easily seen; whether the endomembrane system contributes sound-reflecting boundaries seems to depend on the physiological state of the cells; i.e., in some cells mitochondria are not shown by SAM, although lateral resolution is sufficient, while in others they can be seen.

For soft tissues calculations of impedance differences from the acoustic signal is relatively simple because the contribution of surface waves and Rayleigh waves can be neglected (see for instance Ref. 1, Chap. 9). Intracellular sound velocity and the reflectivity of a cell boundary are closely related. The reflectivity coefficient is determined by impedance differences at the boundary. As previously mentioned, v_c may vary along the acoustical path inside a cell as a consequence of distinct areas filled with cytoskeletal elements of various functional states. Basic assumptions about the physical state of cytoplasm, whether it is treated as a viscoelastic fluid (the widely accepted model) or a quasi-solid, fibrous (or

Figure 3.3. Actin distribution in a Xenopus tadpole endothelial heart cell (XTH-2) compared with sound velocity distribution. (a) Thickness (*continuous line, left ordinate*) and longitudinal sound velocity (*bars, right ordinate*) along the scanning line through the XTH-2 cell marked by arrowheads in (b). In the cell center, no measurements are possible; therefore a gap is left in (a). In fact the slope of the cell is very low; ordinate and abscissa scalings differ by a factor of 10. (b) The SAM image of two living XTH-2 cells taken at 1 GHz (ELSAM®, Leica, Wetzlar, F.R.G.). Scanning line and the intensity distribution along this line are included. The arrowhead marks the range shown in (a). (c) Distribution of F-actin in the same cells shown in (b). The F-actin in the basal cell region was visualized by TRITC-phalloidin fluorescence. The white line indicates the scanning line in (b).

porous) matrix filled with fluid, influence the interpretation that bulk modulus or Young's modulus determines the longitudinal sound velocity according to

$$v_c^2 = \frac{E}{\rho} \tag{1}$$

where v_c is the longitudinal sound velocity in cytoplasm, E is the appropriate modulus of elasticity, and ρ is the density.

Cellular elasticity depends largely on the strain rate; in slow deformation processes, viscous creep may become prominent (e.g., see Ref. 9). In the SAM, we can assume an extremely high strain rate and an extremely small deformation (in the nm range). Thus impedance values exclusively reflect elastic stiffness or Young's modulus, and the viscous properties of cytoplasma are reflected by the attenuation of ultrasound. All these problems concerning cell shape and force distribution, stated theoretically, the question which of the main components of the cytoskeleton affects sound velocity in the cytoplasm, and the modulus of elasticity can be approached experimentally.

3.2.1. Observations on Living Cells in Culture

The XTH-2 cells are a line derived from Xenopus tadpole heart endothelia, which are our pet cell because of their easily handling. According to the method developed by Litniewski and Bereiter–Hahn,[10] which is based on the sum and difference of amplitudes in adjacent constructive and destructive interferences, sound velocity was calculated along the scan line shown in Figs. 3.3(b) and (c) and plotted in Fig. 3.3(a). A typical pattern for cells of this shape arises: The margin on the right [148 μm in (a)] exhibits a relatively high sound velocity. This coincides with an accumulation of F-actin [see Fig. 3.3(c)], which can be assumed to be taut. The following thin area (moving left) is almost devoid of F-actin, and the sound velocity is almost the same as in water (or saline); around 135 μm the scan line passes an actin cable, and sound velocity slightly increases and is high in the ascending part of the dome-shaped central cytoplasm. No flat lamella region is developed at the left side of the cell in contact with a neighbor. Here sound velocity drops almost continuously from the cell center to the periphery. In the very center no interference fringes can be identified, so no measurements are possible [see the gap in Fig. 3.3(a)]. In the lamella zone, sound velocity changes with the distribution of actin fibrils. No comparable relationship could be established for vimentin and microtubules. In the central area (from 130–75 μm), no correlation exists between sound velocity and the actin pattern in the basal region of the cell [see Fig. 3.3(c)]. The subplasmalemmal actomyosin layer can be assumed to be the main structure responsible for the impedance difference at the medium–cell boundary. The reflectivity at this boundary modulates reflections from the cell–substrate interface. The method of calculating impedance differences between two media on the basis of amplitudes in the maxima and minima of the interference fringes is based on the assumption of constant v_c throughout the cytoplasm along the path of longitudinal wave propagation. This requires homogenous distribution of the cytoskeletal elements determining the acoustical properties of cytoplasm. Time-resolved SAM (see Ref. 1, §2.1.3) and confocal laser-scanning microscopy unambiguously showed that this condition is not fulfilled (see Fig. 3.2): The upper and the lower cell surface (surface includes the plasmamembrane with the intimately associated fibrillar layer) exhibit higher impedance than the cytoplasm in between these cortices (see Fig. 3.4).[11,12] The presence of the cortical fibrillar layers itself is a strong indication of inhomogenous properties of cytoplasm. At the thin periphery of the lamellar zone the distinction between an upper and a lower cortical fibrillar layer no longer can be made. Therefore, in this region, variations in sound velocity coincide qualitatively with the distribution of actin fibrils. Contractions in the cortices at the flanks of the central cytoplasm can be assumed to underly the change in slope from the flat area to the dome shaped center. These contractions develop tension in the cortex and thus increase its elasticity which according to Eq. (1) results in an increase in sound velocity.

A simple experiment supports this interpretation. Actomyosin-driven contractions depend on the presence of Ca^{2+} (in the μmolar range). High concentrations of Ca^{2+} (mmolar) induce disassembly of all the fibrillar structures (F-actin and microtubules, collapse of intermediate filaments). These reactions can sequentially be evoked by addition of ionomycin (1μmol/l), a substance which renders the membrane permeable specifically for Ca^{2+}. 30 s after addition of the ionophore, there is an increase in the sound velocity almost everywhere in the cell followed by a decrease to very small values (slightly higher than those of dissolved protein in saline). Actin bundles survived a 4 min treatment only in the very periphery with a velocity of about 1.65 km s^{-1} (see Fig. 3.5).

Figure 3.4. Vero cells grown on soft PVC (almost no reflection of acoustic waves with saline used as a coupling medium) x - z scan (B-scan) through two cells by time-resolved SAM. Strong reflections occur at the upper surface of the cells and at their bottom attached to the plastic. Central parts (containing the nucleus) appear black, indicating a lack of sound reflecting boundaries. (Reproduced from Ref. 12)

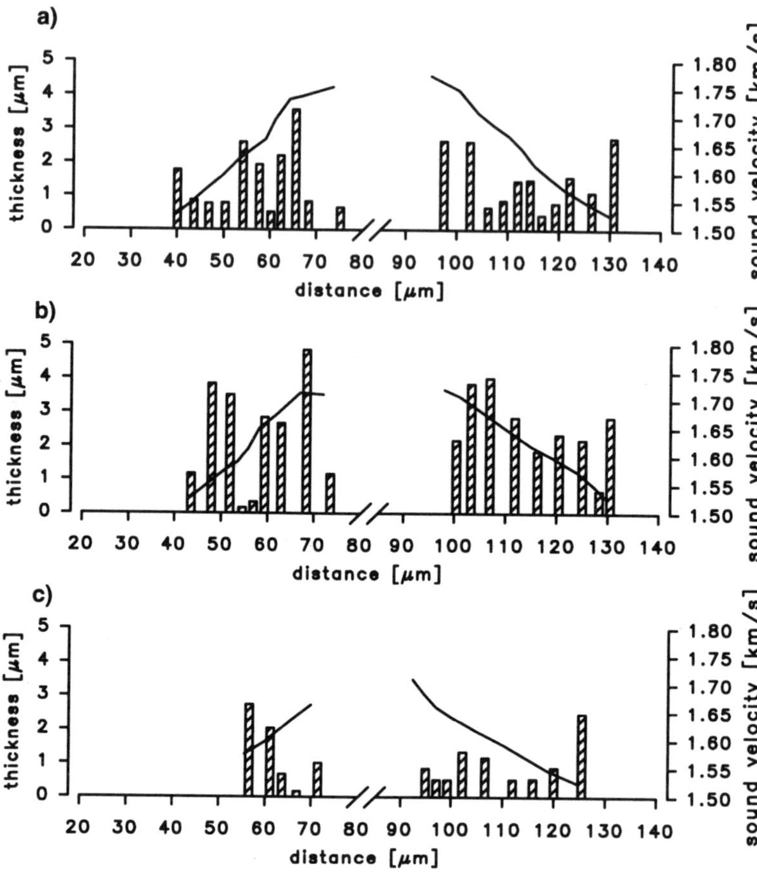

Figure 3.5. Influence of calcium influx in XTH-2 cells mediated by the ionophore ionomycin action on sound velocity. The display corresponds to that in Fig. 3.3. (a) Before adding ionomycin. (b) After a 30-second incubation, an increase of sound velocity occurs that is most prominent in the cell periphery. The thickness of the cell is reduced. (c) After 4 minutes of incubation, sound velocity in the cell body was decreased, whereas in the very periphery, values change only slightly (Ref. 5).

3.2.2. Interpreting SAM Images of Cells Depends on the Model for Cytoplasmic Organization

These observations are highly indicative that in these objects sound velocity reveals mechanical tension produced by contractions of the fibrillar elements of the cytoplasm. Another interpretation is that variations in sound velocity variations reflect changes in the compression modulus, probably due to alterations in the monomer–polymer state of the cytogel. In preliminary measurements of sound velocity (at 500 MHz) during polymerization of actin (1.12–5.6 mg protein per ml), an increase in sound velocity of about 6 ms^{-1} was found. This increase is significant but very small compared to the high velocity values measured in cells.

If a cytoplasmic gel is composed of a fibrillar quasi-solid component (exhibiting elasticity) filled with saline and mono- or oligomeric proteins (exhibiting bulk modulus), the sound velocity v_c in Eq. (1) can be assumed to be composed of two virtual velocities, the velocity in the saline and the velocity in the fibrils. Each is weighted with the volume fraction of the fibrils and the fluid phase. This corresponds to physical treatments of porous media (e.g., see Ref. 13). The diameter of components of the fibrillar layer (described in Ref. 14a as microtrabecular lattice) ranges from 7–30 nm; the gap system in between may be 50–100 nm, far smaller than the wavelength of 1-GHz ultrasound. Another reason why a sound wave of this frequency is not sensitive to boundaries between two components is the hydration shell and a halo of loosely attached molecules around the fibrillar layer that prevent the formation of sharply defined interfaces. These restrictions do not however exclude the introduction of two quasi-different velocities, one related to bulk viscosity and the density of cytosol, the other related to elasticity and the density of the polymeric proteins.

$$v_c = v_s\,(1 - u) + v_f\,u \tag{2}$$

where u is the volume fraction of the fibrillar component.

According to Eq. (2), the virtual sound velocity in the fibrillar phase can be derived and then used for a rough estimation of Young's modulus as revealed from a measured mean sound velocity determined by one of the SAM evaluation methods. Assuming the volume fraction of the fibrillar phase to be 0.1 (this fraction can be determined from electron micrographs), v_s to be 1509 ms^{-1} (the corresponding volume fraction 0.9), and the density of the fibrillar structures to be 1.1 kg dm^{-3}, an elasticity modulus of the fibrillar phase of 7.7×10^2 Nm^{-2} is calculated, corresponding to an overall sound velocity V_c of 1650 ms^{-1} (a value measured typically in cultured cells). The 1900 ms^{-1} lies in the high region of measurements in cultured cells, corresponding to an elastic modulus of 2.7×10^3 Nm^{-2}. Both these values are in the range reported for stresses developed by cells and cellular structures (see Ref. 9). Whether a cell exhibits passive elasticity or develops stresses by active contractions does not make a difference for our calculations.

Calculating Young's modulus (Y) on the basis of a continuum model requires knowledge of the Poisson ratio (σ) of cytoplasm, density (ρ), sound velocity in the cytoplasm (c) (from Ref. 14):

$$Y = g(\mu)\,\rho c^2 \tag{3}$$

where

$$g(\mu) = \left|\frac{(1 + \sigma)(1 - 2\sigma)}{(1 - \sigma)}\right|^2 \tag{4}$$

The Poisson ratio of cytoplasm is not known, and it is difficult to determine. Different ratios are to be expected according to whether the cytoplasm is in a relaxed or contracted state and these ratios may depend on the spatial arrangement of the force-generating elements. Therefore no direct comparison is possible of values obtained by acoustic microscopy with elasticity values derived from various methods for evaluating mechanical properties of cells (e.g., Refs. 15–17), for instance aspiration (e.g., Ref. 18) or poking experiments (see Ref. 19), or most recently, atomic force microscopy (see Refs. 20–22) and ultrasonic force microscopy (see Refs. 23, 24). Thus at the present state of the theory of cytomechanics, the fluid model of cytoplasm does not allow interpretation of SAM data. Similar difficulties arise in interpreting the acoustic properties of suspensions, such as blood, which have been solved by modeling cells as viscoelastic shells enclosing some fluid (see Ref. 25).

3.2.3. *Acoustical Sectioning of Biological Specimens with Time-Resolved SAM*

Time-resolved microscopy was realized several times in the past,[26–29] and it was used to characterize soft and hard biological tissues.[30,31] The short temporal extent of the impulse excitation allows separate echo pulses from the top and bottom of a thin specimen, or from separate interfaces in a layered specimen, to be identified in the received signal; hence the vertical structure in specimens can be resolved. The ultrasound microscope built by Kannigiesser[11,12] extends these earlier developments and opens new applications in biology and medicine. Significant features are the variation of the length of the acoustic pulse over a wide range, high scanning speed (up to three images per second), and most important the ready adaptation to an inverted light microscope. The microscope is essentially a reflection-type SAM of the type realized by Lemons and Quate[32,33] or the time-resolved SAM built at Oxford;[34,30] thus it is a confocal system. Just as a confocal laser-scanning microscope can be used for optical sectioning, such a system provides the possibility of acoustical sectioning, which has been used for instance in waver inspection (e.g., Ref. 35). High reflectivity of the supporting surface combined with low reflectivity of specimens and attenuation by samples normally impede this type of imaging of soft biological samples.

Cells grown on a polyvinyl-chloride surface that almost matches the acoustical impedance of the coupling fluid (reflectivity is 18 dB lower than the reflection of the surface of a polystyrene petri dish) were viewed using ultrasound pulses in the nanosecond range (see Fig. 3.4). This method permitted acoustical sectioning and thus three-dimensional imaging and x/z-scans (B-scans) with a resolution of 1.24 μm in the axial direction (0,54-μm lateral resolution at 1.5 GHz). If echoes in time-resolved SAM are adequately separated in time from the height and position of each maximum in the echo, four crucial parameters can be separated—the difference in time between the reference signal and the reflection from the top layer, the difference in time between the interface of the top surface and the substrate facing surface of a cell, and the relative amplitudes reflected from the top and the substrate interface of the cell. From these parameters, cell thickness, acoustic velocity in the cell, impedance, attenuation, and density can easily be calculated. However in most cases observing cells in culture, signals are too close together to not separate adequately to follow the procedure just outlined. In these cases, expression of signals in the frequency domain and working with their Fourier transforms allow cepstral filtering of the signals and maximum entropy analysis (for details, see Ref. 34), yielding very accurate difference values that can be used to determine further the acoustic properties of the structure under investigation.

The main advantage is that analysis of time-resolved measurements does not require a priori assumptions about the acoustic properties of the specimen. As in methods using phase differences in the acoustical signal, limitations of the method arise from the eveness of the scanning motion and the supporting substrate.

3.2.4. Determining Cell Volume

Determining the volume of cells that are not spherical is very difficult. Suspending spread cells may change their physiological reactions, thus investigations on cell volume regulation or the volume effects of drugs must be based on methods allowing the volume determination of undisturbed cells. The pattern of interference fringes delineates the topography of a cell in SAM images. The difference between the maximum and an adjacent minimum of a fringe corresponds to a thickness difference of $\lambda/4$ (a quarter-wavelength of the longitudinal acoustic wave in cytoplasm). Thus a cell can be partitioned into disks $\lambda/4$ thick, whose area can be determined by planimetry using an image analysis system. Cell volume is given by the sum of the volumes of all these disks.[36] The λ is derived from sound velocity in cytoplasm (v_c), v_c/f (f = frequency). This method was adopted from light reflection microscope measurements.[37,38] Due to longer wavelengths, fewer fringes appear at 1.0 or 1.3 GHz than in light microscope images. If cells are grown on plastic (i.e., polystyrene) surfaces, the contrast is excellent, and fringes are easily identified all over the cell. This clearness of the images makes SAM superior to reflected light pictures, although shorter wavelengths in the light microscope theoretically allow higher measurement accuracy. In both methods, precision is limited by the identification of the different maxima and minima, rather than by step size (thickness of each of the disks) or by small deviations of v_c.

3.2.5. Reliability of Different Methods for SAM Evaluation of the Mechanical Properties of Cytoplasm

Evaluating mechanical properties of cells by SAM inevitably requires some assumptions and simplifications about the ray path (i.e., the angular spectrum, multiple reflections, and surface acoustic waves) and the organization of the specimens (i.e., the number and extension of different thin layers, molecular constituents, isotropy, viscosity, and density). At least four unknowns have to be considered when investigating biological samples on a microscopic scale—thickness, density, sound velocity, and ultrasound attenuation. Model approaches with different degrees of simplification for the quantitative characterization of biological tissue sections have been compared by Akashi and others[(39)] and Kundu and others.[(40)] Uncertainty considerations have been published for laser acoustic microscope measurements of normal and wounded canine skin.[(41)] Briggs[(1)] critically evaluated the limitations of different procedures, I consider a few aspects of cell cultures which appeared after that book. Evaluating the accuracy* of a tissue section measurement can be undertaken by comparing microscopic methods with macroscopic determinations of sound velocity and attenuation of thick slices or whole organs.[(42)] On the subcellular level, this procedure is not possible. In this case, SAM investigations based on different evaluation or imaging procedures can be compared to reveal whether calculated parameters coincide with each other. The following procedures have been developed for evaluating the mechanical properties of cells:

Time-resolved SAM[(12,30,34)]

Frequency modulation method[(7,43)]

Continuous wave acoustic microscopy[(44,45)]

Phase-and-amplitude SAM[(46–48)]

Sum and difference of the amplitude of constructive and destructive interferences combined with positive defocusing[(10)]

$V(z)$-based analysis using
  Large z-defocusing ($\Delta z >$ acoustic wavelength in water)[(10,49–51)]
  Micro defocusing ($\Delta z \leq$ acoustic wavelength in water).[(2,14,40,52)]

Real measurements providing numerical data for sound velocity and attenuation on a subcellular level are rare. Most studies compare model calculations with signals obtained from cells or tissues. However these model calculations do not indicate whether the calculated values are inaccurate or correct (see also Fig. 3.6). No method has gained superiority over others in handling and reliability.

*Accuracy refers to the proximity of the measurement to the true value, precision, the reproducibility of a single measurement.

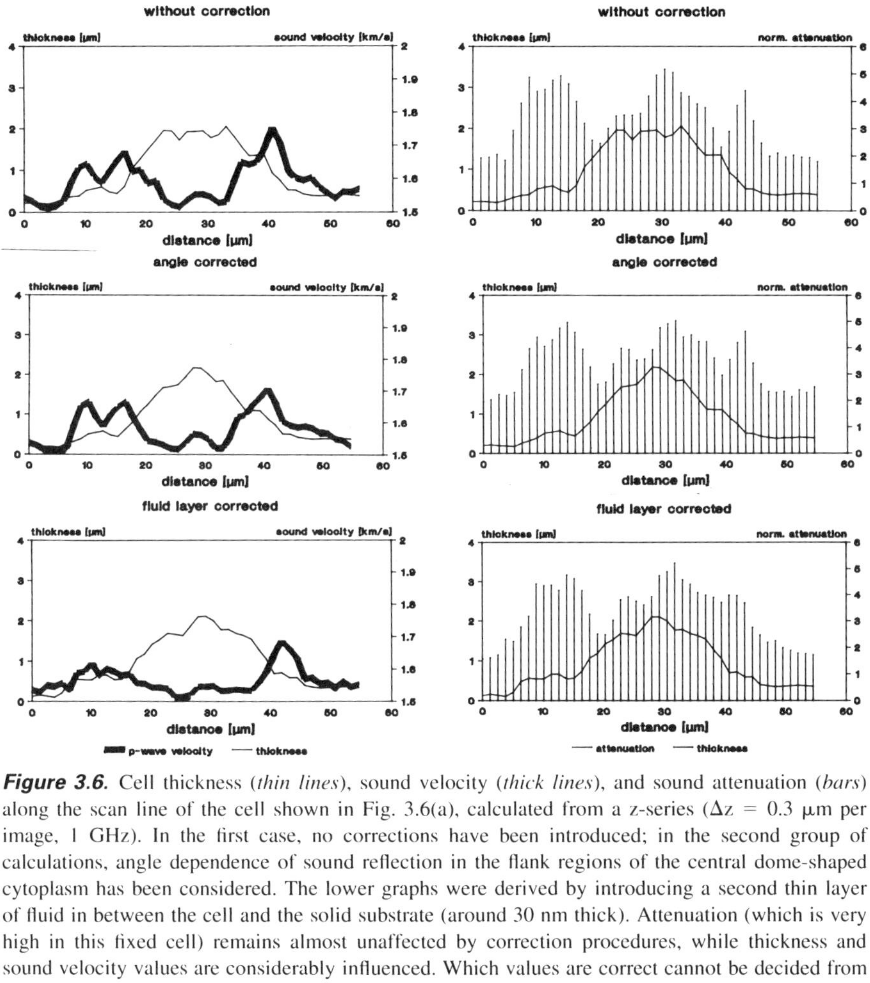

Figure 3.6. Cell thickness (*thin lines*), sound velocity (*thick lines*), and sound attenuation (*bars*) along the scan line of the cell shown in Fig. 3.6(a), calculated from a z-series ($\Delta z = 0.3$ μm per image, 1 GHz). In the first case, no corrections have been introduced; in the second group of calculations, angle dependence of sound reflection in the flank regions of the central dome-shaped cytoplasm has been considered. The lower graphs were derived by introducing a second thin layer of fluid in between the cell and the solid substrate (around 30 nm thick). Attenuation (which is very high in this fixed cell) remains almost unaffected by correction procedures, while thickness and sound velocity values are considerably influenced. Which values are correct cannot be decided from this series only. One method is to compare thickness values calculated from SAM images with those derived from RIC images (see Table 3.1). (from Ref. 40)

In general $V(z)$ procedures including frequency or focus variations [$V(z)$ curves] require several images to be taken before calculations can be made. This limits temporal resolution. Methods based on evaluating interference fringes provide high temporal resolution combined with a loss of spatial resolution, and they do not allow measurements of very thin cytoplasmic layers. Procedures taking advantage of the information in phase and amplitude of the acoustical signal are very sensitive to external disturbances (temperature, scanning plane, and vibrations), which pose great difficulty for practical work. Similar difficulties can arise using time-resolved SAM. On the other hand, if these problems are handled properly, they allow high spatial and temporal resolution. Time-resolved SAM is also a very powerful tool, but only a few observations have been published.

No real comparative studies have been published yet on applying different evaluation or imaging modes to the same cell. [For tissue sections, SLAM and SAM measurements have been compared with each other.][31] Therefore in the case of living cell investigations, values obtained for different specimens in different laboratories must be compared with each other. Considering the significance of cell type, cell shape, and culture density on its mechanical properties, only a range of values can be derived, and no certainty about the accuracy of the measurements exists.

Because of the high motility of living cells (cf. Section 3.2.6.4), comparative studies are restricted to chemically fixed cells or tissue sections. In both cases, it has to be ascertained that no further loss of material to the coupling fluid occurs during the series of measurements. A few examples of the coincidence of SAM images with light reflection microscopic (RIC)[53,97] observations have been published.[54] If applied to the same specimen, the RIC and SAM (Fig. 3.7), allow us to compare thickness measurements obtained by these two procedures totally independent of each other (see Refs. 40, 55). The SAM evaluation was

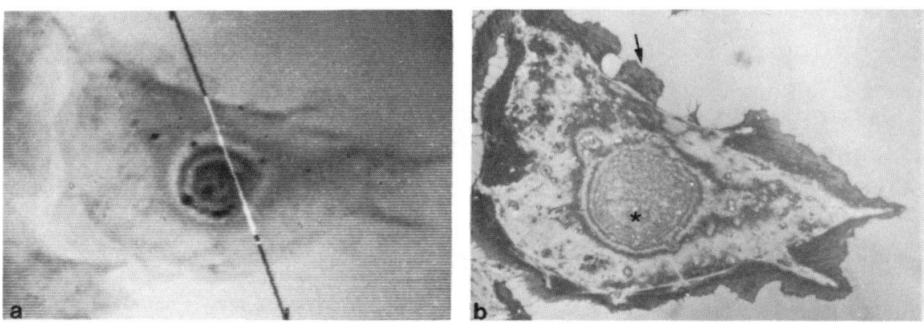

Figure 3.7. The single XTH-2 cell on glass (fixed with glutaraldehyde and paraformaldehyde) visualized with (a) SAM (1 GHz) and with (b) RIC using orange light (580 nm) and conical illumination (mean half-angle of incidence 21°), focus on cell/substrate boundary. In the RIC the very thin marginal lamellae appear dark; in the SAM image, these areas show very low contrast, which could be improved by growing the cells on plastic. As in SAM, RIC images result from interferences of light reflected at the top and the basal cell surface; thus very delicate thickness variations are revealed with good contrast. A general coincidence between the two images exists regarding topography. However many details differ because of different wavelength used and because different physical properties influence image formation. The appearance of granular inclusions (black in the SAM image) is one of the most obvious differences. These are almost invisible in the RIC image. The good contrast of the SAM image (*dark*) along the left margin of the cell has no clear counterpart in the RIC image, which is sensitive to topography, and not to the high elasticity in this region.

Table 3.1. Comparison of Thickness Values at Five Sites of a Cell as Determined from an RIC Image and a SAM Image Series [V(z)-series] either Correcting for the Presence of a Fluid Layer (Thickness 20–60 nm)

Site no.	RIC	SAM without correction	SAM corrected fluid layer
1	0.11	0.13	0.00
2	0.22	0.27	0.18
3	0.22	0.25	0.20
4	0.32	0.35	0.21
5	0.43	0.43	1.07

Source: From Ref. 55.

based on a $V(z)$ series of the cell using six images each with a 0.3-μm focus difference. Calculations were performed either (1) without further corrections according to Kundu and others,[14] (2) including a thin layer of fluid (ca. 30 nm) between the cytoplasm and the solid substratum, or (3) by correcting for influences of slope on the acoustic signal.[40] Whether the presence of such a layer is included in the calculations or not has a significant influence on values for sound velocity in the thin cytoplasmic areas in particular and to a minor extent for attenuation (see Fig. 3.6). In the thin peripheral region where the RIC measurement is very precise, matching values was best without correction factors (see Table 3.1).

3.2.6. Probing Cellular Dynamics

The reaction of cells permeabilized for calcium ions previously reported is an example of the dynamical behavior of cells. This example is helpful in understanding the value of different models of the interaction of cytoplasm with ultrasound. In this section, I present a qualitative approach to cell motility and the underlying mechanics as they become apparent by SAM.

3.2.6.1. Assessing Elasticity and Force Distribution

Let us assume that a cell is connected on one side to a neighboring cell and free on the other side (adhering only to a substrate). By intercellular contacts, these cells can be connected to each other, and the one with a free edge can move toward its neighbor cell [see Fig. 3.8(a)]. The most probable way of achieving this goal involves an asymmetric contraction moving the free cell closer to its neighbor. The cell in Fig. 3.8 moves about 5 μm to the right during a 3-minute interval. It has a gradient of sound velocity increasing from left to right [see Fig. 3.8(b)], revealing a gradient of elasticity. If we agree that cytoplasmic elasticity in these cells results from actomyosin-based contractions, this

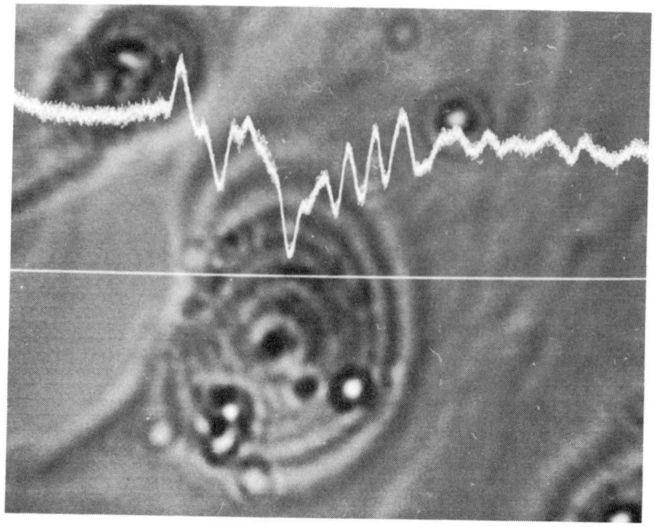

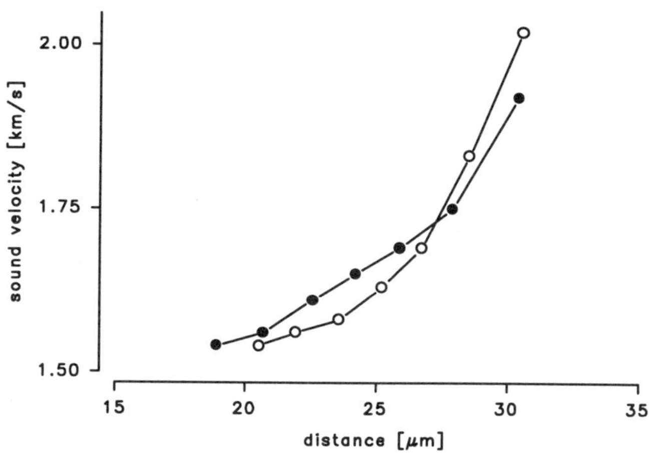

Figure 3.8. The SAM image of an XTH-2 cell on glass, taken at 1 GHz at focus level z = 12 μm (above the glass surface). Its left margin faces a cell-free area, while on its right side, the cell is in contact with another cell (*center, lower right corner*). The graph shows the distribution of sound velocity along the scanning line at time 0 (*open circles*) and 3 minutes later (*filled circles*). During this time interval, the cell moved to the right by about 5 μm. The continuous increase of sound velocity from the left to the right is interpreted to represent increasing mechanical tension in this direction, which is responsible for cell movement. This tension is counteracted by the cell adhesion to the glass. At the end of the short period of displacement, this gradient in tension became diminished. Image width: 100 μm.

gradient of sound velocity represents a gradient of contractile forces underlying the displacement of the cell toward its neighbor on the right.

3.2.6.2. Qualitative Demonstration of Force Distribution

Sound velocity in cells is related to elasticity and contractile forces. Problems in deriving these terms from primary data (velocity) have already been discussed. If biologists wish to understand cell shape and motile phenomena, the gradients and distribution of forces involved are more important in most cases than their exact value. Therefore I and other researchers developed a method of visualizing elasticity distribution in cells using classical image analytical procedures.[56,57] The method is based on considerations underlying calculations of sound velocity and attenuation from the sum and difference of adjacent constructive and destructive interferences. In SAM images of a wedge-shaped thin layer, the intensity differences between the minima and maxima of interference fringes, are closely related to impedance differences between the thin layer and the coupling fluid. They are insensitive to attenuation (see Fig. 3.9). The significant value is the amplitude (or intensity) *difference* between the minima and maxima, not the absolute values. This difference is given by Sobel-filtering images, thereby producing images where increasing grey-level differences in the original are represented by increasing brightness (grey levels). Inversion and the log (or the square) of this first derivative generate images with appropriate brightness and contrast: Large differences in interference fringes appear black on bright backgrounds. In these images, dark zones along interference fringes correspond to areas with high elasticity (see Fig. 3.10).

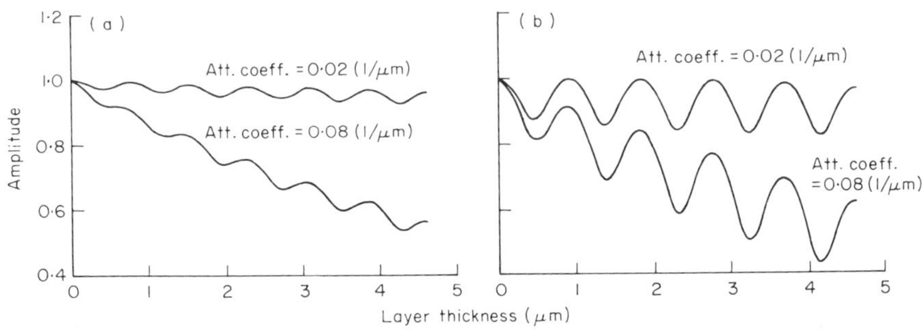

Figure 3.9. Calculating the amplitude of sound reflected from a solid surface covered by a thin layer on glass. The amplitude is plotted as a function of thin layer thickness for two different sound velocities (a) 1550 ms^{-1}, (b) 1850 ms^{-1}, and two attenuation coefficients (in the thin layer). A further assumption was normal incidence of plane waves. Changing the attenuation coefficient alters the average signal, while the variation of velocity in the layer influences the amplitudes of interference maxima and minima. (From Ref. 10)

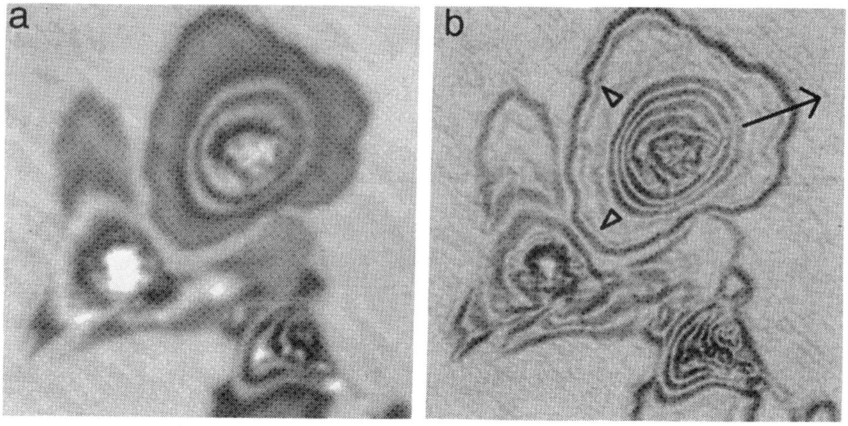

Figure 3.10. Visualization of a subcellular force distribution. (a) Original SAM image (1 GHz) of a Xenopus tadpole epidermis cell in culture on polystyrene (better contrast than on glass). This cell collided with other cells at its lower and left margin, then changed its direction of locomotion, as marked by the arrow in (b). Such a change in locomotion direction should require reorganization of contractile events driving cell motility. This is shown in (b). (b) Results from the first derivation of (a) (Sobel filtering), yielding an intensity-independent measure of contrast; inversion of the filtering result and logarithmic display. Dark zones indicate areas of high contrast caused either by the interference pattern or scattering at edges. (Both phenomena are related to acoustical impedance; see for instance Fig. 3.9). Darker zones appear toward those areas where the cell has made contact previously, while the area where the cell started to extend is relatively bright and of low contrast, indicating a lack of tension (low sound velocity). (This image has been kindly provided by H. Lüers, Cinematic Cell Res. Group, University Frankfurt). Width of single image: 100 μm.

An important advantage of this method is its speed. It needs only single images [not a set of SAM images as in the $V(z)$-based methods], and it can be performed on stored pictures; thus it does not interfere with sampling frequency. For better contrast, cells should be grown on plastic instead of glass. A series of images transformed in the way described provides an easy understanding of cell locomotion caused by the flow of cytoplasm extending a lamella of a keratinocyte in culture (see Fig. 3.11). No comparable method exists for visualizing the mechanical basis of cellular behavior as done by SAM.

3.2.6.3. Investigating Mitosis Using SAM

Mitosis represents one of the most fascinating events studied in cell biology. Mechanical forces are required for all events during this distribution of genetic material to two daughter cells: The nuclear membrane is spread at the end of prophase, then chromosomes are aligned in the equatorial plate, a process guided by the interaction of chromosomes with microtubules, and finally the chromatids separate and move toward the spindle poles. Generation of the driving force and

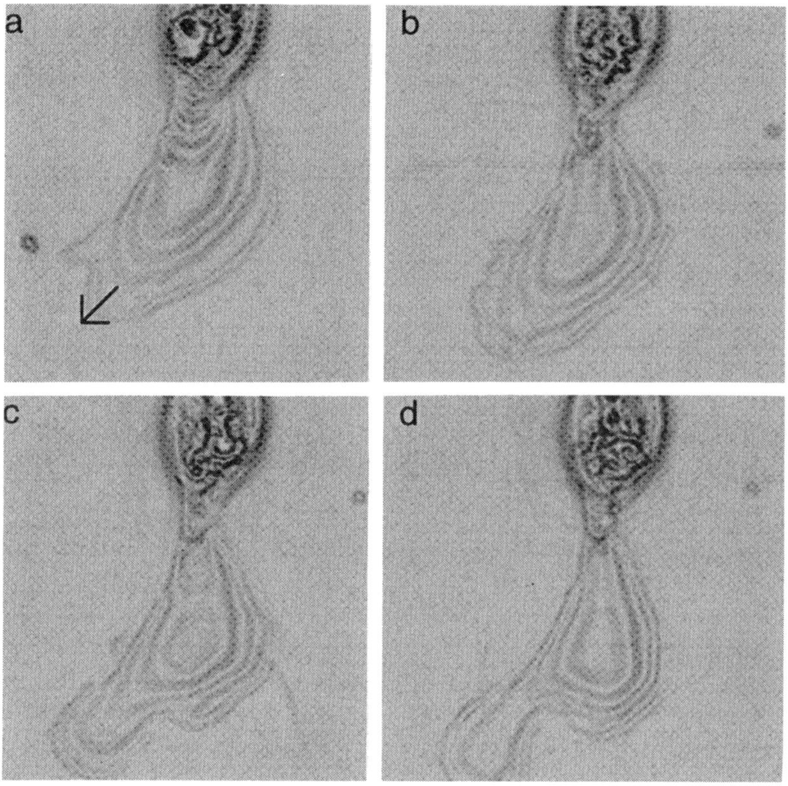

Figure 3.11. A time series showing the extension of the lamella of a Xenopus tadpole epidermis cell (keratinocyte) on plastic. Images have been transformed by the method described in Fig. 3.10; time interval between the single images: 30 s; width of a single image: 62 μm. This series was considered to contribute to our understanding of the mechanism of cell locomotion. One of the main questions is the site of motive force generation in the cell body (or at its periphery) or in the leading lamella. According to the model proposed by the present author, hydraulic pressure generated by contraction of the cortical actomyosin network beneath the membrane provides the force for the extension of the lamella. Following this model, the lamella should be extended at sites where its elasticity is lower than at other positions. Competing models assume extension of the lamella by pressure generated at marginal parts of the lamella, thus at sites of high elasticity (for review see Ref. 57). Fringes representing the contrast function of interferences in orginal SAM images are very weak on the left lower edge of the lamella in (a) and (b), where it is going to become extended, as seen in (c) and (d). The innermost fringe in (b) is also very weak (brightness is close to that in the environment), indicating the mechanical weakness of this area. In the two following stages, an extension of this area and thickness increase occur. Perception of these contrast differences can be improved by pseudocoloring. (Courtesy of H. Lüers, Cinematic Cell Research Group, University Frankfurt/M)

the role played by microtubules are still under debate. Cell division (cytokinesis) is the final event, which in animal cells is achieved by the contraction of a fibrillar circle organized beneath the plasmamembrane around the previous equatorial plane.

Only one investigation using SAM has been reported on this topic.[58] This may be due to difficulties in finding and identifying mitotic cells in the SAM. In the well-spread XTH-2 cells, the mitotic spindle and chromosomes are clearly discerned at 1.3 GHz (see Fig. 3.12). The spindle poles seem to differ in reflecting sound from other parts of the spindle [white arrows in Fig. 3.12(b,c)]. At the beginning of anaphase, chromatids separate [white arrowheads in Fig. 3.12(b,c)]. During the anaphase movement of the chromosomes toward the spindle poles, the attachment zones of the spindle fibers at the chromosomes [area marked by arrows in Fig. 3.12(a)] change their sound reflectivity (or attenuation), which has been interpreted by Lindner and others[58] as a decrease in local elastic modulus caused by microtubule depolymerization. Stabilization of microtubules by taxol however does not change the appearance of this zone. The main physiological significance of the observations is the change in mechanical properties of the spindle fibers in the immediate vicinity of the kinetochores.

Problems interpreting these pictures arise from the rounding of mitotic cells and from the presence of several boundaries—the upper and the lower cell surface, the upper and the lower surface of the mitotic spindle, and the chromosomes in the spindle. Thus, when using ultrasound pulses around 20 ns, interference fringes are generated. Their amplitude varies depending on the plane of focus (e.g., chromosomes appear either dark or bright, depending on focus). The SAM analysis of mitotic events can provide important new insights into the still unsolved problem of how chromosomes are moved; however many difficulties in interpreting images have to be overcome. Time-resolved SAM may be very helpful for investigating mitosis.

3.2.6.4. Motility Phenomena Revealed by Subtracting SAM

Motility designates a change of a structural parameter versus time, which in mathematics is expressed by differentiation. Therefore subtracting scanning acoustic microscopic (SubSAM) images taken at appropriate time intervals visualizes the spatial and temporal changes of mechanical properties of a cell being ascribed to motility.[59] Motility events result in local changes of intensity of the ultrasound signal reflected into the acoustic lens. This signal modulation may be due to changes in elasticity, topography, or attenuation. It is visualized by grey-level differences determined for each pixel in two subsequent images. The mean grey-value change per pixel over a cell therefore provides a measure for total motile activity. Differences increase with time in a cell-type-specific manner and finally reach a saturation value when further changes cancel each other or for

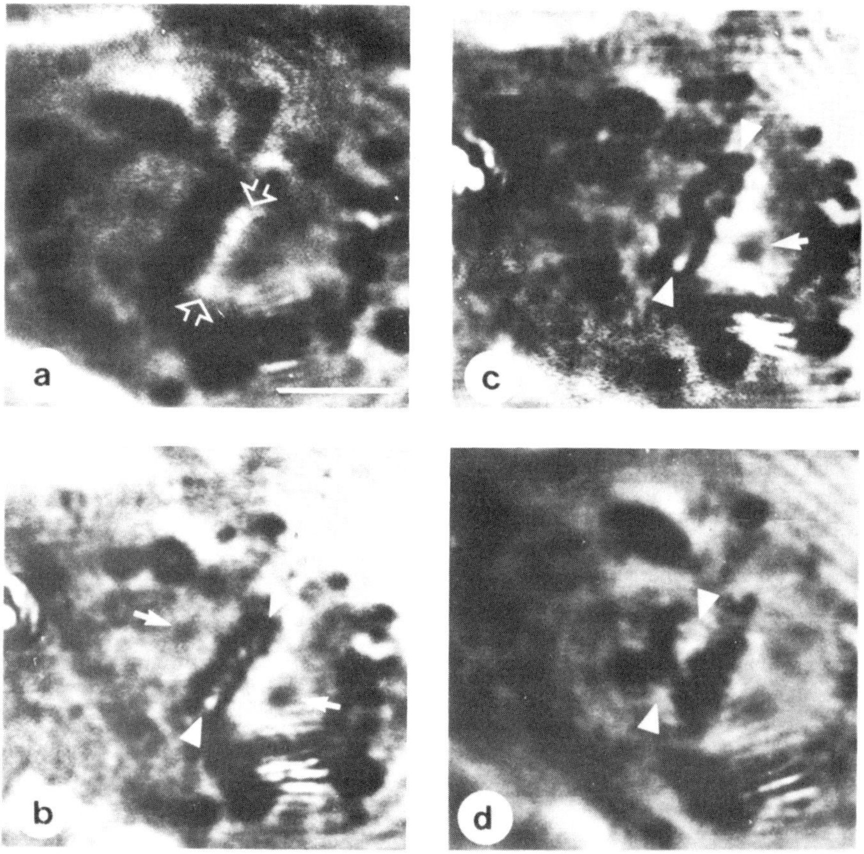

Figure 3.12. The SAM images of the mitotic spindle in XTH-2 cells during the metaphase–anaphase transition (1.3 GHz, ELSAM®; Leica, Wetzlar). (a) Metaphase. Arrows point to a spindle zone in immediate contact with the chromosomes (close to the kinetochores), which are parallel to the equatorial plane and characterized by increased brightness, probably due to enhanced sound reflectivity. Chromosomes are arranged in the equatorial plane and appear dark. (b) Separation of the chromatids appears as a bright zone in the chromosomal mass (*white arrowheads in b–d*). This cleft broadens in (c) and (d) during the progression of anaphase. The white arrow in (c) points to one of the spindle poles. (From Ref. 58) Bar: 10 μm.

instance when a vacuole moves away from its starting position by more than its own diameter. A series of difference images obtained from a very malignant sarcoma line[60] gives an idea of the existence and plasticity of the cytoplasmic domains of motility becoming apparent by this method (see Fig. 3.13).

Absolute differences of grey levels of corresponding pixels in images of a specimen taken at time t_0 and t_1 are calculated. In video-based image analysis,

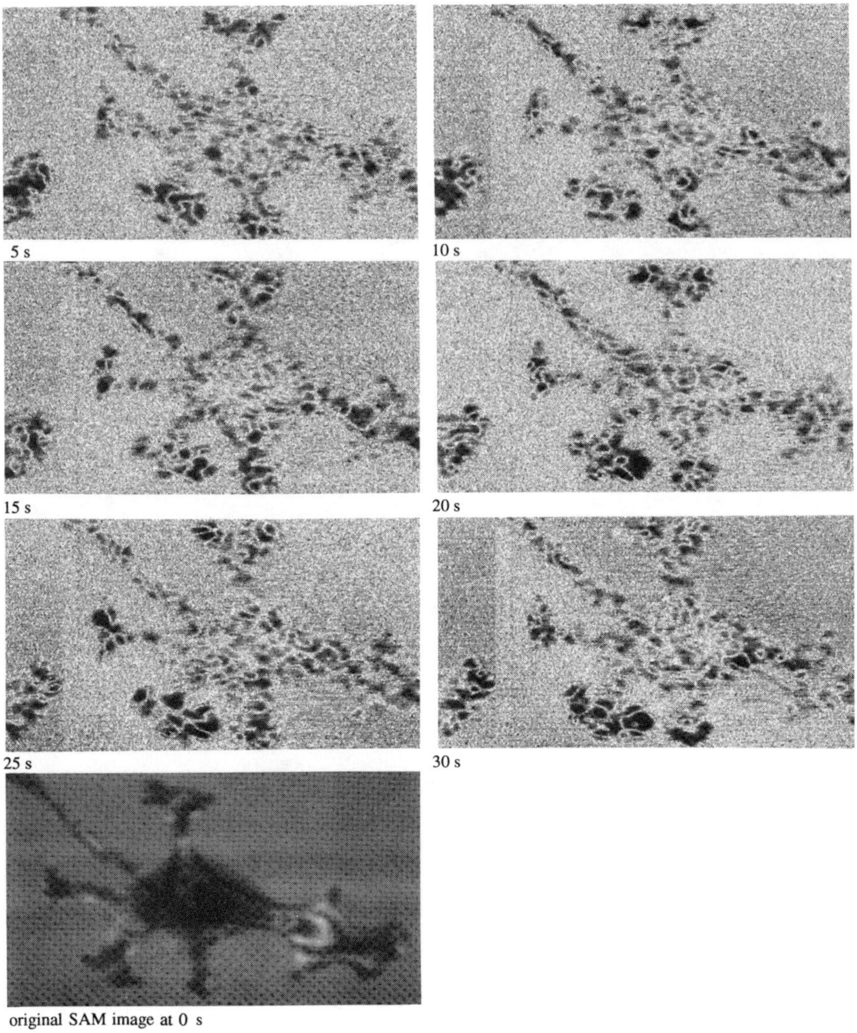

Figure 3.13. A series of SubSAM images of an A870 cell, a highly metastatic sarcoma cell line. Eight SAM pictures have been taken at 1 GHz with a 5-second time difference. The series shows absolute grey-level difference in subsequent images. These difference values have been enhanced; large differences are dark (high motility); bright areas indicate that no difference was detected. The highest motile activity is found in the lamellar areas of cell processes. Sharply delineated motile areas that change their shape and position (nucleating domains) appear. In this example, such domains are found all over the cell. One original SAM image of the cell is shown at the bottom of the left column.

no grey values smaller than nought can be handled; therefore in difference images, all negative grey values are represented by their positive notation. Because of the very small differences arising in a short time interval ($\geq$ 5 s), difference images must be enhanced. This is performed first by inverting the images (bright areas become dark, and vice versa) followed either by log filtering or forming the square of each pixel value.

The SubSAM approach enables researchers to unveil cytoplasmic motions otherwise hidden in the cell. The sensitivity of this method is very high: If we assume a minimum signal change of three grey-level units (to exceed noise) and a total dynamic range of the image of 150 grey levels, changes corresponding to topographical deviations of $\lambda/200$ (e.g., 7.5 nm at 1 GHz) are revealed. Whether these are local alterations of thickness indeed or local changes in sound velocity cause the grey-level difference remains unresolved. These considerations show that for quantitation of motile activities, the dynamic range of the images must be known, and it should be high and reproducible.

The SubSAM led to the recognition of dynamic but distinct domains as elements of cell motility.[59] Nucleating domains (see Fig. 3.13) are sharply delineated areas of motility close to each other or embedded in a larger zone of quiescence. The second group (wavelike domains) corresponds to the typical ruffling motions usually found at the edge of the cells. The third type revealed large-area cell dislocations due to the scanning motion of the acoustic lens. This back and forth displacement (oscillating domains) was found only in well-spread and almost immobile cells. These domains indicate considerable interaction of the fluid shear introduced by the scanning motion of the acoustic lens. In the future, this principle may be developed for systematically mapping elastic properties of cells and sheets of cells.

The value of SubSAM is its two-dimensional analysis of early and rapid motile responses of a cell to various stimuli (for instance, chemoattractants, drugs, or cell contacts) and its characterization of the behavior of different cell types. The SubSAM opens new areas for acoustic microscopy in cell biology.

3.3. Probing Microspheres and Freshly Isolated Outer Hair Cells

The scanning movement of the acoustical lens causes great difficulties when investigating single cells or groups of cells not attached to a solid surface. This may either be overcome by coating a glass surface with poly-lysine, laminin, or other substances increasing adhesivity. Another possibility is embedding in agar or agarose gel of a low melting point. Covering cells with a thin layer of polymers like formvar or mylar disturbs the acoustic image at least at frequencies in the

GHz range; at lower frequencies (5 MHz), even a 3-μm-thick polyethylene foil does not disturb the acoustical image of tissue sections.[61–63]

Problems of immobilization-impeded investigations refer to research of outer hair cells of the guinea pig cochlea that I tried to perform in collaboration with Drs. Brownell and Shehata from the Department of Otolaryngology (Baltimore, Maryland). These cells can be induced to contract along their long axis by electrical stimulation. The mechanism of force generation is still unclear.[64,65] After embedding in a thin layer of agar, high reflectivity (and thus stiffness) was found in the cuticular plate and the microvilli emanating from this plate, while only minor sound echoes were produced in the periphery of the cell.[66] This result coincided with our observations (unpublished), which in addition revealed extremely high values for acoustic attenuation (up to four times that of water).

At lower frequencies of ultrasound, which do not allow resolution at the subcellular level, greater distances from the acoustic lens to the specimen are used. Thus microspheres (spherical aggregates of tumour cells) fixed at the tip of a conical hole in an agar gel can be investigated by SAM (80 MHz). When focused on the equatorial plane of a microsphere, central necrotic cells can easily be distinguished from cortical viable cells by higher backscatter (see Fig. 3.14). Damage of cells in the intermediate, hypoxic layers induced by addition of 2-nitroimidazoles was followed by a time series of SAM images.[67] It was fascinating to follow the dynamics of progressing cell damage with this noninvasive

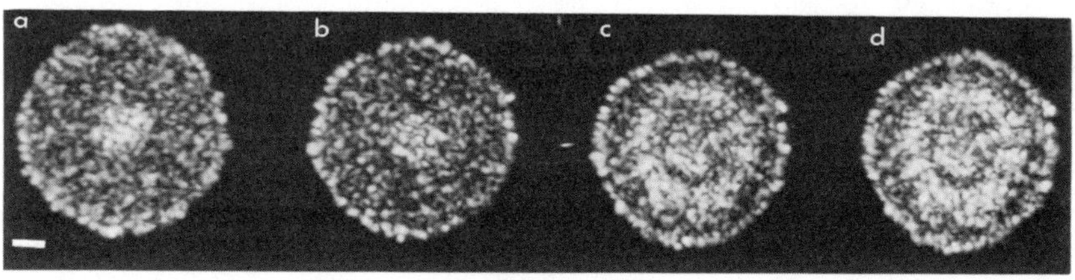

Figure 3.14. The C-scan images of multicellular spheroids composed of EMT6/To cells taken at 80 Mhz. The transducer is moved vertically to bring its focus to the desired depth in the spheroid. At this sectional plane, an x/y scan is performed, and the backscattered signal arriving in the preset time interval is used to generate the image. The spheroid is placed at the bottom of a conical groove in agarose. This spheroid has been treated with 5-mM 1-methyl-2-nitroimidazole (INO_2) for (a) 0, (b) 1, (c) 5 h, and (d) 6 hours. At low oxygen concentration (in hypoxic cells), the nitroso reduction product (INO) is toxic at micromolar concentrations. Depletion of cellular sulphhydryls and subsequent calcium influx may be the immediate reason for cell death. The core of necrotic cells appears brighter than the outer shells of living cells, as can be seen by increased backscatter. After 5 hour incubation, the necrotic core is almost unchanged, but the shell of hypoxic cells underwent severe damage, which is shown by increased backscatter. This reaction is more pronounced after 6 hours. In histological sections, these cells appear shrunken, and they exhibit condensed and collapse chromatin. Some cell damage was also observed in the outermost aerobic portion (for further description, see Ref. 67). (Courtesy of Dr. A. M. Rauth, Ontario Cancer Institute, Princess Margaret Hospital, Toronto). Bar: 100 μm.

method, although the reason for the increased signal during cell death is not understood. The authors assume that Ca^{2+} induced disassembly of cytoskeletal elements, with induction of nuclear pycnosis the reason for enhanced sound reflection. Indeed dead cells display a less homogeneous cytoplasm (because of multiple boundaries in SAM) than living cells.

3.3.1. Probing Tissue Sections

Images of unstained tissue sections show substantial contrast in the acoustic microscope, indicating considerable variations in the elastic properties and material density among tissue components. Investigations on a microscopic scale can improve the feasibility of ultrasound diagnostic imaging for evaluating pathologic conditions. The theoretical basis of clinical ultrasound diagnostics can be improved by comparing large-range A- or B-scans with SAM histological examinations (e.g., Refs. 68, 69). In addition to classical staining methods, information gained from SAM inspection can close the gap in our understanding of the ultrasound imaging of organs. The main difficulty for comparative *in vivo* studies and examinations of histological sections is the great difference between ultrasound frequencies that have to be used for different scales of investigation. While sound velocity in tissues is independent of frequency over a wide range, attenuation increases significantly with increasing frequency.

The interaction of biological tissues with ultrasound has been investigated preferentially in the MHz range (10–500 MHz) using either SAM or SLAM. In most cases, cell aggregates and connective tissue elements have been identified without looking at cellular or subcellular details. The prominent appearance of extracellular matrix components in SAM pictures contrasts with the light microscope appearance of tissue sections emphasizing cellular components of a tissue (see Fig. 3.15). Mechanical properties of tissues in organs are primarily determined by water content, properties of the connective tissue, hydraulic pressure in blood vessels and the interstitium, cell aggregates and their interaction rather than by the mechanics of single cells; therefore an overall view at relatively low magnification gives the proper impression of a tissue's mechanical properties.

Van der Steen investigated the influence of formalin (4% buffered solution) on the acoustic properties (at 5 MHz) of rabbit liver. Sections from fixed material, stained with haematoxylin eosin or unstained, were compared with unfixed slices. Formalin-fixed thin sections were measured before and after a freeze–thawing cycle.(61–63,69,70) Formalin fixation decreases sound velocity in rabbit liver by 0.9%,(63) which is in the range of earlier observations reporting 1.5%(71) and 0.6%,(72) and attenuation increases by 30%. All the steps of paraffin embedding have a great influence on acoustic parameters; sound velocity is reduced, and paraffin has an extremely high attenuation as compared to tissue. Thus remnants of paraffin may be responsible for the high attenuation measured in deparaffinized

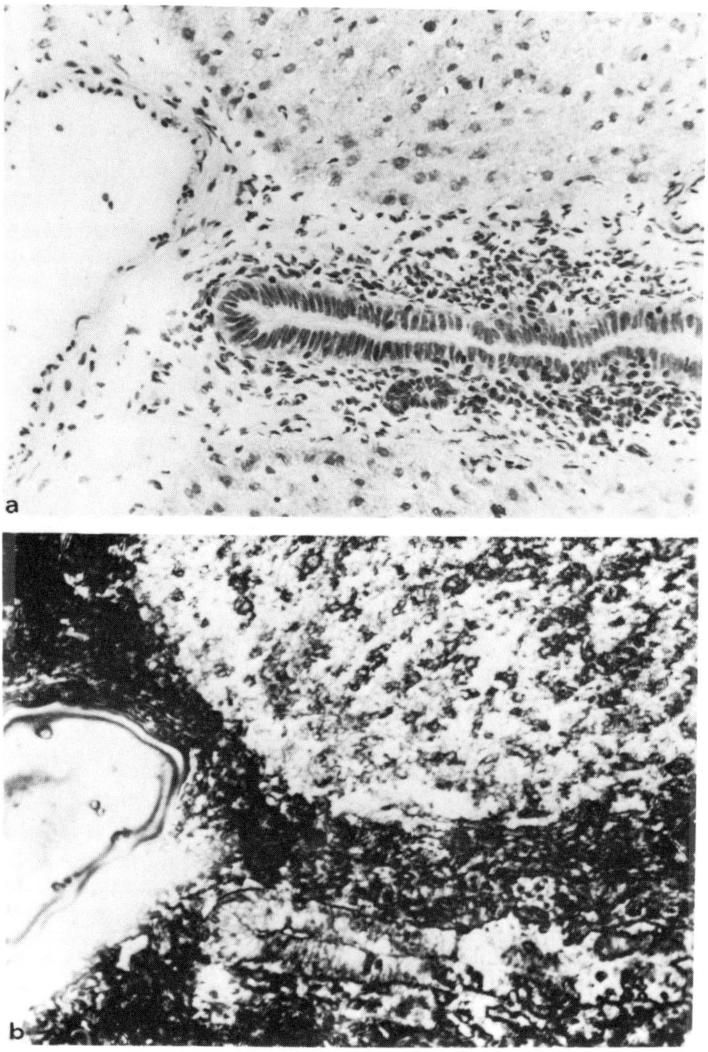

sections. Tissues embedded in paraffin or Epon (unpublished observation) are not suited for SAM investigation. Staining had a major influence only on attenuation. The freeze–thawing cycle had no significant influence on the properties of prefixed sections. The general appearance of frozen liver sections viewed at 1.2 GHz was better after fixating sections with alcohol than after formalin fixation.

Experiences with tissue sections cannot be fully transferred to investigating cells in culture. In sections many cells are opened, and emphasis is on the whole tissue ensemble, not subcellular details. The latter depend on active cellular processes that cannot be reduced to the distribution of material. The best preservation of the original acoustic velocity was found after air drying cells and rewetting them with a buffer (unpublished observations). Formalin treatment, and more extensively fixation with glutaraldehyde, changed sound velocity and attenuation in an unpredictable manner when evaluated at subcellular resolution. Overall shape however was preserved best with a glutaraldehyde/paraformaldehyde fixation (Karnovski's fixation).

3.3.2. Soft Tissues

Skin, liver, skeletal muscle, and heart are the most intensively studied organs.(31,70) In mammalian skin sections, the corneal layer, basal and suprabasal epithelial layers, and the dermal connective tissue are easily recognized.(73,74) Focusing on top of a section revealed different reflectances related to impedance differences in those regions. More recently healing canine skin wounds has been followed using SLAM (100 MHz/10–40 MHz).(75) Ultrasonic velocity and attenuation coefficient values were obtained for frozen sections of normal skin and for wound tissue (9, 20, 34, and 49 days after wounding by incision). In general there appeared to be an increase in both speed and attenuation coefficient with increasing wound age. Good correlation was found between collagen concentration and speed and attenuation coefficient as well. Speed and attenuation

Figure 3.15. Formalin-fixed cryosection through rabbit liver, hematoxylin-eosin stained. (a) Light microscope transmission view. (b) The SAM image of the same section (1.2 GHz), double-transmission image. The section was placed on glass, so sound waves penetrated the tissue twice. In (a) the most prominent feature is the epithelium lining a bile duct; part of a cross-sectioned vein is seen on the left, including three leucocytes (appearing as dark spots) in the lumen. The major part of this section contains connective tissue; the nuclei of fibrocytes are the dominant structures. In the SAM image (b), the bile duct epithelium appears relatively bright (low attenuation); nuclei are invisible, while the basal lamina of the duct epithelium is clearly delineated. Fibrous (collagenous) tissue surrounding the bile duct yields high contrast due to the high attenuation as compared to other tissue components. The wall of the blood vessel also gives good contrast. (from Ref. 63) coefficient were negatively correlated with water content.[74] No significant differences in acoustic wave propagation were found between sections parallel to the epidermis and those oriented perpendicular to the epidermis. These observations fit well with the general concept that protein concentration is the main determinant for acoustic properties in tissues.[76,77] Tissue attenuation in particular depends on collagen concentration, while backscattering in medical ultrasound probing does not seem to be related to sound velocity.[69] In human epidermis however dry mass determined by microinterferometry does not correlate well with attenuation because an additional structural component seems to be involved;[73] the same conclusion was drawn from comparisons of attenuation and dry mass in subcellular domains of living cells. These findings seem to be contradictory. The most probable explanation is the order of magnitude of the field in the sections used for the correlation. Olerud and others[75] relate an average attenuation to a total collagen content in a given tissue zone, while data from Bereiter–Hahn and others[73] are based on measurements of very small volumes where structural and acoustic inhomogeneity is very high.

The same question, whether ultrasonic propagation properties of tissue can be modeled as functions of the constituents, has been tackled by O'Brien and others,[78] studying sections from normal rat liver and fatty livers induced by orotic acid diet. With SLAM no correlation between ultrasonic speed and protein concentration (% wet weight) was found in normal and fatty livers, while ultrasonic speed correlated positively with water content and negatively with fat content. The ultrasonic attenuation coefficient barely showed a relation to one of these three tissue constituents. Recent measurements of attenuation of human fetal liver prior and after birth point to a positive correlation between the acoustic attenuation coefficient measured at 5 MHz and the glycogen content of the liver.[79] In contrast to proteins, glycogen is deposited as granular material and thus may increase scattering. The structural component of ultrasound attenuation is also exemplified by a significant anisotropy of canine myocardium.[6]

Another interesting application of SAM in histology may be in revealing age-related changes in acoustic properties.[80,81] Most recently ocular tissues have been investigated by van der Steen.[70] He calculated for example an increase in longitudinal sound velocity of 7.7×10^2 ms^{-1} per year for human lenses. Attenuation and sound velocities for human and porcine eye tissues have been determined at 20 Mhz. The highest sound velocities were found in the sclera (1597 $\pm$ 20.3 ms^{-1} in human eyes, 1654 $\pm$ 17.8 ms^{-1} in porcine sclera); the lens exhibited the highest attenuation in both species.

High-resolution echography (22 MHz) can close the gap between histology, acoustic microscopy, and echography. Echo strength and echo heterogeneity of small skin lesions and skin tumors correlate well with cellularity and collagen heterogeneity, respectively.[82]

3.4. Bone and Other Collagenous Tissues

Connective tissue derivatives are made from extracellular matrix material. They exhibit a wide range of mechanical properties related to their function in the organism; i.e., tendon is optimized to transduce tension, dermal cutis layers act as pillows, and articular cartilage functions like a lubricated shock absorber. In bone the extracellular matrix is stiffened by mineralization, which is a prerequisite for fulfilling its function as a skeleton. All these structures owe their specific mechanical properties to the distinct orientation of collagen fibrils and cross-linking with each other and interaction with an almost amorphous matrix of acid glycoproteins and glucosaminoglycans. These macromolecular organizations provide tensile strength or restrict the flux of water through the composite layers.[83,84] Knowing the relationship between mineralization and collagen orientation, the influence of amorphous matrix constituents, and ultrasound propagation are of considerable interest for medical diagnosis and evaluating race horses. One of the main problems in the clinical use of ultrasonic measurements of bone is how to use results to predict such mechanical properties as elastic modulus and strength.[85,86] Some difficulties arise from the porous structure of bone, which limits the value of general ultrasound investigation, but this is overcome by applying real microscopic techniques.

Mineralization is easily revealed by acoustic microscopy, and extracting the mineral from frozen sections can be followed immediately by SAM.[3] At least over a certain range of demineralization, the speed of ultrasound propagation in bovine bone is linearly correlated with its mineral content.[87] In equine cortical bone ultrasound speed also has a positive linear relationship when compared with bone specific gravity.[88] On the other hand, newly formed periosteal bone enclosing an implant reveals a significantly reduced acoustic impedance when compared with areas distinct from the implant.[89] This difference is attributed to imperfect mineralization. This may also be the reason for the increasing speed of surface acoustic waves in adult human bones.[90]

The role of collagen arrangement and the matrix in between the collagen fibrils are more difficult to assess. While ultrasonic speed (longitudinal wave) along the long axis of purified collagen is considerably faster than across the long axis (3,64 km/s and 2.94 km/s, respectively),[91] the situation is reversed in cartilage where a higher speed of longitudinal wave propagation is found across the long axis of the collagen fibrils than at random orientations of the collagen fibrils.[92] Thus sound velocity determined with the SLAM clearly reveals differences in collagen orientation in cartilage (see Fig. 3.16). The apparent difference in anisotropic sound propagation in purified collagen I and cartilage is difficult to explain. It may be due to differences in the macromolecular organization of the material used in both these studies. Mayev[91] investigated dry collagen (Type

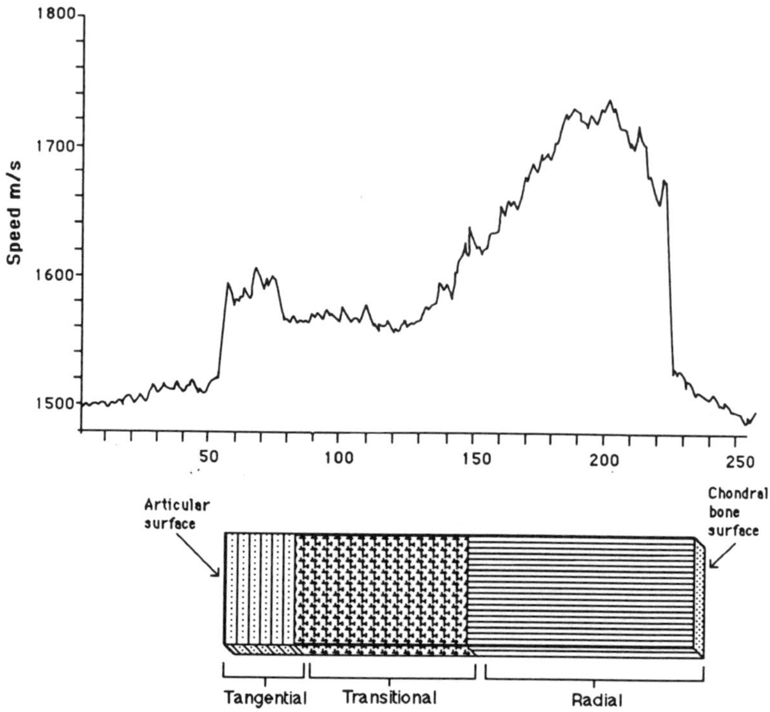

Figure 3.16. A profile of longitudinal wave speed as a function of tissue depth for a slice of normal cartilage (from the stifle of a 1–2-year-old steer) cut perpendicular to the articular surface. Analysis was performed using a SLAM. Below the speed profile, the predominant orientations of collagen fibrils from articular to chondral surfaces are shown. The greatest speed is seen in the (radial) zone where sound waves are propagating perpendicular to (across) the long axis of the collagen fibrils. The speed is lower at the articular surface (along/across) where propagation is more parallel to the preferred fibril orientation. (Fibrils to tend to be random in the two-dimensional plane of the tangential zone near the articular surface.) Speed is lowest in the transitional zone (random collagen orientation); a gradual transition occurs from one preferred orientation to the other. (from Ref. 92)

I), while Agemura and coworkers[92] studied fresh cartilage, which is made up primarily of Type II collagen and contains a large amount of bound water that may be engaged in sound propagation. Collagen II is made up of thinner fibrils than collagen I, and it is cross-linked to the proteoglycan-rich matrix via collagen IX. Disruption of the intermolecular cross-links, which reduces tensile strength by a factor of 100, has almost no influence on ultrasonic speed, but it markedly increases the attenuation coefficient. Removing proteoglycans and glucosaminoglycans and other noncollagen proteins did not change acoustic attenuation.

Therefore we have to agree that ''the mechanisms responsible for acoustic wave propagation in highly organized biological materials are still not well understood. However, it is also clear that careful consideration must be given to orientation as well as to the fundamental intermolecular structure and composition.''[92]

3.5. Model Investigations of Gels and Membranes

We have learned how difficult interpreting SAM images is. One way of improving our understanding of the mechanisms responsible for acoustic wave propagation in biological material is to study isolated components, such as extracellular matrix components, model membranes, or gels prepared from cytoskeletal elements. Initial experiments performed in my laboratory in collaboration with Prof. Grill from the physics department are encouraging. Acoustic parameters change significantly and in a predictable manner during the polymerization of actin and tubulin. These polymerizations often are induced by changing the ionic strength of the saline. Mixing freshly introduced, highly concentrated salt solutions can be delayed by several minutes even when the mixture is agitated mechanically. Because of the high influence of salt concentration on sound attenuation and propagation,[93,94] these mixing phenomena may obscure polymerization-based acoustic changes.

Polyvinylalcohol (PVA) gels exposed to a series of freezing and thawing cycles turn into a strong rubberlike elastic mass. Investigating changes in mechanical properties accompanying this conversation by acoustic imaging at 80–320 MHz and in parallel with a torsion rheometer showed that elastic modulus of this porous gel did not increase in parallel with ultrasound velocity. The latter reached a stable value after about ten freezing cycles, while Young's modulus still increased during additional freezing cycles.[98]

Membrane properties in particular can be revealed by investigating model systems. One fruitful approach was developed by Bianco and others,[49] who collected multilayers of omega-tricosenoic acid with the Langmuir–Blodgett technique on silanized glass. From $V(z)$ curves, leaky surface acoustic wave velocity was determined. The linear decrease with increasing lipid layer thickness (28–220 nm) found experimentally was explained by model calculations.

3.6. Conclusions

Acoustic *microscopy* applications in biology and medicine are still in their infancy. However methods have been developed to allow this technique to visualize the mechanical properties of hard tissues and the motile activities of cells, which are not revealed by any other method with the same ease and resolution. Using easily available image analytical methods and combining these possibilities with SAM provide an excellent tool for studying the motility of cells attached to a substrate. Cellular reactions to receptor binding or any other external stimulus are immediately apparent in SAM images. This tool awaits its use in developmental biology, membrane research, and cancer research, as well. The SAM images of tissue sections emphasize connective tissue elements, a feature totally different from traditional histology, which concentrates on cellular components and very often overlooks the significance of extracellular matrix structures in maintaining and controlling cell activities. The shape of an organism or a body is also primarily determined by extracellular matrix.

Finally, an interesting application has to be mentioned, that of acoustical tweezers,[95] which like optical tweezers[96] allow us to manoeuvre laterally a trapped object. In addition movement in the axial direction can be achieved by tuning the frequency.

A map with large white areas lies in front of us if acoustic microscopy is extended. Improved analytical methods and investigations of model gels seem to be the most important goals for further research. Frequency-specific interactions of ultrasound with biological material are another field where we lack knowledge. The increasing sensitivity of acoustic microscopes for measuring very small acoustic path deviations is interesting in macromolecular research; however phase-sensitive imaging,[46,47] which considerably increases resolution on the z-axis is difficult to handle because of its high sensitivity to environmental influences.

Acknowledgments

Results are included in Chapter 3 that have not been previously published. This research was supported by the Gesellschaft der Freunde und Förderer der Johann Wolfgang Goethe Universität and the Deutsche Forschungsgemeinschaft (Be 423/13, 16). Thanks are due to all colleagues who contributed photographs to be reproduced.

References

1. Briggs, A. (1992). *Acoustic Microscopy,* Monographs on the Physics and Chemistry of Materials (R. Brook, P. B. Hirsch, C. J. Humphreys, N. F. Mott, eds.). Clarendon Press, Oxford, UK.
2. Hildebrand, J. A. and Rugar, D. (1984). Measurement of cellular elastic properties by acoustic microscopy. *J. Microsc.* **134**, 245–60.
3. Bereiter–Hahn, J., Litniewski, J., Hillmann, K., Krapohl, A., Zylberberg, L. (1988). What can scanning acoustic microscopy tell about animal cells and tissues? *Acoust. Imaging* **17**, 27–38.
4. Lüers, H., Bereiter–Hahn, J., Litniewski, J. (1992). In: *Acoustical Imaging,* vol. 19 (H. Ermert and H. P. Harjes, eds.) Plenum Press, New York, pp. 511–16.

5. Lüers, H., Hillmann, K., Litniewski, J., Bereiter-Hahn, J. (1992). Acoustic microscopy of cultured cells: distribution of forces and cytoskeletal elements. *Cell Biophys.* **18**, 279–93.
6. Mottley, J. G. and Miller, J. G. (1990). Anisotropy of the ultrasonic attenuation in soft tissues: measurements *in vitro*. *J. Acoust. Soc. Am.* **88**, 1203–10.
7. Akashi, N., Kushibiki, J., Chubachi, N., Dunn, F. (1989). Considerations for quantitative characterization of biological tissues by scanning acoustic microscopy. *Acoust. Imaging* **17**, 183–91.
8. Litniewski, J. and Berieter-Hahn, J. (1991). In: *Acoustical Imaging*, vol. 19 (H. Ermert and H. P. Harjes, eds.), pp. 535–38. Plenum Press, New York.
9. Bereiter-Hahn, J. (1987). In: *Cytomechanics* (J. Bereiter-Hahn, O. R. Anderson, W. E. Reif, eds.), pp. 3–30. Springer, Heidelberg; New York, Berlin.
10. Litniewski, J. and Bereiter-Hahn, J. (1990). Measurements of cells in culture by scanning acoustic microscopy. *J. Microsc.* **158**, 95–107.
11. Kanngiesser, H. (1990). Ultraschallmikrokopie von schwach reflektierenden biologischen Strukturen. Dissertation techn. Wissenschaften, Zürich.
12. Kanngiesser, H. and Anliker, M. (1992). In: *Acoustical Imaging*, vol. 19 (H. Ermert and H. P. Harjes, eds.), pp. 517–22. Plenum Press, New York.
13. de Boer, R., Ehlers, W., Kowalski, S., Plischka, J. (1991). Porous media, a survey of different approaches. *Forschungsber. Fachereich Bauwesen* **54**, 153 pp.
14. Kundu, T., Bereiter-Hahn, J., Hillmann, K. (1991). Measuring elastic properties of cells by evaluation of scanning acoustic microscopy V (Z) values using simplex algorithm. *Biophys. J.* **59**, 1194–1207.
14a. Wolosewick, J. J., Porter, K. R. (1979). Microtrabecular lattice of the cytoplasmic ground substance. *J. Cell Biol.* **82**, 114–39.
15. Felder, S. and Elson, E. L. (1990). Mechanics of fibroblast locomotion—quantitative analysis of forces and motions at the leading lamellas of fibroblasts. *J. Cell Biol.* **111**, 2513–26.
16. Simon, S. I. and Schmid-Schönbein, G. W. (1990). Cytoplasmic strains and strain rates in motile polymorphonuclear leukocytes. *Biophys. J.* **58**, 319–32.
16a. Schmid-Schönbein, G. W., Woo, S. L. Y., Zweifach, B. W. (1986). *Frontiers in Biomechanics*, Springer, New York, Heidelberg, Berlin.
17. Janmey, P. A. (1991). A torsion pendulum for measurement of the viscoelasticity of biopolymers and its application to actin networks. *J. Biochem. and Biophys. Methods* **22**, 41–53.
17a. Trombitas, K., Pollack, G. H., Wright, J., Wang, K. (1993). Elastic properties of titin filaments demonstrated using a freeze-break technique. *Cell Motil. Cytoskeleton* **24**, 274–83.
18. Hiramoto, Y. (1987). In: *Cytomechanics* (J. Bereiter-Hahn, O. R. Anderson, W. E. Reif, eds.), pp. 31–46. Springer, Heidelberg, New York, Berlin.
19. Elson, E. L. (1988). Cellular mechanics as an indicator of cytoskeletal structure and function. *Annu. Rev. Biophys. Chem.* **17**, 397–430.
20. Butt, H. J., Wolff, E. K., Gould, S. A., Northern, B. D., Peterson, C. M., Hansma, P. K. (1990). Imaging cells with the atomic force microscope. *J. Struct. Biol.* **105**, 54–61.
21. Henderson, E., Haydon, P. G., Sakaguchi, D. S. (1992). Actin filament dynamics in living glial cells imaged by atomic force microscopy. *Science* **257**, 1944–46.
22. Horber, J. K. H., Haberle, W. Ohnesorge, F., Binnig, G., Liebich, H. G., Czerny, C. P., Mahnel, H., Mayr, A. (1992). Investigation of living cells in the nanometer regime with the scanning force microscope. *Scanning Microsc.* **6**, 919–30.
23. Kolosov, O., Ogiso, H., and Yamanaka, K. (1993). Ultrasonic force microscope (UFM)—a new approach to an NDE on nanometer scale. Part II. Experimental investigation of nonlinear viscoelastic properties of various materials using UFM. Spring meeting of the Mechanical Engineering Laboratory, AIST, MITI, Tsukuba, Japan.
24. Kolosov, O., Ogiso, H., and Yamanaka, K. (1993). Ultrasonic force microscope (UFM)—a new approach to an NDE on nanometer scale. Part I. Nonlinear detection of ultrasonic vibration in

an AFM. Theory and experiment. Spring meeting of the Mechanical Engineering Laboratory, AIST, MITI, Tsukuba, Japan.
25. Zinin, P. V. (1992). Theoretical analysis of sound attenuation mechanisms in blood and in the erythrocyte suspensions. *Ultrason.* **30**, 26–34.
26. Sinton, A. M., Briggs, G. A. D., Tsukahara, Y. (1989). Time-resolved acoustic microscopy of polymer coatings. *Acoust. Imaging* **17**, 87–95.
27. Weaver, R. L., Sachse, W., Niu, L. (1989). Transient ultrasonic waves in a viscoelastic plate. *Theory. J. Acoust. Soc. Am.* **85**, 2255–61.
28. Weaver, R. L., Sachse, W., Niu, L. (1989). Transient ultrasonic waves in a viscoelastic plate: applications to materials characterization. *J. Acoust. Soc. Am.* **85**, 2262–67.
29. Yamanaka, K. (1983). Surface acoustic wave measurements using an impulsive converging beam. *J. Appl. Phys.* **54**, 4323–29.
30. Daft, C. M. and Briggs, G. A. (1989). The elastic microstructure of various tissues. *J. Acoust. Soc. Am.* **85**, 416–22.
31. Daft, C. M., Briggs, G. A., O'Brien, W. D., Jr. (1989). Frequency dependence of tissue attenuation measured by acoustic microscopy. *J. Acoust. Soc. Am.* **85**, 2194–2201.
32. Lemons, R. A. and Quate, C. F. (1974). Acoustic microscope—scanning version. *Appl. Phys. Lett.* **24**, 163–65.
33. Quate, C. F. (1979). The acoustic microscope. *Scientif. Am.* **241**, 62–70.
34. Briggs, G. A. D., Wang, J., Gundle, R. (1993). Quantitative acoustic microscopy of individual living human cells. *J. Microsc.* **172**, 3–12.
35. Attal, J., Saied, A., Saurel, J. M., Ly, C. C. (1989). Acoustic microscopy: deep focusing inside materials at gigahertz frequencies. *Acoust. Imaging* **17**, 121–30.
36. Hegner, S. and Bereiter–Hahn, J. (1993). In: Cell and Tissue Culture Models in Dermatological Research (A. Bernd, J. Bereiter–Hahn, F. Hevert, H. Holzmann, eds.), pp. 58–66, Springer, New York, Heidelberg, Vienna.
37. Bereiter–Hahn, J., Strohmeier, R., Beck, K. (1983). Determination of the thickness profile of cells with the reflection contrast microscope. *Scient. Techn. Inf.* **8**, 125–50.
38. Strohmeier, R., Bereiter–Hahn, J. (1987). Hydrostatic pressure in epidermal cells is dependent on Ca-mediated contractions. *J. Cell Sci.* **88,** 631–40.
39. Akashi, N., Kushibiki, J., Chubachi, N. (1989). Quantitative characterization of biological tissues by acoustic microscopy-effects of multiple reflection and viscosity, Technical Report of the IECE 43.80 Ev. 43, 767–75.
40. Kundu, T., Bereiter–Hahn, J., Hillmann, K. (1992). Calculating acoustical properties of cells: influence of surface topography and liquid layer between cell and substrate. *J. Acoust. Soc. Am.* **91**, 3008–17.
41. Steiger, D. L., O'Brien, W. D., Jr., Olerud, J. E., Riederer–Henderson, M. A., Odland, G. F. (1988). Measurement uncertainty assessment of the scanning laser acoustic microscope and application to canine skin wound, *IEEE Trans. Ultrasonics, Ferroelectr. and Frequency* **35**, 741–48.
42. Weisser, G., Fein, M., Zuna, I., Lorenz, A., Lorenz, W. J. (1992). In: *Acoustical Imaging,* vol. 19 (H. Ermert and H. P. Harjes, eds.), pp. 409–414. Plenum Press, New York.
43. Okawai, H., Tanaka, M., Dunn, F., Chubachi, N., Honda, K. (1989). In: *Acoustical Imaging* (H. Shimizu, N. Chubachi, J. Kushibiki, eds.), pp. 193–201. Plenum Press, New York.
44. Kulik, A., Gemaud, G., Sathish, S. (1989). Continuous wave reflection scanning acoustic microscope (SAMCRUW). *Acoust. Imaging* **17**, 71–78.
45. Kulik, A., Richard, P., Sathish, S., Gremaud, G. (1992). In: *Acoustical Imaging,* vol. 19 (H. Ermert and H. P. Harjes, eds.), pp. 697–702, Plenum Press, New York.
46. Hillmann, K., Bereiter–Hahn, J., Grill, W. (1994). Determination of ultrasonic attenuation in small samples of solid material by scanning acoustic microscopy with phase contrast. *J. of Alloys and Compounds* (in press).

47. Daft, C. M., Weaver, J. M., Briggs, G. A. (1985). Phase contrast imaging of tissue in the scanning acoustic microscope. *J. Microsc.* **139**, RP3–RP4.
48. Liang, K., Bennet, D. D., Khuri-Yakub, B. T., Kino, G. S. (1985). Precise phase measurements with the acoustic microscope. *IEEE Trans. Sonics Ultrason.* **32**, 266.
49. Bianco, B., Cambiaso, A., Paradiso, R., Tommasi, T. (1992). Experimental and theoretical surface acoustic wave analysis of thin-film lipid multilayers. *Appl. Phys. Lett.* **61**, 402–404.
50. Tommasi, T., Cambiaso, A., Buzzoni, G., Grattarola, M., Bianco, B. (1992). In: *Acoustical Imaging,* vol. 19 (H. Ermert and H. P. Harjes, eds.), pp. 529–44, Plenum Press, New York.
51. Yu, Z. and Boseck, S. (1992). In: *Acoustical Imaging,* vol. 19. (H. Ermert and H. P. Harjes, eds.), pp. 617–22.
52. Chubachi, N., Kamai, H., Samonyia, T., Wakahara, T. (1992). In: *Acoustical Imaging,* vol. 19 (H. Ermert and H. P. Harjes, eds.), pp. 685–90.
53. Gingell, D. and Todd, I. (1979). Interference reflection microscopy: a quantitative theory for image interpretation and its application to cell-substratum separation neasurement. *Biophys. J.* **26**, 507–26.
54. Bereiter–Hahn, J. (1987). In: *Scanning Imaging Technology* (T. Wilson and I. Balk, eds.) SPIE, Washington, vol. 809: 162–65.
55. Bereiter–Hahn, J., Berghofer, F., Kundu, T., Penzkofer, C., Hillmann, K. (1992). In: *Acousto-Optics and Acoustic Microscopy,* vol. 140 (S. M. Graceweski and T. Kundu, eds.), pp. 71–80, Am. Soc. Mech. Engineers, vol. 140, pp. 71–80.
56. Bereiter–Hahn, J. and Lüers, H. (1990). Shape changes and force distribution in locomoting cells. Investigations with reflected light and acoustic microscopy. *Eur. J. Cell Biol.* **53**; *Suppl.* **31**, 85.
57. Bereiter–Hahn, J. and Lüers, H. (1994). In: *Mechanics of Actively Locomoting Cells* (N. Akkas, ed.), ASI series **84**, Springer, Heidelberg, New York, Berlin, pp. 181–230.
58. Lindner, A., Winkelhaus, S., Hauser, M. (1992). In: *Acoustical Imaging,* vol. 19 (H. Ermert and H. P. Harjes, eds.), pp. 523–28, Plenum Press, New York.
59. Vesely, P., Lüers, H., Riehle, M., Bereiter–Hahn, J. (1994). Subtraction-scanning acoustic microscopy reveals motility domains in cells *in vitro. Cell Motil. Cytoskeleton* **29**, 231–240.
60. Veselý, P., Chaloupková, A., Urbanec, P., Urbancová, H. Bohá, L., Krchnáková, E., Franc, F., Sprincl, L., Vousden, K., Moss, R., Heaysmann, J., Dilly, P. N. (1987). Patterns of *in vitro* behaviour characterizing cells of spontaneously metastasizing K2M rat sarcoma. *Folia Biol. (Praha)* **33**, 307–24.
61. van der Steen, A. F. W., Cuypers, M. H. M., Thijssen, J. M., de Wilde, P. C. M. (1991). Influence of histochemical preparation on acoustic parameters of liver tissue, a 5-MHz study. *Ultrasound Med. & Biol.* **17**, 879–91.
62. van der Steen, A. F. W., Cuypers, M. H. M., Thijssen, J. M., Ebben, G. P. J., de Wilde, P. C. M. (1992). In: *Acoustical Imaging,* vol. 19 (H. Ermert and H. P. Harjes, eds.), pp. 529–34, Plenum Press, New York.
63. van der Steen, A. F. W., Thijssen, J. M., Ebben, G. P. J., de Wilde, P. C. M. (1992). Effects of tissue-processing techniques in acoustic (1.2 GHz) and light microscopy. *Histochem. J.* **97**, 195–99.
64. Brownell, W. E. (1986). In: *Auditory Frequency Selectivity* (B. C. Moore, C. D. Patterson, eds.), Plenum Press, New York, pp. 109–118.
65. Brownell, W. E., Bader, C. R., Bertrand, D., de Ribaupierre, Y. (1985). Evoked mechanical responses of isolated cochlear outer hair cells. *Science* **227**, 194–96.
66. Zenner, H. P., Gitter, A. H., Rudert, M., Ernst, A. (1992). Stiffness, compliance, elasticity, and force generation of outer hair cells. *Acta Otolaryngol. (Stockh)* **112**, 248–53.
67. Bérubé, L. R., Harasiewicz, K., Foster, F. S., Dobrowsky, E., Sherar, M. D., Rauth, A. M. (1992). Use of a high-frequency ultrasound microscope to image the action of 2-nitroimidazoles in multicellular spheroids. *Brit. J. Cancer* **65**, 633–40.

68. Coleman, D. J., Silverman, R. H., Rondeau, M. J., Lizzi, F. L., Mclean, J. W., Jakobiec, F. A. (1990). Correlations of acoustic tissue typing of malignant melanoma and histopathologic features as a predictor of death. *Am. J. Ophthalmol.* **110**, 380–88.
69. van der Steen, A. F. W., Thijssen, J. M., van der Laak, J. A. W. M., de Wilde, P. C. M., Ebben, G. P. J. (1993). Quantitative correlation of acoustical and light microscopy. *Proc. IEEE London.*
70. van der Steen, A. F. W. (1993). *Akoestische biomicroscopie*, Benda BV, Nijmegen.
71. Bamber, J. C., Hill, C. R., King, J. A., Dunn, F. (1979). Ultrasonic propagation through fixed and unfixed tissues. *Ultrasound Med. & Biol.* **5**, 159–65.
72. Kremkau, F. W., Barnes, R. W., McGraw, C. P. (1981). Ultrasonic attenuation and propagation speed in normal human brain. *J. Acoust. Soc. Am.* **70**, 29–38.
73. Bereiter–Hahn, J. and Buhles, N. (1987). In: *Imaging and Visual Documentation in Medicine* (K. Wamsteker, U. Jonas, G. Veen, V. Waes, eds.), pp. 537–43. Excerpta Medica, Amsterdam.
74. Kolosov, O., Levin, V. M., Mayev, R. G., Senjushkina, T. A. (1987). The use of acoustic microscopy for biological tissue characterization. *Ultrasound Med. & Biol.* **13**, 477–83.
75. Olerud, J. E., O'Brien, W. D., Jr., Riederer–Henderson, M. A., Steiger, D. L., Debel, J. R., Odland, G. F. (1990). Correlation of tissue constituents with the acoustic properties of skin and wound. *Ultrasound Med. & Biol.* **16**, 55–64.
76. Dunn, F., Edmonds, P. D., Fry, W. J. (1969). In: *Biomedical Engineering* (H. Schwan, ed.), pp. 205–332, McGraw-Hill, New York.
77. Dunn, F. and O'Brien, W. D., Jr. (1978). *Ultrasonic Absorption and Dispersion in Ultrasound: Its Application in Medicine and Biology* (F. J. Fry, ed.), pp. 393–439, Elsevier, Amsterdam.
78. O'Brien, W. D., Jr., Erdman, J. W., Jr., Hebner, T. B. (1988). Ultrasonic propagation properties (a-100 MHz) in excessively fatty rat liver. *J. Acoust. Soc. Am.* **83**, 1159–66.
79. Carson, P. L., Chiang, E. H., Rubin, J. M., Meyer, C. R., Andersen, H. F., Marks, T. I. (1991). Pre- to postnatal reduction in ultrasound attenuation coefficient of the liver. *Invest. Radiol.* **26**, 8–12.
80. Afschrift, M., Cuvelier, C., Ringoir, S., Barbier, F. (1987). Influence of pathological state on the acoustic attentuation coefficient slope of liver. *Ultrasound Med. & Biol.* **13**, 135–39.
81. Hartman, P. J., Oosterveld, B. J., Thijssen, J. M., Rosenbusch, G. J. E. (1991). Variability of quantitative echographic parameters of the liver: intra- and interindividual spread, temporal and age-related effects. *Ultrasound in Med. & Biol.* **17**, 857–67.
82. Bamber, J. C., Harland, C. C., Gusterson, B. A., Mortimer, P. S. (1992). In: *Acoustical Imaging*, vol. 19 (H. Ermert and H. P. Harjes, eds.), pp. 369–74, Plenum Press, New York.
83. Comper, W. D., Zamparo, O. (1990). Hydrodynamic properties of connective tissue polysaccharides. *Biochem. J.* **269**, 561–64.
84. Suh, J. K. and Woo, S. L.-Y. (1992). Negative hydrostatic pressure in articular cartilage extracellular matrix under cyclic compressive load, *Adv. Bioengineering* **22**, 199–202.
85. Ashman, R. B. and Rho, J. Y. (1990). Use of a transmission ultrasonoic technique for the *in vitro* evaluation of bone ingrowth. *J. Biomech.* **23**, 941–43.
86. Martin, R. B. (1991). Determinants of the mechanical properties of bones. *J. Biomech.* **24**, 79–88.
87. Tavakoli, M. B. and Evans, J. A. (1991). Dependence of the velocity and attenuation of ultrasound in bone on the mineral content. *Phys. Med. Biol.* **36**, 1529–37.
88. McCarthy, R. N., Jeffcott, L. B., McCartney, R. N. (1990). Ultrasound speed in equine cortical bone—effects of orientation, density, porosity, and temperature. *J. Biomech.* **23**, 1139–43.
89. Shie, S.-J., Zimmermann, M. C., Parsons, J. R., Langrana, N. (1992). Scanning acoustic microscopy and the structural adaptation of bone. *Adv. Bioengineering* **22**, 127–29.
90. Hein, H.-J., Czurratis, P., Bernstein, A., Boseck, S. (1993). Possibilities and problems of structure characterization of human bone. *Techn. & Health Care,* 279–80.
91. Mayev, R. G. (1988). Scanning acoustic microscopy of polymeric materials and biological substances. *Arch. Acoust.* **13**, 13–43.

92. Agemura, D. H., O'Brien, W., Jr., Olerud, J. E., Chun, L. E., Eyre, D. E. (1990). Ultrasonic propagation properties of articular cartilage at 100 MHz. *J. Acoust. Soc. Am.* **87**, 1786–91.
93. Dahint, R., Grunze, M., Josse, F., Andle, J. C. (1992). Probing of strong and weak electrolytes with acoustic wave fields. *Sens. and Actuators* **9**, 155–62.
94. Dahint, R., Shana, Z. A., Josse, F., Riedel, S. A., Grunze, M. (1993). Identification of metal ion solutions using acoustic plate mode devices and pattern recognition. *IEEE Trans. Ultrasonics, Ferroelectrics, and Freq. Control* **40**, 114–20.
95. Wu, J. R. (1991). Acoustical tweezers. *J. Acoust. Soc. Am.* **89**, 2140–43.
96. Chubachi, N., Kushibiki, J., Sannomiya, T., Akashi, N., Tanaka, M., Okawai, H., Dunn, F. (1987). Scanning acoustic microscope for quantitative characterization of biological tissues. *Acoust. Imaging* **16**, 1–9.
96. Liang, H., Wright, W. H., Wei, H., Berns, M. W. (1991). Micromanipulation of mitotic chromosomes in PTK2 cells using laser-induced optical forces (optical tweezers). *Exp. Cell Res.* **197**, 21–35.
97. Bereiter–Hahn, J., Fox, C. H., Thorell, B. (1979). Quantitative reflection contrast microscopy of living cells. *J. Cell Biol.* **82**, 767–79.
98. Kolosov, O., Suzuki, M., and Yamanaka, K. (1993c). Microscale evaluation of the local viscoelastic properties of polymer gel for artificial muscles using acoustic microscopy. Sixth Polymer Science Symposium on Molecular Gel Resolution, Tokyo, pp. 62–63.

4

Lens Geometries for Quantitative Acoustic Microscopy

Abdullah Atalar, Hayrettin Köymen, Ayhan Bozkurt, and Göksenin Yaralioğlu

4.1. Introduction

The purpose of the first Lemons-Quate acoustic microscope[1] was to image the surfaces of materials or biological cells with a high resolution. Unfortunately, competition with the optical microscope was only partially successful due to the high degree of absorption in the liquid-coupling medium at high frequencies. Increasing the resolution beyond optical limits was possible with the use of hot water[2] or cryogenic liquids,[3] at the cost of operational difficulty and system complexity. Meanwhile it was shown that the acoustic microscope can generate information that has no counterpart in the optical world.[4] The presence of leaky waves resulted in an interference mechanism known as $V(z)$ curves. The $V(z)$ method involves recording the reflected signal amplitude from an acoustic lens as a function of distance between the lens and the object. This recorded signal is shown to depend on elastic parameters of the object material. After underlying processes are well understood, new lens geometries or signal-processing electronics are designed to emphasize the advantage of the acoustic lens. In any case,

A. Atalar, H. Köymen, A. Bozkurt, and G. Yaralioğlu • Electrical and Electronics Engineering Department, Bilkent University, Ankara, Turkey

Advances in Acoustic Microscopy, Volume 1, edited by Andrew Briggs.
Plenum Press, New York, 1995.

the aim has been to increase the quantitative characterization ability of the microscope.

An acoustic microscope is typically operated in the frequency range of 10–2000 MHz for different purposes, such as detecting microcracks, adhesion problems, imaging biological specimens, microstructures, delaminations, residual stresses, measuring anisotropy, film thickness, SAW velocity, and attenuation.[5] The low-frequency end is suitable for higher penetration depth, whereas high frequencies are used to achieve high resolution. The $V(z)$ response of the microscope is affected by variations in leaky wave velocities. Since these leaky waves probe the subsurface of the material under investigation, any change in material elastic constants, density, bonding strength, etc., cause a perturbation in the velocity of the leaky wave. The problem of characterizing and interpreting images becomes more difficult for multilayered materials, since a family of leaky waves is excited. To investigate the contribution of the leaky wave to the acoustic microscope response, an angular spectrum approach[6] is typically employed in analysis and simulation studies. Ray representation developed by Bertoni and coworkers[7,8,9] is a less accurate method compared to this approach, but it provides physical insight into underlying mechanisms.

In Chapter 4 we first present different lens geometries, starting with the conventional lens. Possible lens configurations with increased leaky wave sensitivity are shown. A Lamb wave lens with its inherent advantages for characterization is described in some detail. Then, lenses with directional characteristics, such as line-focus-beam (LFB) lens, are briefly described. Other lenses designed for this purpose and in particular slit-aperture and V-groove lenses are presented. We compare different signal-processing schemes with respect to sensitivity to elastic parameters and surface topography. Capabilities of different configurations for subsurface penetration and quantitative characterization ability are given. The $V(f)$ measurement technique is described. Chapter 4 also discusses the accuracy of leaky wave velocity measurement using the $V(z)$ method.

4.2. Acoustic Lens Geometries

4.2.1. Conventional Spherical Lens

Very little has changed in spherical acoustic lens design since its first introduction.[10] A high-performance lens (see Fig. 4.1) can be designed by finding an optimum combination of cavity size, operating frequency, opening angle, antireflection layer, transducer size, and buffer rod length. An important aspect of the design is diffraction effects in the buffer rod when determining transducer size. A solution minimizing diffraction loss can be found in Ref. [11]. The conventional lens has circular symmetry, and hence it is not expected to show sensitivity

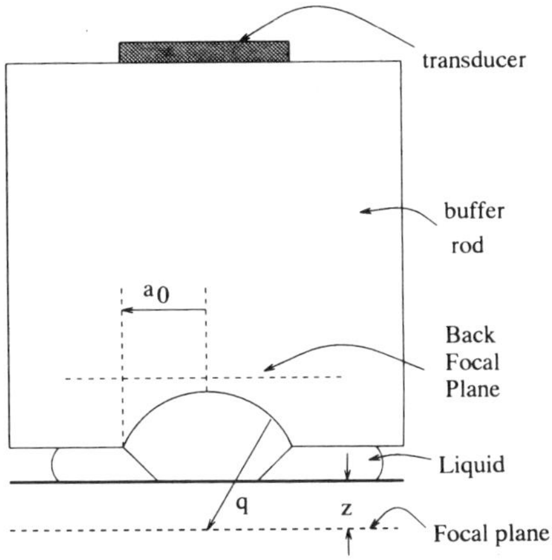

Figure 4.1. Geometry of a spherical lens.

to the orientation of a particular surface of a crystalline object. However, since different surfaces of a crystal may have different reflection characteristics, the conventional lens can differentiate between grains on a polycrystalline object and generate a considerable contrast in acoustic micrographs. This sensitivity can be improved by a suitable selection of the transducer size and shape.[12] The resolution of images is determined mainly by the operating frequency and the opening angle of the lens. Since tilting the lens axis with respect to the object surface is equivalent to increasing the opening angle of the lens, it is possible to obtain an increased resolution in one direction, but this method is not very popular.

In a multilayered solid structure, many leaky modes can be generated. A spherical acoustic microscope lens excites all these modes as long as they lie within the opening angle of the lens. The presence of many simultaneous modes clutters $V(z)$ curves with such objects and makes interpretation difficult. Nevertheless the spherical acoustic lens is still the most preferred lens design for imaging because of its good resolution performance.

To be able to compare lenses with different geometries, we write the specular reference component of the lens response in the following form:

$$V_G(z)\cos(2kz - \omega t) \tag{1}$$

and the leaky wave component in the form

$$V_R(z)\cos(2kz\cos\theta_c - \omega t) \tag{2}$$

where θ_c is the critical angle; k is the wave number in the liquid medium; ω is the carrier frequency; z is the distance between the focal plane and the object surface, increasing as separation between the lens and object increases; $V_G(z)$

and $V_R(z)$ are the normalized z variation of the specular geometric and leaky wave parts.[8] While $V_G(z)$ has a maximum at focus point, $z = 0$, $V_R(z)$ has a maximum at some negative z. For substantial interference between these two components, nonzero values of the functions $V_G(z)$ and $V_R(z)$ should overlap in a wide range of z values. Obviously it is desirable to have large values for both.

For a spherical lens, we can find the specular part by assuming a unity reflection coefficient and uniform insonification in the $V(z)$ integral

$$V_G(z) = \int_0^{a0} rR\left(\frac{r}{q}\right)[u_1^+(r)]^2\, P^2(r) \exp\left[i2kz\sqrt{1-\left(\frac{r}{q}\right)^2}\right] dr \qquad (3)$$

$$= \int_0^{a0} r \exp\left[i2kz\sqrt{1-\left(\frac{r}{q}\right)^2}\right] dr$$

where R is the reflection coefficient at the liquid–object interface as a function of the sine of the incidence angle; u_1^+ is the acoustic field at the back focal plane of the lens; P is the pupil function of the lens, which includes the transmission coefficient of the antireflection layer; q is the focal length of the lens; and a_0 is the lens pupil radius. With paraxial approximation the integral becomes

$$V_G(z) \approx \exp(2kz) \int_0^{a0} r \exp\left(\frac{-ikzr^2}{q^2}\right) dr \qquad (4)$$

$$= \left(\frac{a_0^2}{2}\right) \exp\left[i2kz\left(\frac{1-a_0^2}{4q^2}\right)\right] \sin\frac{(kza_0^2/2q^2)}{(kza_0^2/2q^2)}$$

Therefore $V_G(z)$ decays very rapidly as a *sinc* function of defocus distance, in the form $1/|z|$. The leaky wave part can be found by substituting $(k_p^2 - k_0^2)/(k_x^2 - k_p^2)$ for the reflection coefficient $R(k_x/k)$,[(8)] where k_p and k_0 are the leaky wave pole and zero

$$V_R(z) = \int_0^{R} r\frac{k_p^2 - k_0^2}{k^2(r/q)^2 - k_p^2}[u_1^+(r)]^2P^2(r) \exp\left[i2kz\sqrt{1-\left(\frac{r}{q^2}\right)}\right] dr \qquad (5)$$

Figure 4.2 depicts numerically calculated V_G and V_R for a spherical lens at 1 GHz, with unity insonification and pupil function. The $V_G(z)$ is like a *sinc* function as expected. It drops to −35 dB level at $z = -30$ μm. When aluminum is chosen as object material, the leaky wave part $V_R(z)$ is −7.3 dB at its maximum. When sapphire is the object material, $V_R(z)$ has a peak of −26 dB. Since the signals are either very low or drop very rapidly, the conventional lens is not the best lens for characterization purposes.

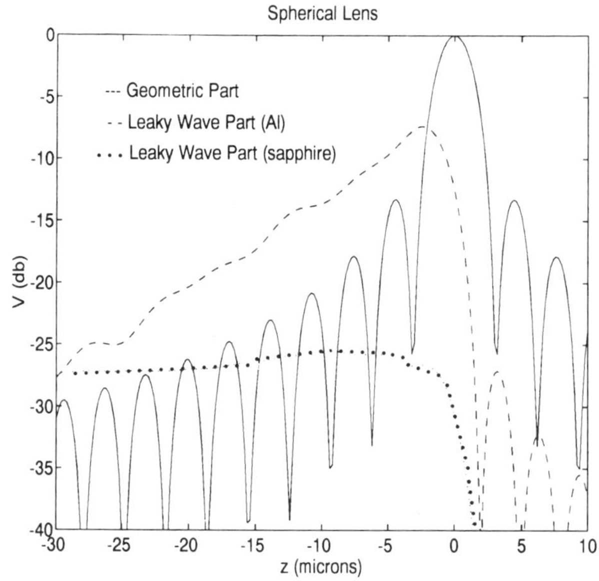

Figure 4.2. Calculated geometrical and leaky wave parts of $V(z)$ for a spherical lens (frequency, $f = 1$ GHz; focal length, $q = 115$ μm; pupil radius, $a_0 = 75$ μm). Leaky wave parts are shown for aluminum and sapphire.

4.2.2. Lamb Wave Lens

The Lamb wave lens (see Figure 4.3) was introduced to increase the leaky wave content of the acoustic lens response.[13] It excites only one of the possible leaky modes at a time by generating conical wave fronts at a fixed incidence angle.[14] Rays emerging from the central flat portion of the Lamb wave lens are incident normally onto the object surface, and they return to the lens to provide the reference $V_G(z)\cos(2kz - \omega t)$ term given above. Since the critical angle of a layered material depends on frequency, it is possible to excite the leaky modes selectively by matching the fixed incidence angle with the corresponding critical angle at a particular frequency. Since most of the incident rays are at or near the critical angle, the conversion efficiency to the leaky mode is much higher than that of the conventional lens. Moreover the reference signal is higher than the reference produced by a conventional lens, and this signal does not decrease appreciably when the lens is defocused; in other words, $V_G(z)$ function is nearly uniform.

The Lamb wave lens can be used for imaging, since it provides a reasonable focusing performance. Furthermore subsurface images have a well-defined coni-

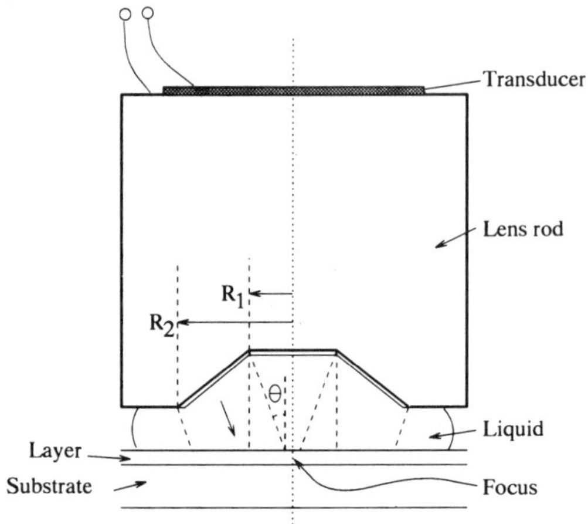

Figure 4.3. The geometry of the Lamb wave lens.

cal-focusing resolution, while corresponding images resolution obtained by the conventional lens is spoiled by aberrations due to refraction at the interfaces. We must note that the alignment of the Lamb wave lens with the object surface is difficult, since the flat part of the Lamb wave lens must be parallel to the object surface.

The $V(z)$ measurement resulting from a center-blocked Lamb wave lens gives an interference-free curve (see Fig. 4.4). The peak at $z = 0$ is due to specularly reflected rays. The second peak around $z = -4$ mm is created by leaky waves. The minimum at $z = -2.5$ mm is due to destructive interference between reflected and leaky waves, which are π out of phase. If waves incident from the central portion of the Lamb wave lens are allowed to overlap leaky waves, an interference mechanism similar to the conventional lens results. For this purpose, the center of the lens should not be blocked. As depicted in Fig. 4.4, a $V(z)$ curve with a good number of fringes is easily obtained. For large positive z values, no interference occurs. There the only contribution is from the normal specular component (see Fig. 4.5). At points near $z = 0$, interference results from the superposition of the normal specular component and the oblique specular reflection. The fringe period is determined by the cone angle of the Lamb wave lens, and it carries no or little information about the object. For more negative z values however, interference occurs between normal specularly reflected rays and leaky waves. In this case, periodicity is determined by the leaky wave velocity of the particular mode. Figure 4.6 shows the calculated geometric and leaky wave parts of $V(z)$ for a Lamb wave lens for aluminum and sapphire. As compared to a spherical lens, the leaky wave part is 6 dB greater for aluminum and 20 dB greater for sapphire. Figure 4.7 depicts the $V(z)$ variation for negative z values for a layered material when the c_{44} of the layer is perturbed by 1%. A

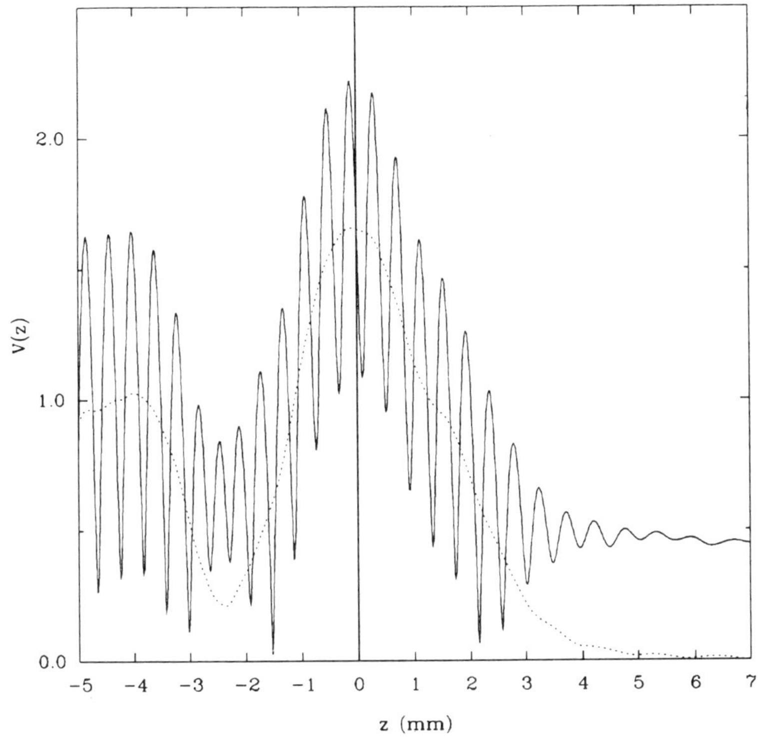

Figure 4.4. Lamb wave lens $V(z)$ curves (*center blocked and not blocked*) for 0.6-mm copper on steel (f = 9.6 MHz). Material parameters: Copper: v_l = 5010 m/s, v_s = 2270 m/s, ρ = 8.93 g/cm^3; steel: V_l = 5900 m/s, v_s = 3200 m/s, ρ = 7.9 g/cm^3. *Dashed line,* center blocked; *solid line*, center not blocked.

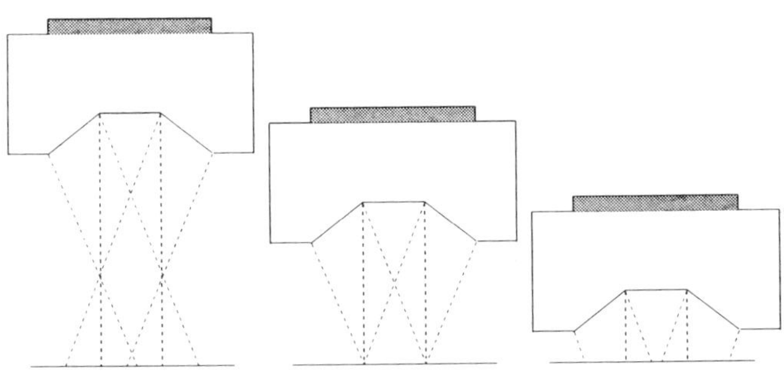

Figure 4.5. Beams generated by Lamb wave lens under different object positions.

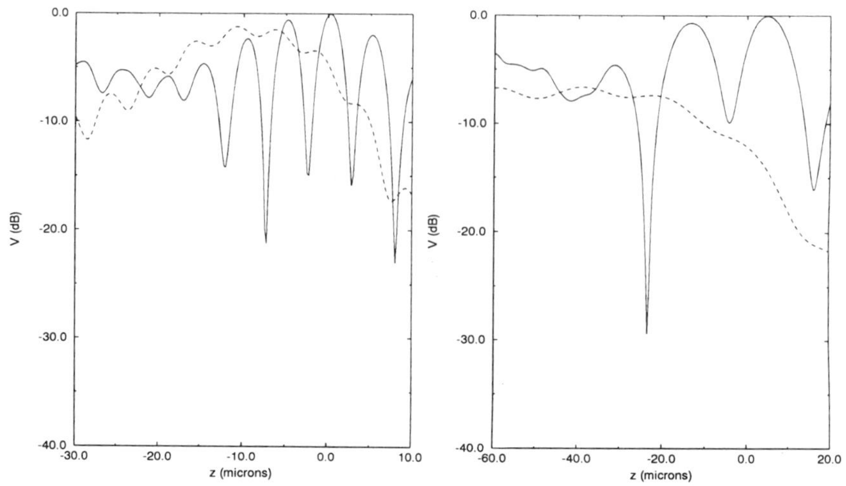

Figure 4.6. Calculated geometric and leaky wave parts of $V(z)$ for Lamb wave lenses designed for aluminum and sapphire objects (frequency, $f = 1$ GHz). For the first lens, $R_1 = 15$ μm, $R_2 = 30$ μm. For the second lens, $R_1 = 20$ μm, $R_2 = 40$ μm. *Dashed line*, leaky wave part; *solid line*, geometric part.

0.5% perturbation in the shear wave velocity is readily detectable. With a 40-dB signal-to-noise ratio, it is possible to detect a velocity perturbation of 0.0075% in the layer material. With the same signal-to-noise ratio, a conventional lens with an envelope detector can detect only a 0.12% velocity perturbation.

4.2.3. Directional Lenses with a Noncircular Transducer

To obtain orientation dependence, the circular symmetry of the lens must somehow be suppressed. It can be done on the transducer side by a noncircular-shaped bowtie,[15] butterfly,[16] rectangular,[17] two separate,[18] or shear wave generating transducers[19] (see Fig. 4.8). The shape of the transducer may emphasize certain directions, even though diffraction effects in the buffer rod tend to smooth out directivity. A shear wave transducer has a polarization direction, and hence circular symmetry does not exist. Moreover with shear wave excitation instead of longitudinal waves, specular reflection contribution at the output can be minimized. With such a lens, there are no incident rays normal to the object surface.[20] When such a lens is moved out of focus, the only contribution to the output is from leaky waves. However the output signal is rather dull, since there is no interference. All of these lenses have good resolution performance, and they can be used as imaging lenses. But diffraction in the lens rod limits their use as accurate characterization tools. Differential phase contrast lens[21,22] can also be considered in this class.

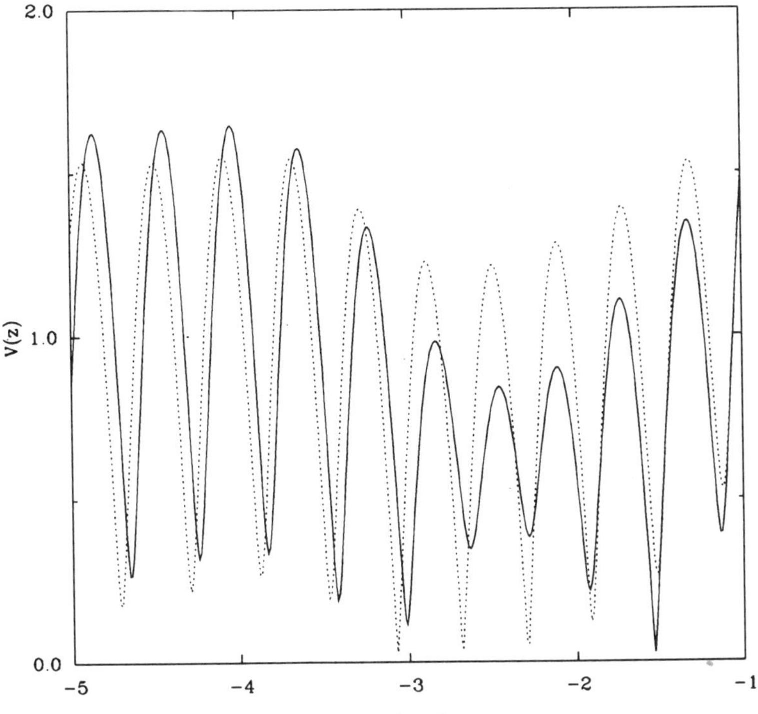

Figure 4.7. Lamb wave lens $V(z)$ curves for copper on steel (*solid*) and for a copper layer c_{44} perturbed by 1% (*dashed*) (f = 9.6 MHz).

4.2.4. Line Focus Beam Lens

Kushibiki and coworkers proposed and used an LFB acoustic lens. The virtues and limitations of this lens are extensively described in the literature.[23–26] Such a lens is formed by a cylindrical cavity in the lens rod (see Fig. 4.9). Generated rays hit the object surface in the same azimuthal plane, so leaky waves

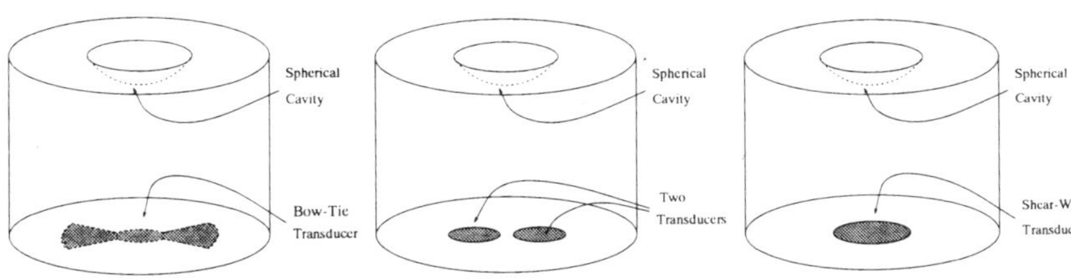

Figure 4.8. Bow-tie, two-transducer, and shear wave transducer lenses.

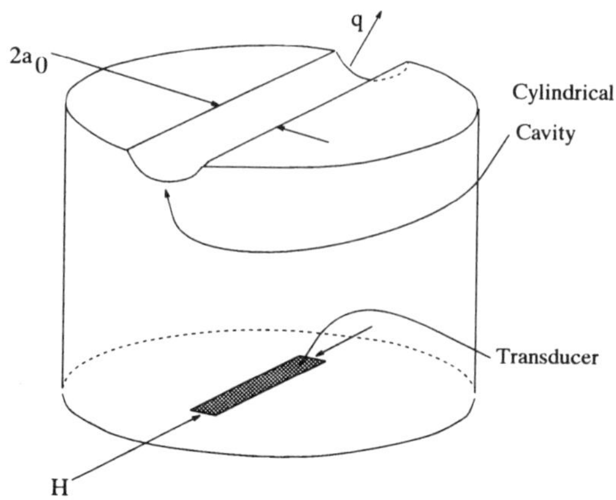

Figure 4.9. FB lens.

in that plane affect the response of the lens. This lens has poor resolution in one direction, and it cannot be used for imaging, but it has very good directional sensitivity, and it is suitable for characterizing anisotropic crystals.[27] Aligning the lens with the surface is more difficult compared to the conventional lens because the line focus must be parallel to the object surface. To compare the LFB lens with the conventional lens, we must determine the specular and leaky wave amplitudes. An expression for the geometric part $V_G(z)$ can be written as

$$V_G(z) = \int_0^{a0} R\left(\frac{r}{q}\right)[u_1^+(r)]^2 P^2(r) \exp\left[i2kz\sqrt{1-\left(\frac{r}{q}\right)^2}\right] dr \tag{6}$$

$$= \int_0^{a0} \exp\left[i2kz\sqrt{1-\left(\frac{r}{q}\right)^2}\right] dr$$

A paraxial approximation shows that $V_G(z)$ decays as $1/\sqrt{-z}$, and hence more slowly than the spherical case. The leaky wave part can be estimated from

$$V_R(z) = \int_0^{a0} \frac{k_p^2 - k_0^2}{k^2(r/q)^2 - k_p^2}[u_1^+(r)]^2 P^2(r) \exp\left[i2kz\sqrt{1-\left(\frac{r}{q}\right)^2}\right] dr \tag{7}$$

Note that these formulas differ from spherical lens formulas solely in the factor of r inside the integral. That factor exists in the spherical lens case due to the circular nature of the geometry. Figure 4.10 shows calculated $V_G(z)$ and $V_R(z)$ for an LFB lens with uniform insonification. The leaky wave part is about 3 dB

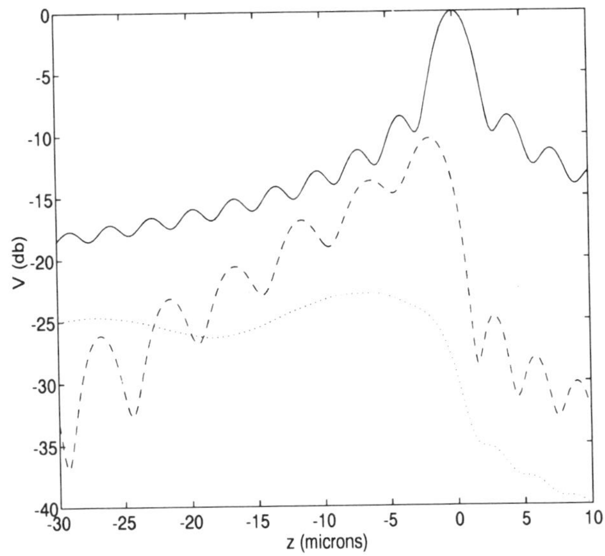

Figure 4.10. Calculated geometrical (solid line) and leaky wave parts of $V(z)$ for an LFB lens (f = 1 GHz, q = 115 μ, a_0 = 75 μm). Leaky wave parts are shown for aluminum (*dashed line*) and sapphire (*dotted line*).

smaller than the spherical case, as expected from geometrical considerations.* But, the geometric part is larger. The oscillations in the leaky wave part can be attributed to diffraction effects. LFB lens generates extended $V(z)$ interference and gives excellent characterization performance.

4.2.5. Slit-Aperture Lens

Lenses providing directionality in the transducer shape suffer from a diffraction effect in the buffer rod and provide only limited directionality, not sufficient for accurate measurement. If directionality is introduced in the lens pupil plane, diffraction in the lens rod plays no role. A slit aperture can be formed easily, and it provides sufficient directionality.[28] Figure 4.11 shows the geometry of the slit-aperture lens. It consists of a conventional acoustic lens with a specially designed aperture in front. The slit is formed by parallel edges of two absorbing sheets. The distance between the sheets determines the slit width. The directional

*The fraction of incident power that can be coupled to the leaky waves is $2\pi a_0\Delta/\pi a_0^2 = 2\Delta/a_0$ for the spherical lens and $2H\Delta/2Ha_0 = \Delta/a_0$ for the LFB lens, where Δ is the width of the narrow strip in the pupil function responsible for leaky wave excitation. Here the lens opening angle is chosen to be just sufficient to excite leaky waves.

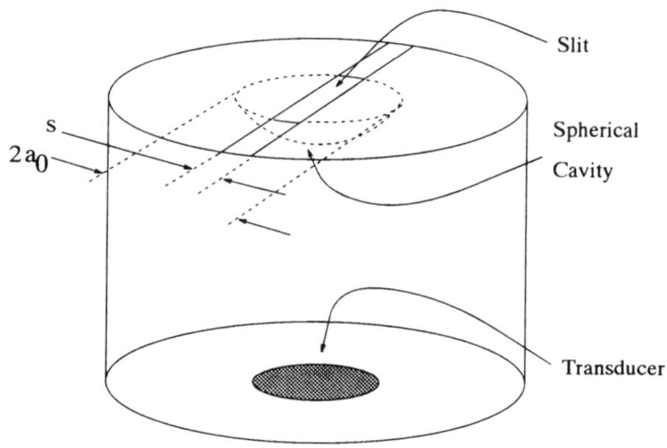

Figure 4.11. Geometry of the slit-aperture lens.

knife-edge-aperture lens[29] is essentially equivalent to this lens when the knife edge blocks less than half of the pupil. The parameter s is then twice the distance between the knife edge and the pupil center.

A diffraction-corrected ray theory approach for the slit lens is given in Ref. (30). Although the ray theory provides very valuable physical insight into the problem, it cannot handle the problem correctly near the focus point, and it gives only an approximate solution. Here we give an expression for the slit-aperture lens $V(z)$ using the angular spectrum approach.[31]

$$V(z) = \int_0^{a0} R_e\left(\frac{r}{q}\right)[u_1^+(r)]^2 \exp\left[i2kz\sqrt{1-\left(\frac{r}{q}\right)^2}\right] dr \tag{8}$$

where

$$R_e\left(\frac{r}{q}\right) = r\int_0^{2\pi} R\left(\frac{r}{q}, \phi\right) P^2(r, \phi)\, d\phi \tag{9}$$

Here R_e is the average reflection coefficient of an anisotropic material as a function of the sine of the incidence angle. For a slit lens, the pupil function can be written as

$$P(x, y) = \begin{cases} 1 & \text{if } |x| < s/2 \quad \text{and} \quad x^2 + y^2 \leq a_0 \\ 0 & \text{otherwise} \end{cases} \tag{10}$$

where s is the slit width and a_0 is the lens aperture radius. Note that the apodization effect caused by the matching layer on the lens can be included in the function $u_1^+()$.

Calculating the reflection coefficient at a liquid–anisotropic solid interface is not a trivial procedure, since an analytic expression does not exist. Therefore numerical methods are used to calculate it.[32–34] In its most general form, the anisotropic solid is represented by its 21 elastic constants, so materials with arbitrary orientation can be handled by transforming the stiffness matrix by multiplying with Bond matrices.

We have calculated the $V(z)$ response of a slit-aperture lens with the following parameters: Operation frequency of 200 MHz, lens cavity radius of 500 μm, lens aperture radius a_0 = 435 μm, transducer radius a_T = 435 μm, lens buffer rod length L = 10mm. Slit width s was kept as a variable parameter. Various single-crystal solids are considered. Figure 4.12 shows the calculated $V(z)$ curves for a GaAs substrate (001) surface. Different curves are for different orientations of slit with respect to crystal symmetry axes. For a slit width equal to one-tenth of the lens aperture, the loss in the signal level is about 15 dB. Figure 4.13 depicts a similar set of $V(z)$ curves for the (111) surface of GaAs. Calculations are repeated for different slit widths. As the slit width is increased, the resulting $V(z)$ curves converge to each other for different orientations of an anisotropic crystal. As the slit width is reduced, there is a loss in the signal output level, but different orientations have different responses.

Figure 4.14 shows the results of simulations for varying slit widths for a GaAs (111) surface. In the same figure, the actual SAW velocity is indicated as a function of propagation direction. For slit widths less than 5% of the aperture size, the extracted velocities follow the actual velocity curve closely. But as slit size increases, the error between the estimated and actual velocities increases very rapidly. When the slit size is greater than 20% of the aperture size, estimates are quite unreliable.

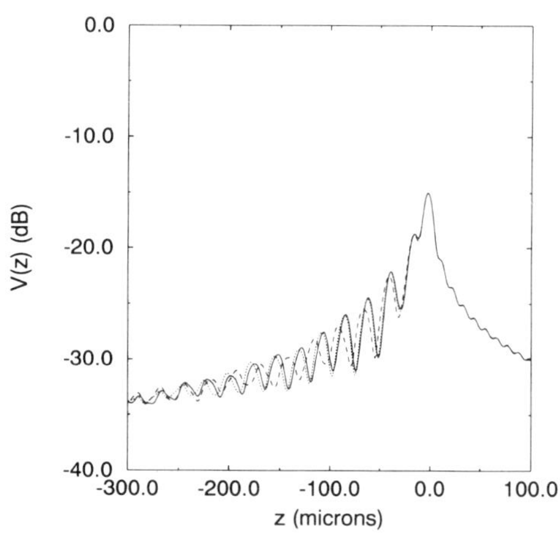

Figure 4.12. Calculated $V(z)$ curves for GaAs (001) surface at various directions with a slit-aperture lens. $S = 0.1a$; solid line, 0 degrees; dotted line, 15 degrees; dashed line, 30 degrees.

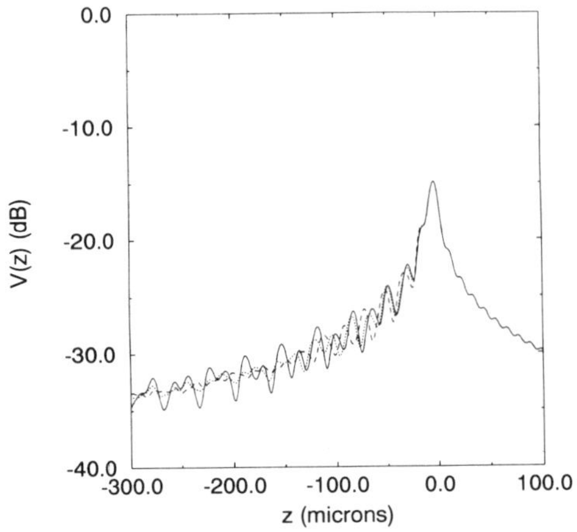

Figure 4.13. Calculated $V(z)$ curves for a GaAs (111) surface at various directions with a slit-aperture lens. $S = 0.1a_0$; solid line, 0 degrees; dotted line, 10 degrees; dashed line, 20 degrees.

4.2.6. V-Groove Lens

The V-groove lens is a combination of an LFB lens and the Lamb wave lens.[35] It inherits the high efficiency of the Lamb wave lens, while providing the high directionality of the LFB lens. The V-groove lens differs from the LFB lens in the way the refracting element of the lens is fabricated. Unlike the cylindrical cavity of the LFB lens, the V-groove lens has a V-shaped groove with a flattened bottom, as shown in Fig. 4.15. Essentially, the relationship between an LFB lens and a conventional lens is the same as the relationship between the V-groove lens

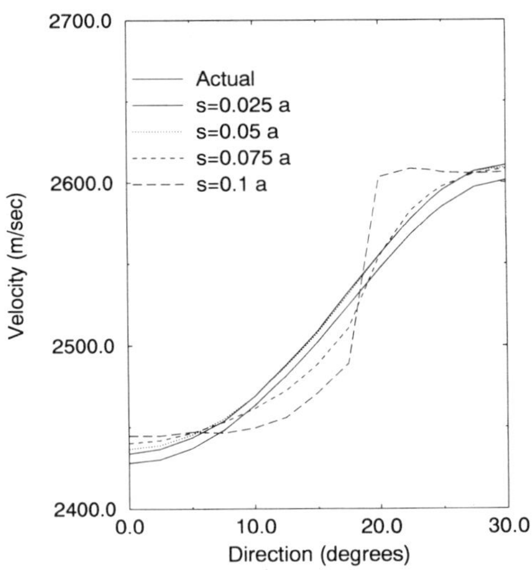

Figure 4.14. Actual and estimated velocities on a GaAs (111) surface as a function of direction under varying slit widths.

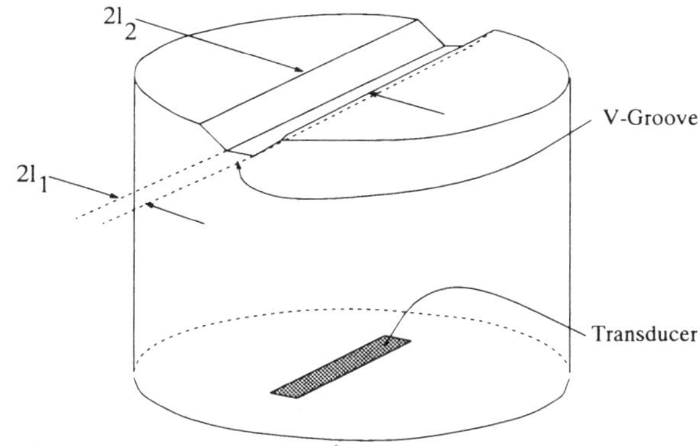

Figure 4.15. Geometry of the *V*-groove lens.

and the Lamb wave lens.[13] In other words, leaky waves are selectively excited on the surface of the object when a V-groove lens is used, while in the case of an LFB lens, all modes are excited due to its wide angular spectrum.

The transducer is designed to insonify a substantial portion of the groove with a minimal waste of power elsewhere. The flat bottom part does not cause refraction, and thus a part of the incident beam insonifies the object surface at normal incidence. Symmetrical sides of the groove cause a refraction, and hence two symmetrical beams insonify the object surface at the same incidence angle. The interference of the refracted beams, which encounters leaky wave modes on the object surface, with the reference beam resulting from the specular reflection of the normally incident beam produces the $V(z)$. A good match between the median direction of the refracted beam and the critical angle for the object improves measurement accuracy.

Simulating the performance of a lens involves propagating acoustic waves between the transducer and the refracting element. The wave front is then propagated through the refracting element using ray theory. The wave front is reflected from the object surface on propagation in liquid. This analysis is similar to the one developed for the Lamb wave lens,[13] except for the circular symmetry. While the circular symmetry of the Lamb wave lens allows the use of a fast Hankel transform for propagation purposes, the propagation problem in a V-groove lens requires the more costly two-dimensional Fourier transform (FFT).

Lens alignment is achieved easily by maximizing the signal from the flat part of the lens, since the maximum signal is reached when the object surface is perfectly parallel to the flat part of the V-groove. A simulated $V(z)$ curve for a V-groove lens with the (001) surface ([100] propagation) of silicon as the reflector is depicted in Fig. 4.16 with experimental measurements.

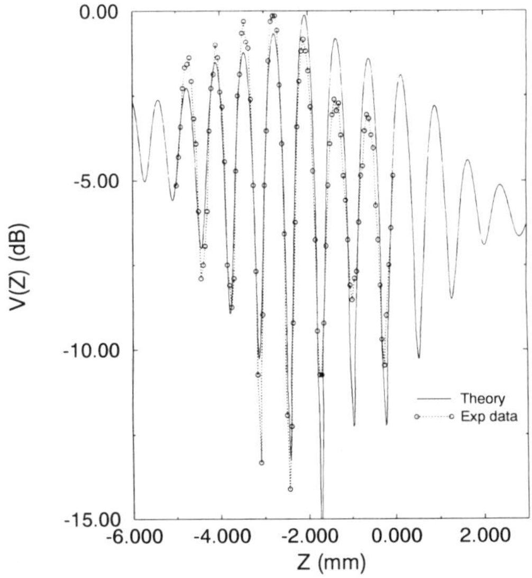

Figure 4.16. Calculated (*solid line*) and measured (*dotted line*) $V(z)$ values along [100] direction on (001) surface of Si at $f = 25$ MHz for a V-groove lens.

To compare the performance of the V-groove lens with other lenses, geometric and leaky wave parts of a received signal are calculated for typical lenses. Figure 4.17 is a plot of the results. Obviously the geometric part is much greater than either the LFB or the spherical lens. The leaky wave part is about 12 dB greater than the LFB lens.

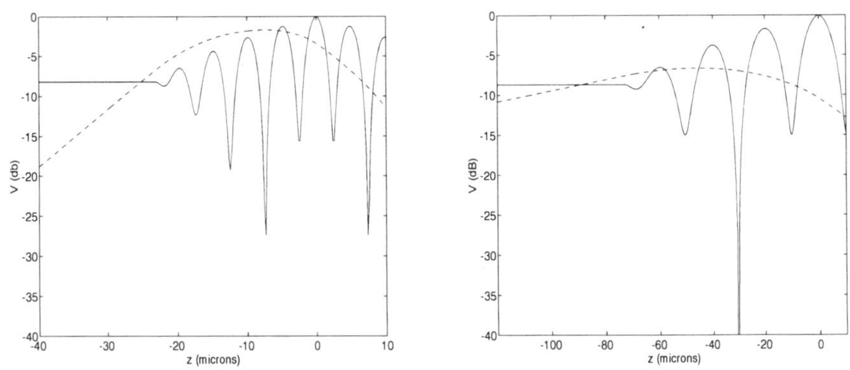

Figure 4.17. Calculated geometrical (*solid line*) and leaky wave (*dashed line*) parts of $V(z)$ for V-groove lenses designed for aluminum (right) and sapphire (left) objects ($f = 1$ GHz). (a) $l_1 = 14$ μm, $l_2 = 36$ μm. (b) $l_1 = 22$ μm, $l_2 = 60$ μm. Diffraction effects are neglected.

4.3. Comparing Signal Processing Electronics

In this section we compare various signal-processing systems presently in use for acoustic microscopy. To be able to compare them quantitatively, we define a figure of merit and calculate it for all systems under similar conditions. A white gaussian noise $n(t)$ with noise power W is considered. We find the response of the particular configuration with respect to variations in the critical angle and surface topography.

4.3.1. Conventional Envelope Detection System

The conventional acoustic microscope system shown in Fig. 4.18 uses an envelope detector to obtain the output signal. Such a detector can have no sensitivity to the phase of the received signal. This is not a great disadvantage, since the microscope output signal is the result of an interference anyway. The low-pass, filtered linear envelope detector output can be written as

$$V(z) = LPF\{|V_G(z)\cos(2kz - \omega t) + V_R(z)\cos(2kz\cos\theta_c - \omega t) + n(t)|\} \qquad (11)$$

where $n(t)$ denotes the noise and LPF [.] the low pass filtering operation. A typical $V(z)$ curve is shown in Fig. 4.19. Quantitative object parameters can be deduced either by an inversion integral applied to the $V(z)$ data[36] or by Fourier transformation of data points to determine the characteristic periodicity.[24]

4.3.2. Amplitude and Phase-Measuring Synchronous Detection System

It is possible to detect the phase of an acoustic microscope output voltage with a phase-sensitive receiver system.[37] In such a system, the received signal is multiplied with a reference signal obtained from the carrier (see Fig. 4.20).

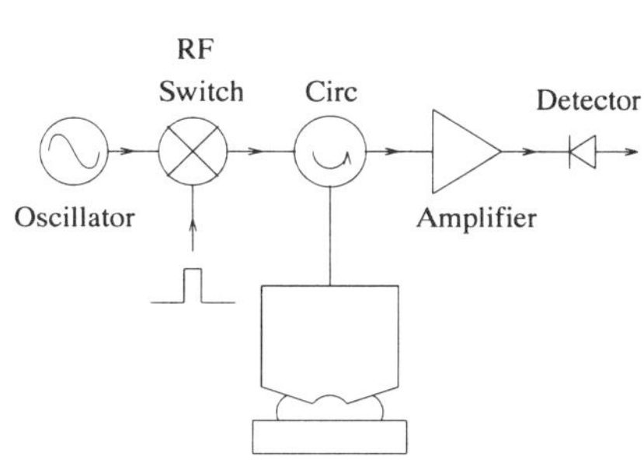

Figure 4.18. A conventional acoustic microscope system.

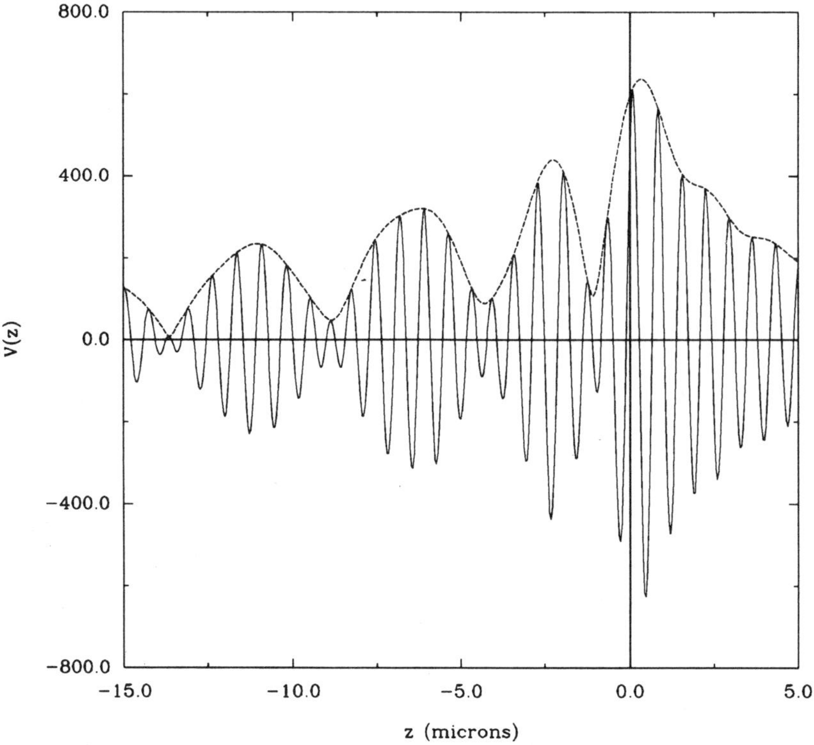

Figure 4.19. $V(z)$ curves for aluminum for conventional (*dotted line*) and phase-measuring (*solid line*) systems ($f = 1100$ MHz).

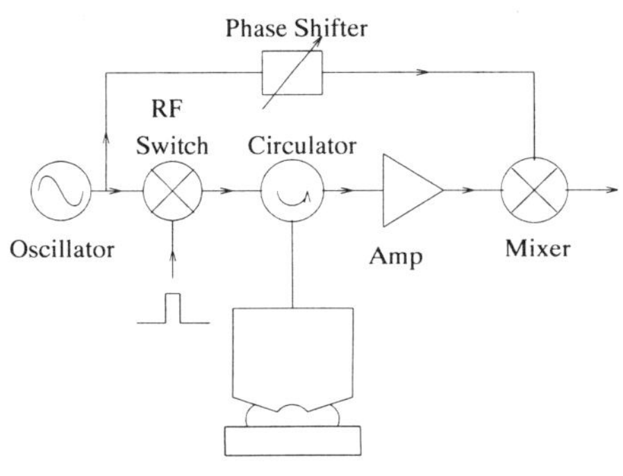

Figure 4.20. A phase-measuring acoustic microscope system.

Low-pass, filtered mixer output gives a voltage highly sensitive to the surface topography of the object material. This signal can be expressed as

$$V(z) = V_G(z)\cos(2kz + \psi) + V_R(z)\cos(2kz\cos\theta_c + \psi) + n_c(t)\cos\psi \qquad (12)$$

where $n_c(t)$ is the in-phase component of the noise and ψ is the phase shift of the carrier. The output voltage varies sinusoidally with distance with an envelope that is the same as that obtained from the conventional system (see Fig. 4.19). If ψ is changed in small steps and the output signal is measured, it is possible to find the complex amplitude of the signal. With suitable averaging methods, it is possible to detect surface height variations less than 1/500 of the wavelength[38] or to measure residual stresses.[39]

4.3.3. Conventional System with Added Carrier

A conventional system with a reduced on/off ratio in the pulse generation electronics may result in a phase-sensitive system without the extra cost of phase-measuring electronics. In this case, carrier leakage causes a pseudo-phase-sensitive system. This leakage can be introduced intentionally in a controlled manner (see Fig. 4.21). After detection the output signal contains components that vary sinusoidally as a function of z, the lens object separation in Fig. 4.22. The resulting envelope detector output is written as

$$V(z) = LPF\{|V_G(z)\cos(2kz - \omega t) + V_R(z)\cos(2kz\cos\theta_c - \omega t) + C\cos(\psi - \omega t) + n(t)|\} \qquad (13)$$

where C is the magnitude of the leak component and ψ is its phase. If the amount of leakage can be varied, a maximum sensitivity can be reached.

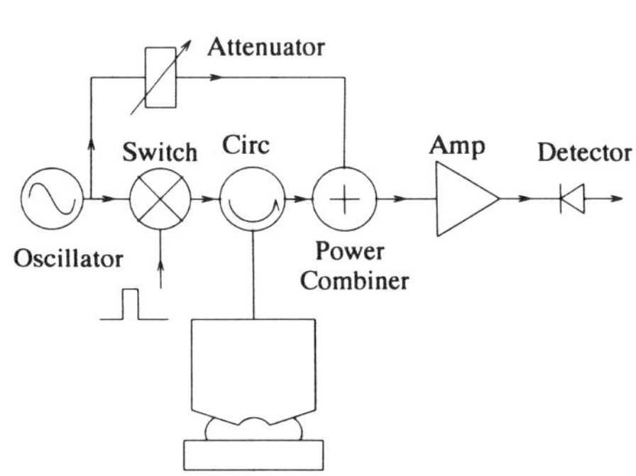

Figure 4.21. A conventional system with some added carrier.

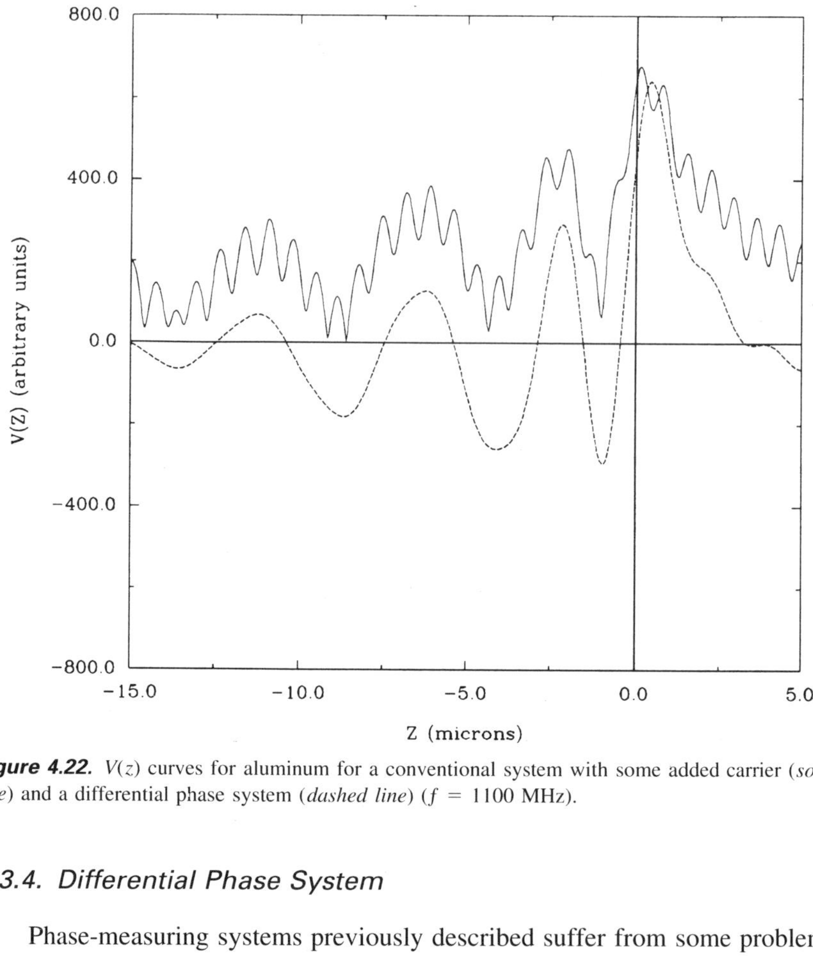

Figure 4.22. $V(z)$ curves for aluminum for a conventional system with some added carrier (*solid line*) and a differential phase system (*dashed line*) (f = 1100 MHz).

4.3.4. Differential Phase System

Phase-measuring systems previously described suffer from some problems that reduce measurement accuracy. Such external factors as vibration of the mechanical-scanning stage or temperature fluctuations in the coupling medium result in phase errors. These phase errors reduce the accuracy of quantitative measurements. To isolate the external factors, a reference signal that undergoes the same phase fluctuation may be used. This reference may be the specular reflection component. The procedure is to separate the specular and leaky wave contributions in the time domain and to multiply these two signals after proper delay (Fig. 4.23). Low-pass filtered mixer output can be given as

$$V = V_G(z)V_R(z)\cos[2kz(1 - \cos\theta_c)] + noise \tag{14}$$

This setup provides excellent sensitivity to material properties, since it can measure small variations in the phase of leaky waves. In addition to having a

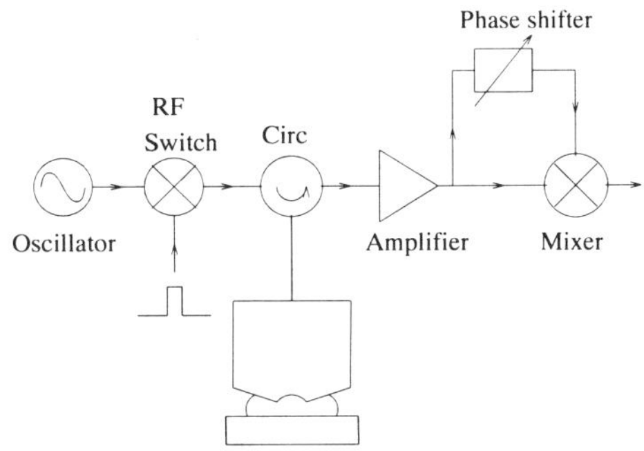

Figure 4.23. A differential phase-measuring acoustic microscope.

reduced sensitivity to external factors, it is insensitive to small surface height variations. As shown in Fig. 4.22, the output signal is predominantly determined by object-elastic properties. The greatest difficulty is to separate these two signals in the time domain. For this purpose, lenses must have wide bandwidths to pass short pulses, and a substantial defocus is necessary, causing a loss in resolution. Special geometries involving shear wave or mixed-mode transducers can be employed to obtain separated pulses without a great resolution sacrifice.[40]

4.3.5. Comparing Systems

We can compare the performance of these schemes in the presence of noise for either critical angle measurement or surface topography. In the first case, we define a figure of merit FM_θ as

$$FM_\theta = \frac{(\partial V/\partial\theta_c)^2}{(N_0 * SNR_i)} \tag{15}$$

where SNR_i is the signal-to-noise ratio defined as total power input to the lens divided by noise power W. The N_0 is the output noise power; N_0 may contain signal cross noise terms depending on the processing scheme employed, and in general it is not equal to W. This figure of merit indicates how sensitive the system is as a function of critical angle at a given defocus. Moreover it includes the effects of the presence of noise. In the second case, we define another figure of merit, FM_z, which shows the sensitivity with respect to surface height z.

$$FM_z = \frac{[(1)\partial V/2k(\partial z)]^2}{(N_0 * SNR_i)} \tag{16}$$

We assume a unity total input power with a sufficiently high signal-to-noise ratio. In the carrier-added system, the added carrier amplitude is much higher than other signal components. The choice of ϕ and z is made such that expressions are simplified and figure of merit values are maximized. The results are presented in Table 4.1 in terms of a normalized geometric part V_G, a leaky wave part V_L, and the critical angle θ_c. Inspecting the FM_θ formulas indicates that critical angle sensitivity grows as $|kz|$ is increased. Moreover higher critical angles result in more sensitive systems. The synchronous system is the best. The carrier-added system approaches the performance of the synchronous system if the added carrier is sufficiently large. The envelope detector system is the worst. In all cases, increasing the geometric part and the leaky wave part is rewarding. Formulas for FM_z show that the synchronous system also gives the best z sensitivity. The least sensitive system is again the envelope detection scheme.

We used these formulas and typical V_G and V_L values for different lens geometries. Tables 4.2 and 4.3 give typical values assuming a large signal-to-noise ratio at the input. This assumption makes sure that the envelope detector operates above the threshold.

Results show that the best system for critical angle measurement is the synchronous detection scheme. All other systems have similar but lower figures of merit. The conventional envelope detection system performance deteriorates according to the expression given in Table 4.1 as the input signal-to-noise ratio falls below the detector threshold. Moreover the nonlinearity of the detector diode may introduce some errors in quantitative measurements. For low signal levels, dropping below the threshold can be avoided by increasing the added carrier amplitude. Synchronous detector and differential phase systems, on the other hand, maintain the same figure of merit even for small input signal levels. The differential phase system is inferior to the synchronous detector, since what may be considered as the carrier is also corrupted with noise. As already mentioned, the differential phase system gives the highest performance configuration because of its relative insensitivity to external factors.

As far as measuring surface topography is concerned, the synchronous system is the most sensitive by a large margin. It is followed by the carrier-added envelope detector scheme. The other two methods are relatively insensitive to surface topography. In those systems, reduced leaky wave contribution results in diminished z sensitivity.

Although the synchronous detection scheme gives the best figures of merit, implementation cost is high. The envelope detector with carrier-added provides

Table 4.1. Figures of Merits of Different Signal-Processing Systems with Respect to Critical Angle and Topography

System	FM_θ (Critical angle figure of merit)	FM_z (Topography figure of merit)
Envelop detection	$V_G^2V_L^2(2kz\sin\theta_c)^2/(V_G+V_L)^2$	$V_G^2V_L^2(1-\cos\theta_c)^2/(V_G+V_L)^2$
Carrier added	$< V_G^2(2kz\sin\theta_c)^2$	$< (V_G+V_L\cos\theta_c)^2$
Synchronous	$V_G^2(2kz\sin\theta_c)^2$	$(V_G+V_L\cos\theta_c)^2$
Differential phase	$V_G^2V_L^2(2kz\sin\theta_c)^2/(V_G^2+V_L^2)$	$V_G^2V_L^2(1-\cos\theta_c)^2/(V_G^2+V_L^2)$

Table 4.2. Typical Critical Angle Figure of Merit Values for $kz = 27\pi$ and $\theta_c = 30$[a]

	FM_θ			
System	Spherical[b]	Lamb[c]	LFB[d]	V-Groove[e]
Envelope detection	0.034	0.84	0.14	1
Synchronous	0.84	3.4	0.41	5.2
Differential phase	0.049	1.7	0.28	2

[a]These are typical values when aluminum is used as the object.
[b]For a spherical lens: $V_G = 0.05$, $V_L = 0.2$.
[c]For a Lamb wave lens: $V_G = 0.4$, $V_L = 0.4$.
[d]For an LFB lens: $V_G = 0.2$, $V_L = 0.14$.
[e]For a V-groove lens: $V_G = 0.4$, $V_L = 0.5$.

a significant improvement over the conventional system with little or no cost. Expressions given for FM_θ indicate that a large defocus factor kz is preferable. In all cases, figures of merit improve sharply with increased leaky wave signal amplitude. Therefore lens configurations maximizing this amplitude have higher figure of merit regardless of the detection scheme employed. However when the leaky wave signal is comparable with the geometrical signal, analysis of $V(z)$, based on a linear approximation can no longer be used.[(41)]

4.4. The V(f) Characterization Method

In Section 4.4 we describe an alternative to the $V(z)$ characterization technique. This method is applicable to acoustic lens systems with restricted angular coverage and a reasonable frequency bandwidth. The output voltage of a Lamb wave or a V-groove lens recorded as a function of frequency results in a unique curve [$V(f)$].[(14)] Excited leaky modes reveal themselves as peaks in the $V(f)$ curve at frequencies where the critical angle of the mode matches the fixed incidence angle of the lens. Since the excitation frequency of leaky modes is highly dependent on elastic parameters of the layers and the bond quality at interfaces, peak positions in the $V(f)$ curve are very sensitive to these parameters.

Table 4.3. Typical Topography Figure of Merit Values for $kz = 27\pi$ and $\theta_c = 30$

	FM_z	
System	Spherical	Lamb
Envelope detection	0.000028	0.00071
Synchronous	0.049	0.56
Differential phase	0.000042	0.0014

A typical $V(f)$ curve for a center-blocked Lamb wave lens is shown in Fig. 4.24. If the center is not blocked, a more interesting $V(f)$ curve results. As depicted in the same figure, this curve has an interference pattern that results in high sensitivity. Since a signal is now present even in the absence of leaky waves, a background level always exists. This background level can be avoided if the specular and leaky wave components are separated and mixed with each other. Separation can be achieved in the time domain. As plotted in the same figure, this pattern is free from background while maintaining high sensitivity.

For the purpose of investigating the performance of the Lamb wave lens in a layered structure, a sample composed of silver and copper layers on a steel substrate is considered. Dispersion characteristics of this structure, with perfect bonds at the interfaces, is depicted in Fig. 4.25. The vertical axis shows the incidence angle in a water medium for which the best excitation of a particular mode is achieved. It can be observed that two modes can be excited at approximately 35° in the frequency range of 4–9 MHz.

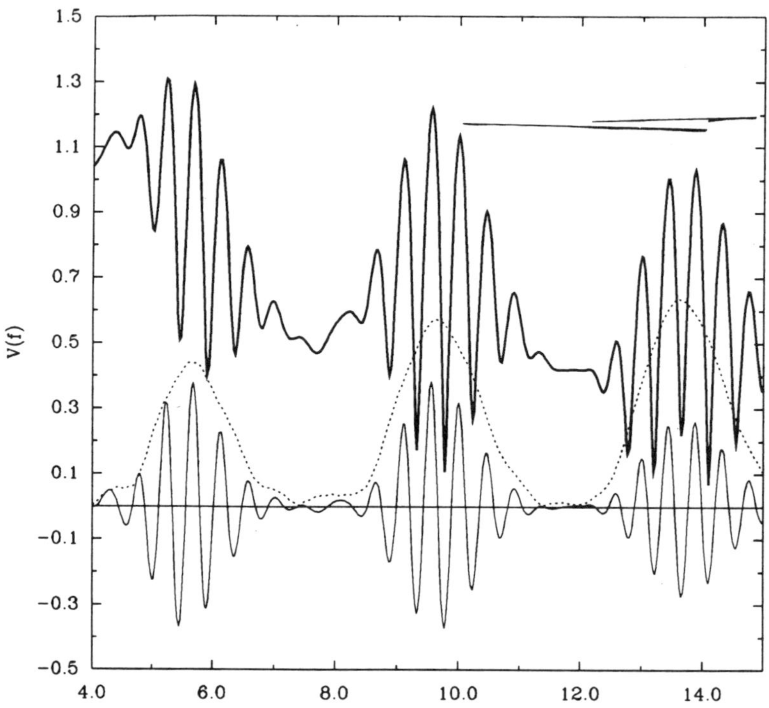

Figure 4.24. $V(f)$ curves for 0.6-mm copper on steel using different Lamb wave systems. *Dotted line,* center blocked; *heavy line,* center not blocked; *thin line,* center and side mixed.

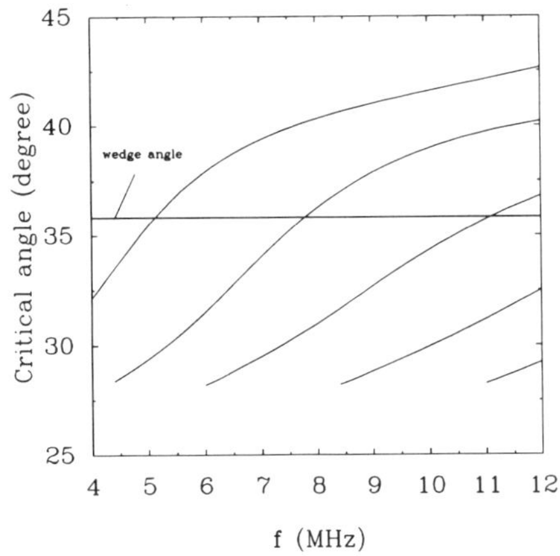

Figure 4.25. Dispersion characteristics of modes in a multilayered structure composed of 90 microns of silver and 505 microns of copper on steel substrate. A 35.8° wedge angle shows the angle of incident waves generated by the Lamb wave lens used in experiments.

A better understanding of the reflection phenomenon in layered materials can be achieved by decomposing the reflection coefficient R into two parts[8]: A surface reflection coefficient R_s and a subsurface reflection coefficient R_u. The R_s is due to the interface between the liquid and the topmost layer, excluding effects of leaky waves and subsurface layers. This part is found by considering a half-space made up of the same material as the topmost layer, with the leaky wave component suppressed by setting phase variation in this reflection coefficient to zero[42]

$$R_s = |R_T|$$

where R_T is the reflection coefficient between the liquid and the half-space made up of the topmost layer material. The subsurface reflection coefficient is found by subtracting the surface reflection coefficient from the original, $R_u = R - R_s$. The subsurface reflection coefficient includes effects of leaky waves as well as subsurface specular reflections. Figure 4.26 depicts the magnitude of the calculated subsurface reflection coefficient for the material just described. Leaky waves, such as leaky Rayleigh, leaky Lamb waves, and lateral waves, emerge as peaks at corresponding incidence angles. The magnitude of the peaks can be greater than unity but always less than or equal to 2. This is due to opposing phases of the surface and subsurface reflection coefficients. Peaks arise from Lamb and lateral waves. The width of the peaks indicates the angular spread within which the mode can be efficiently excited. This in turn determines the aperture size of the Lamb wave lens for the most efficient excitation.

Peaks in the $V(f)$ curve are related to particular Lamb wave modes. We may suspect that each of these modes is predominantly present in one of the layers of the multilayered structure. To determine the validity of this hypothesis,

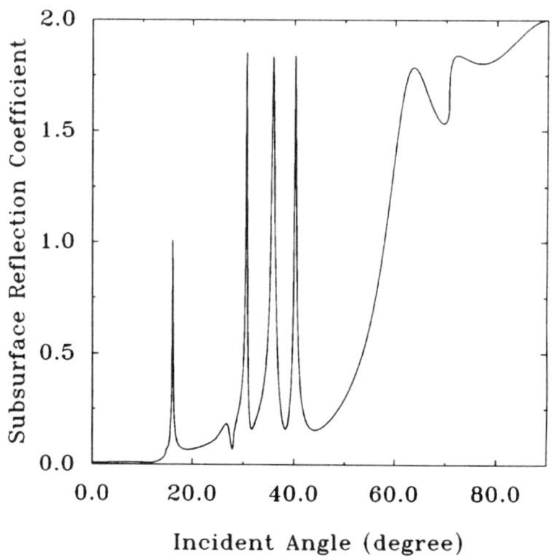

Figure 4.26. A subsurface reflection coefficient for the multilayered structure composed of 90-μm Ag and 505 μm of Cu on a steel substrate at 7.8 MHz.

we calculated the lateral component of the Poynting vector for plane wave excitation for various critical angles. The poynting vector is a measure of energy flow. Figure 4.27 shows the results for a multilayered structure under two different excitation angles. There is a concentration of acoustic energy along particular interfaces. Hence different interfaces are responsible for modes at these two angles. Once this relationship is ascertained, images obtained from these modes can be attributed to those interfaces. Therefore the Lamb wave lens has the inherent ability to generate selective interface images.

The effectiveness of a Lamb wave lens using the $V(f)$ technique for quantitative characterization can be evaluated for the sample described in the previous section. Figure 4.28 depicts the calculated $V(f)$ curve and measurements for the case when both bonds are perfect. Measurements are performed by using a Lamb wave lens that has the same wedge angle of 45.9°. A very good agreement is obtained between experiment and theory. The $V(f)$ measurements and calculated curve for a region with a disbond in the upper interface is given in Fig. 4.29. A slight shift can be observed in the curve as expected. The effect of a disbond in lower interface on V(f) can be observed in Fig. 4.30. The central mode is now reduced to 7 MHz.

We may exploit the sensitivity of the Lamb wave lens by generating a *peak frequency image* (PFI), which is obtained by mapping the peak frequency of a particular leaky mode at every point in the form of an image. The PFI provides information directly about the spatial variation of parameters of the layers and the bond quality at interfaces. Considering that each layer and interface in a multilayered material affect these modes in a different manner, the effect of each parameter at a desired layer and interface can be selectively observed and

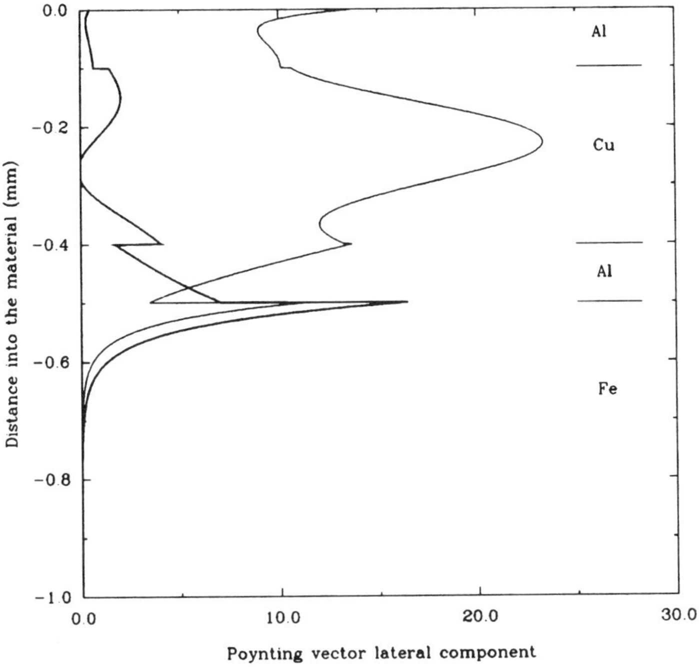

Figure 4.27. The lateral poynting vector component as a function of distance in the three-layer material. $f = 7$ MHz; *thin line,* theta = 28.4; *heavy line,* theta = 32.6.

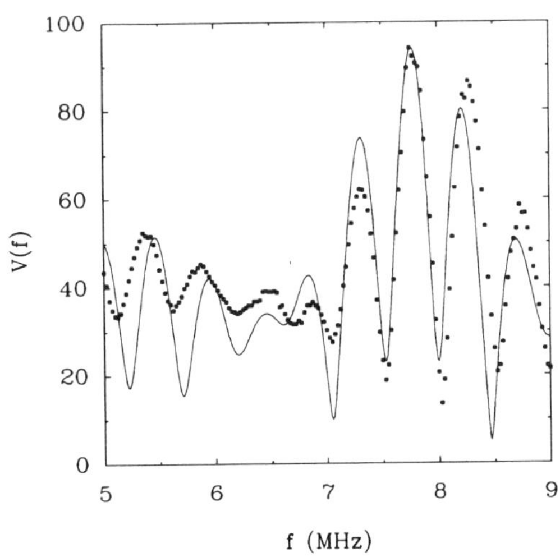

Figure 4.28. Measured (*squares*) and calculated (*line*) $V(f)$ curves (good bond) for the Ag–Cu–steel sample.

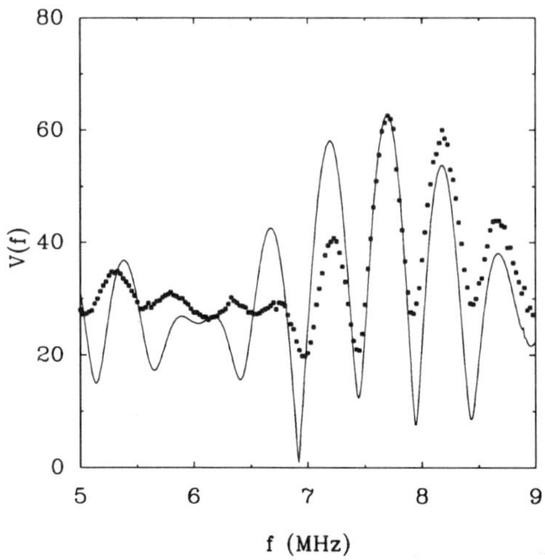

Figure 4.29. Measured (*squares*) and calculated (*line*) $V(f)$ curves (disbond at upper interface) for the Ag–Cu–steel sample.

characterized by the Lamb wave lens. The PFI images can be obtained very fast because there is no need to scan the object in the z direction, an operation that is inherently very slow. This mode of imaging is preferable, since it reduces the ambiguity that may arise in conventional imaging, where changing z may cause contrast reversal. Figure 4.31 shows a PFI image obtained from a Lamb wave lens for a copper layer on steel where a slippery bond (disbond)[43] is intentionally induced in a restricted region. This region shows up with a high contrast in the image.

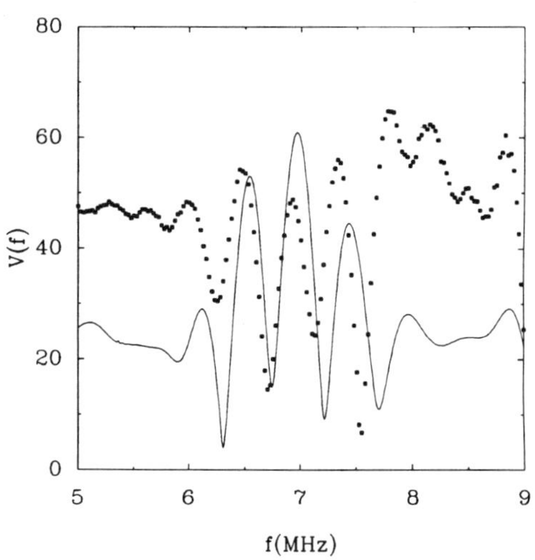

Figure 4.30. Measured (*squares*) and calculated (*line*) $V(f)$ curves (disbond at lower interface) for the Ag–Cu–steel sample.

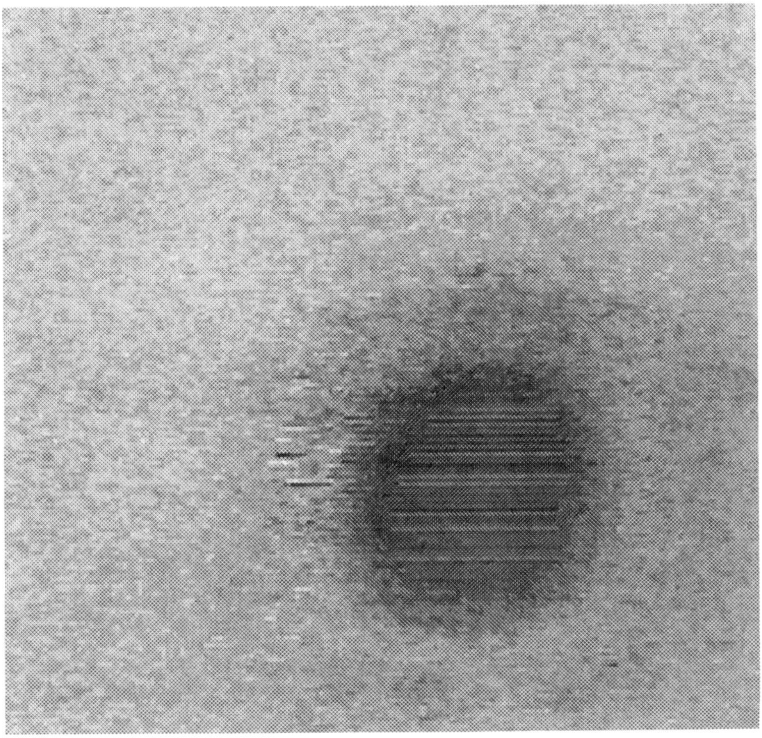

Figure 4.31. The peak frequency image of a 0.6-mm thick copper layer on steel obtained by a Lamb wave lens. Frequency range is 7.2–8.7 MHz. The interface contains an intentionally generated disbond region. Image dimensions are 20 mm by 20 mm.

4.5. Accuracy of Velocity Measurement Using the V(z) Method

The LFB, slit-aperture and V-groove lenses are primarily used to determine leaky wave velocities on anisotropic material surfaces. Spherical and Lamb wave lenses can also be employed for similar purposes when the material is isotropic. The $V(z)$ curves obtained by these lenses are analyzed and velocities are found. The procedure for extracting velocity information[41] from $V(z)$ can be summarized as follows

- Obtain $V(z)$ for the object.
- Obtain $V_{ref}(z)$ for the same object with its reflection phase zeroed. Experimentally $V_{ref}(z)$ can be obtained by using Pb as the object.
- Find $V^2(z) - V^2_{ref}(z)$. If the leaky wave content is small, $V(z) - V_{ref}(z)$ can be used instead.[23]
- Pad data with zeroes.

- Apply FFT to find the period of oscillation.
- Determine velocity from the period.

To deduce the absolute accuracy of this method, we performed a series of simulations for LFB and V-groove lenses. As the object material, we used a number of isotropic crystals whose elastic constants are known. From the known constants of crystal and liquid medium first, we determined the leaky surface wave velocity. This involves calculating complex poles of the reflection coefficient. Calculated Rayleigh velocity values are shown in the first column of Tables 4.4 and 4.5. Then we tried to find the same velocity using the $V(z)$ method. The $V(z)$ responses are calculated from the known elastic parameters of the materials and the known geometry of the lenses. These responses are then input into the velocity extraction algorithm using the FFT method.[(23)] Calculations are made under various assumptions. First uniform insonification of the lens is assumed. Then the actual field pattern calculated by using diffraction effects in the buffer rod is used. Finally the reference material is chosen as Pb rather than the phase-zeroed reflection coefficient. Different assumptions result in slightly different estimates. Extracted velocities are then compared with velocities computed directly from elastic parameters. Error values are tabulated in the same tables. Obviously our simulations do not include effects of mechanical and measurement inaccuracy in the z direction, object surface alignment error, electrical noise and temperature change that normally occur in an actual experiment. In that sense, our accuracy estimations are optimistic.

Results indicate that we may have an error on the order of 1% in either direction. The errors that we find here are somewhat less than the experimental absolute errors reported by Kushibiki and others[(44)], as expected. However, our findings show that errors cannot be corrected simply by a bias argument based on measurements from standard specimens. The error figures reported in Ref. 44 are very close to each other, since materials under investigation are very similar to each other with similar leaky wave velocities. Since the direction of

Table 4.4. Absolute Errors Using LFB Lens for Different Materials under Different Assumptions[a]

Material	Actual	Uniform field	Error	Ideal reference	Error	Pb reference	Error
Aluminum	2858.6	2861.1	0.08%	2863.6	0.17%	2862.0	0.12%
Chromium	3656.7	3660.0	0.09%	3669.3	0.34%	3665.0	0.22%
Alumina	5679.0	5696.6	0.30%	5706.4	0.48%	5608.2	−1.25%
Silicon carbide	6809.5	6850.1	0.60%	6884.8	1.11%	6672.4	−2.01%

[a]Actual velocities (first column) are compared with $V(z)$-extracted velocities under a uniform field (second column), real field with ideal reference reflector (fourth column), real field with Pb as reference reflector (sixth column) assumptions.

Table 4.5. Absolute Errors Using V-Groove Lens for Different Materials under Different Assumptions[a]

Material	Actual	Uniform field	Error	Ideal reference	Error	Pb reference	Error
Aluminum	2858.6	2854.3	−0.15%	2848.8	−0.34%	2846.3	−0.43%
Chromium	3656.7	3654.0	−0.07%	3650.3	−0.18%	3642.9	−0.38%
Alumina	5679.0	5665.6	−0.24%	5636.3	−0.75%	5659.3	−0.35%
Silicon carbide	6809.5	6810.5	0.01%	6792.3	−0.25%	6741.7	−0.98%

[a]Actual velocities (first column) are compared with $V(z)$ extracted velocities under uniform field (second column), real field with ideal reference reflector (fourth column), real field with Pb as reference reflector (sixth column) assumptions.

error changes for varying leaky wave velocities, a simple multiplicative error correction is obviously not possible.

The reason for the existence of this systematic error can be attributed to asymmetry in the contribution of the leaky wave pole-zero pair. This asymmetry is especially pronounced for Lamb wave or surface-skimming mode phase transitions,[45] but it exists even for the simple Rayleigh wave mode. Hence error levels given in Table 4.5 can be considered as the lower bound for the absolute error.

A correction factor can be applied for each measurement if an error simulation is performed for the lens and material under consideration. The correction factor is not fixed, but rather depends on sample parameters.

Table 4.6 summarizes simulation results for a slit lens with $s = 0.1a$. Various single crystals with different cuts along various propagation directions are considered. Table 4.6 shows the percentage difference between the $V(z)$-extracted velocity and actual velocity as absolute error. Simulations are done for a slit lens with $s = 0.1a$. For most materials, the error is about 1%. Repeatability of experiments can be much better for each of the techniques than the values given in Table 4.6.

Table 4.6. Calculated Wave Velocities versus Actual Velocities with Slit-Aperture Lens ($s = 0.1a$)

Material	Cut	Direction	Actual V_r	Extracted V_r	Error (%)
Al	011	0	2973.5	3008.1	1.16
Al	111	0	2841.2	2869.5	1
GaAs	011	0	2813.7	2831.8	0.64
GaAs	111	0	2433.8	2443.6	0.41
Si	011	0	5010.4	4994.1	−0.33
Si	011	10	4991.5	4970.3	−0.43
Si	011	20	4936.5	4900.8	−0.72
Si	011	30	4850.0	4812.7	−0.77

4.6. Conclusions

We compared various lens designs and signal-processing systems used in acoustic microscopy. The conventional spherical lens offers the best resolution performance, but it has an inferior characterization ability. The circular symmetry of this lens makes only isotropic material characterization possible. An emphasized leaky wave contribution and hence a higher figure of merit can be obtained by a Lamb wave lens at the expense of a limited range of measurable object velocity. It is possible to measure perturbations in the object parameters at least an order of magnitude smaller than possible with conventional lens. The Lamb wave lens has an additional advantage: Possible modes can be separated by proper choice of frequency and/or cone angle.

The LFB lens has a directional sensitivity, hence it can be used to characterize anisotropic substrates. But it has poor resolution in one direction and a slightly lower figure of merit, since the leaky wave component is smaller in amplitude compared to the conventional lens.

The slit-aperture lens is a compromise between characterization accuracy and good resolution. Unlike the LFB lens, it can be used as an imaging lens while providing a directionality for velocity measurement. The usefulness of this lens is limited by its low signal-to-noise ratio.

The directional properties of a V-groove lens is comparable to those of a LFB lens, while its leaky wave excitation efficiency is as good as the Lamb wave lens. Since its excitation angle is fixed, a given V-groove lens can be used only for a limited range of velocities. For best results, a matching V-groove lens must be used for each material.

The comparison of signal-processing systems is based on a quantitative measure in the form of a figure of merit. In particular sensitivity with respect to object parameters and object surface topography is considered. The sensitivity to object parameters is in general tightly connected to the level of the leaky wave signal component. Phase measuring systems are by far the best systems in terms of sensitivity, although they are difficult to implement. Adding the carrier signal to the received signal before the envelope detection is a simple modification, and it provides a considerable improvement in the topography measurement and detector threshold extension in the conventional system. An increased immunity to external perturbations can be attained with a differential-phase system, provided specular and leaky wave signals can be separated from each other in the time domain.

Figure of merit values indicate that limited angular coverage lenses, such as Lamb wave and V-groove lenses, provide better characterization performance at the cost of reduced versatility.

An alternative characterization method is discussed. The method requires recording the output voltage as a function of input frequency. The resulting curve, called $V(f)$, is also highly object-dependent. A new mode of imaging, PFI, suitable for Lamb wave lens, is presented, which reduces the interpretation difficulty of conventional imaging.

The velocity determination method using $V(z)$ has inherent systematic errors. Error is due to the nonsymmetric nature of reflection coefficient phase transitions. The error can be as high as 1% depending on the object material, lens insonification, and reference material choice. A correction factor can be chosen for each particular case after careful simulation of the problem.

Acknowledgments

This work is supported by the Turkish Scientific and Research Council, TUBITAK.

References

1. Lemons, R. A. and Quate, C. F. (1974). Acoustic microscopy, scanning version. *Appl. Phys. Lett.* **24,** 163–65.
2. Hadimioğlu, B. and Quate, C. F. (1984). Water acoustic microscopy at suboptical wavelengths. *Appl. Phys. Lett.* **43,** 1006–1007.
3. Hadimioğlu, B. and Foster, J. S. (1984). Advances in superfluid helium acoustic microscopy. *J. Appl. Phys.* **56,** 1976–80.
4. Atalar, A. Quate, C. F., Wickramasinghe, H. K. (1977). Phase imaging with the acoustic microscope. *Appl. Phys. Lett.* **31,** 791–93.
5. Briggs, A. (1992). *Acoustic Microscopy.* Oxford University Press, Oxford.
6. Atalar, A. (1978). An angular spectrum approach to contrast in reflection acoustic microscopy. *J. Appl. Phys.* **49,** 5130–39.
7. Parmon, W. and Bertoni, H. L. (1979). Ray interpretation of the material signature in the acoustic microscope. *Electron Lett.* **15,** 681–686.
8. Bertoni, H. L. (1984). Ray-optical evaluation of V(Z) in the reflection acoustic microscope. *IEEE Trans. Sonics Ultrason.* **31,** 105–16.
9. Chan, K. H. and Bertoni, H. L. (1977). Ray representation of longitudinal waves in acoustic microscopy. *IEEE Trans. Ultrason. Ferroelect. and Freq. Control* **38,** 27–34.
10. Quate, C. F., Atalar, A., Wickramasinghe, H. K. (1979). Acoustic microscopy with mechanical scanning—a review. *Proc. IEEE* **67,** 1092–1114.
11. Atalar, A. (1988). A fast method of calculating diffraction loss between two facing transducers. *IEEE Trans. Ultrason. Ferroelect. Freq. Control* **35,** 612–18.
12. Atalar, A. (1987). In *Proceedings of IEEE 1987 Ultrasonics Symposium,* IEEE Press, New York, pp. 791–94.
13. Atalar, A. and Köymen, H. (1989). In: *Proceedings of IEEE 1989 Ultrasonics Symposium,* IEEE Press, New York, pp. 813–16.
14. Atalar, A., Köymen, H., Değertekin, L. (1990). In: *Proceedings of 1990 Ultrasonics Symposium,* IEEE Press, New York, pp. 359–62.

15. Davids, D. A., and Bertoni, H. L. (1986). In: *Proceedings of IEEE 1986 Ultrasonics Symposium*, IEEE Press, New York, pp. 735–40.
16. Kanai, H., Chubachi, N., Sannomiya, T. (1992). Microdefocusing method for measuring acoustic properties using acoustic microscope. *IEEE Trans. Ultrason. Ferro. Freq. Cont.* **39,** 643–52.
17. Sannomiya, T., Kushibiki, J., Chubachi, N., Matsuno, K., Suganuma, R., Shinozaki, Y. (1992). In: *IEEE Ultrasonic Proceedings* 731–34.
18. Hildebrand, J. A., and Lam, L. K. (1983). Directional acoustic microscopy for observation of elastic anisotropy. *Appl. Phys. Lett.* **42,** 413–15.
19. Khuri-Yakub, B. T. and Chou, C.-H. (1986). In: *Proceedings of IEEE 1986 Ultrasonics Symposium,* IEEE Press, New York, pp. 741–44.
20. Chou, C.-H. and Khuri-Yakub, B. T. (1987). In: *Proceedings of IEEE 1987 Ultrasonics Symposium,* IEEE Press, New York, pp. 813–16.
21. Routh, H. F., Pusateri, T. L., Nikoonahad, M. (1989). In: *Proceedings of IEEE 1989 Ultrasonics Symposium,* IEEE Press, New York, pp. 817–20.
22. Routh, H. F., Sivers, E. A., Bertoni, H. L., Khuri-Yakub, B. T., Waters, D. D. In: *Proceedings of 1990 Ultrasonics Symposium,* IEEE Press, New York, pp. 931–36.
23. Kushibiki, J., and Chubachi, N. (1985). Material characterization by line-focus-beam acoustic microscope. *IEEE Trans. Sonics Ultrason.* **32,** 189–212.
24. Kushibiki, J., Ueda, T., Chubachi, N. (1987). In: *Proceedings of IEEE 1987 Ultrasonics Symposium,* IEEE Press, New York, pp. 817–21.
25. Kushibiki, J., Takahashi, H., Kobayashi, T., Chubachi, N. (1991). Quantitative evaluation of elastic properties of $LiTaO_3$ crystals by line-focus-beam acoustic microscopy. *Appl. Phys. Lett.* **58,** 893–95.
26. Kushibiki, J., Takahashi, H., Kobayashi, T., Chubachi, N. (1991). Characterization of $LiNbO_3$ crystals by line-focus-beam acoustic microscopy. *Appl. Phys. Lett.* **58,** 2622–24.
27. Kushibiki, J. and Chubachi, N. Application of LFB acoustic microscope to film thickness measurement. *Electron Lett.* **23,** 652–54.
28. Davids, D. A., Wu, P. Y., Chizhik, D. (1989). Restricted aperture acoustic microscope lens for Rayleigh wave imaging. *Appl. Phys. Lett.* **54,** 1639–41.
29. Kolosov, O. V. and Yamanaka, K. (1994). Adjustable acoustic knife edge for anisotropic and dark field acoustic imaging. *Jpn. J. Appl. Phys.* **33,** 329–33.
30. Chizhik, D. and Davids, D. A. (1992). Applications of diffraction-corrected ray theory to the slot lens in acoustic microscopy. *J. Acoust. Soc. Am.* **92,** 3291–3301.
31. Atalar, A., Ishikawa, I., Ogura, Y., Tomita, K. (1993). In: *IEEE Ultrasonic Proceedings,* IEEE Press, New York.
32. Somekh, M. G., Briggs, G. A. D., Ilett, C. (1984). The effect of elastic anisotropy on contrast in the scanning acoustic microscope. *Phil. Mag. A* **49,** 179.
33. Arikan, O., Teletar, E., Atalar, A. (1989). Reflection coefficient null of acoustic waves at a liquid–anisotropic–solid interface. *J. Acoust. Soc. Am.* **85,** 1–10.
34. Nayfeh, A. H. (1991). Elastic wave reflection from liquid-anisotropic substrate interfaces. *WAMOD* **14,** 55–67.
35. Bozkurt, A., Yaralioğlu, G., Atalar, A., Koymen, H. (1993). In: *IEEE Ultrasonic Proceedings,* IEEE Press, New York, pp. 583–586.
36. Yu, Z. and Boseck, S. (1990). In: *Acoustical Imaging,* vol. 12. (H. Ermert and H.-P. Harjes, eds.). Plenum Press, New York.
37. Reinholdtsen, P. A., Chou, C.-H., Khuri-Yakub, B. T. (1987). In: *Proceedings of IEEE 1987 Ultrasonics Symposium,* IEEE Press, New York, pp. 807–11.
38. Khuri-Yakub, B. T., Reinholdtsen, P., Chou, C.-H., Parent, P., Cinbis, C. In: *Acoustical Imaging,* vol. 17. (H. Shimizu, N. Chubachi, and J. Kushibiki, eds.). Plenum Press, New York, 583–586.

39. Meeks, S. W., Peter, D., Horne, D., Young, K., Novotny, V. (1989). In: *Proceedings of IEEE Ultrasonics Symposium,* IEEE Press, New York, pp. 809–12.
40. Chou, C.-H., Hsieh, C. P., Khuri-Yakub, B. T. (1990). In: *Proceedings of 1990 Ultrasonics Symposium,* IEEE Press, New York, pp. 887–90.
41. Briggs, A. (1992). Acoustic microscopy—a summary. *Rep. Prog. Phys.* **55,** 851–909.
42. Chou, C.-H., Khuri-Yakub, B. T., Kino, G. S. (1984). Lens design for acoustic microscopy. *IEEE Trans. Ultrason. Ferroelect. Freq. Control,* **35,** 464–69.
43. Hansen, P. B. and Bjørnø, L. (1989). In: *Proceedings of IEEE 1989 Ultrasonics Symposium,* IEEE Press, New York, 1125–1128.
44. Kushibiki, J., Wakahara, T., Kobayashi, T., Chubachi, N. (1992). In: *IEEE Ultrasonic Proceedings,* IEEE Press, New York, pp. 719–22.
45. Tsukahara, Y., Neron, C., Jen, C. K., Kushibiki, J. (1993). In: *IEEE Ultrasonic Proceedings,* IEEE Press, New York, 593–598.

5

Measuring Thin-Film Elastic Constants by Line-Focus Acoustic Microscopy

Jan D. Achenbach, Jin O. Kim, and Yung-Chun Lee

5.1. Overview

Determining the elastic constants of anisotropic films deposited on anisotropic substrates from $V(z)$ measurements obtained by using a line-focus acoustic microscope is discussed in Chapter 5. The procedure has three essential components: (1) measuring the $V(z)$ curve as a function of direction of wave mode propagation in the thin-film/substrate system at fixed frequency and/or as a function of the frequency or film thickness for a fixed direction, (2) developing a theoretical measurement model for parametric studies of $V(z)$ curves, and (3) obtaining elastic constants by systematically comparing wave-mode velocities obtained from the theoretical model and $V(z)$ measurements. Examples primarily concern transition metal nitride films and superlattice films used as hard protective coatings for softer surfaces. Results are presented for several thin-film/substrate configurations. Advantages of the method as well as remaining problems that require further investigation are discussed.

JAN D. ACHENBACH, JIN O. KIM, AND YUNG-CHUN LEE • Center for Quality Engineering and Failure Prevention, Northwestern University, Evanston, Illinois 60208
Advances in Acoustic Microscopy, Volume 1, edited by Andrew Briggs.
Plenum Press, New York, 1995.

5.2. Introduction

The quantitative mode of acoustic microscopy has proven to be a very useful technique for determining elastic constants, the thickness, and even the mass density of a thin film deposited on an elastic substrate. Determining these quantities is generally based on measuring the $V(z)$ curve, which is a record of the modulus of the measured voltage as a function of the distance z between the focus of the lens and the specimen surface.

The potential for quantitative measurements by using the $V(z)$ curve was apparently first recognized by Weglein.[1] For a point-focus acoustic microscope Weglein[2] proposed to determine film thickness by using $V(z)$-derived information on velocity variation with normalized film thickness of the lowest mode of wave propagation (the leaky Rayleigh mode) in a thin-film/substrate system.

As discussed in Chapter 5, the velocities and the attenuation of modes of elastic wave propagation in a film/substrate configuration can be obtained from the periodic variation and the decay of the $V(z)$ curve. If the mass density is known, elastic constants can subsequently be obtained by using an appropriate technique to best fit theoretical velocities obtained from a $V(z)$ measurement model to velocities obtained from the measured $V(z)$ curve.

Both a point-focus beam and a line-focus beam have been used to measure $V(z)$ curves. A point-focus beam excites wave modes propagating in radial directions on the specimen, and hence only the mean value of elastic properties around the axis of the lens can be measured. This restricts the application of point-focus acoustic microscopy to characterizing isotropic materials. A line-focus lens on the other hand generates wave modes propagating in a specific direction, namely, normal to the focal line, and thus it is particularly useful for measuring velocities and hence elastic constants in anisotropic materials. These differences between quantitative point-focus and line-focus acoustic microscopy (LFAM) are illustrated in Fig. 5.1. Chapter 5 is exclusively concerned with LFAM.

5.2.1. Terminology

The terminology surface acoustic wave (SAW) used in Chapter 5 refers to wave motion along the surface of bare solids, fluid-loaded bodies, and in thin-film/substrate configurations. For the latter case, it is assumed that the wave motion penetrates the substrate but decays exponentially with depth. Thus in this chapter, SAWs include Rayleigh waves, surface-skimming longitudinal waves, leaky Rayleigh waves, pseudosurface waves, and wave modes in thin-film/substrate systems. The latter are sometimes referred to as generalized Lamb modes, and for isotropic thin-film/substrate systems the second mode is known as the Sezawa wave.

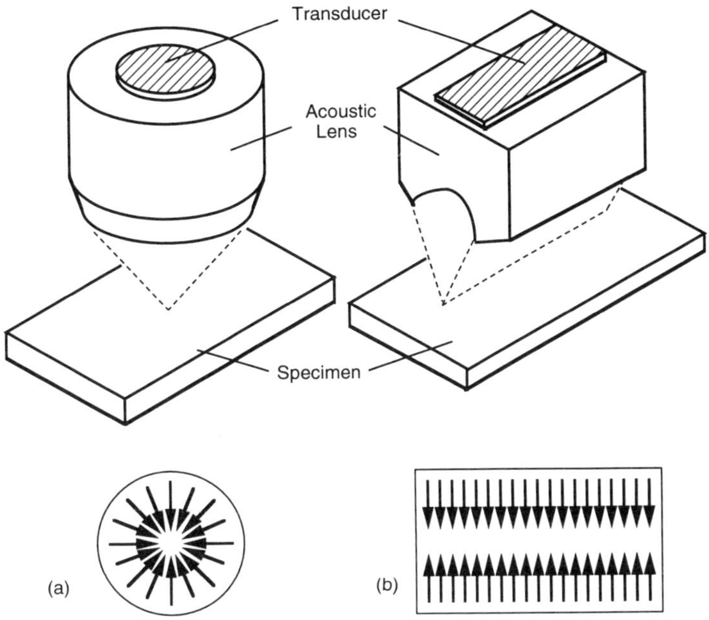

Figure 5.1. Acoustic lenses: (a) point-focus lens, (b) line-focus lens.

5.2.2. Thin Films

There are many present and potential applications of configurations consisting of a thin film deposited on a substrate. Among these we are primarily concerned with thin films deposited to improve the hardness and/or thermal properties of surfaces. Thin-film technology does however also include high T_c superconductor films, films for magnetic recordings, superlattices, and films for band gap engineering and quantum devices.

Diamond film has a number of remarkable properties: It is the hardest substance known, and it has a higher modulus of elasticity than any other material. When free of impurities, it has one of the highest resistivities. It also combines a very high thermal conductivity with a low thermal expansion coefficient to yield high resistance to thermal shock. Lastly diamond is very resistant to chemical attack.

Transition metal nitride films are commonly used as hard protective coatings for softer surfaces. Superlattice films, including TiN/NbN, TiN/VN, and TiN/VNbN have however been shown to exhibit much higher hardness than homogeneous nitride films.

When a thin layer is deposited on a substrate, mechanical properties of the thin layer depend on the deposition method; hence they should be determined by measuring the actual thin-layer/substrate configuration. Since the layer may be very thin, this is not a trivial task but one requiring sophisticated equipment and techniques.

5.2.3. Line-Focus Acoustic Microscopy

An excellent exposition of acoustic microscopy can be found in the book by Briggs.[3] Here we briefly summarize some relevant aspects of the operation of a line-focus acoustic microscope which consists of four main components: the acoustic probe, the tone-burst-mode measurement system for transmitting and receiving electrical signals, the mechanical system for alignments and movements to record $V(z)$ curves, and a computer for controlling the system and processing the recorded waveforms. Figure 5.2 shows the configuration of the acoustic probe and the specimen. A ZnO-film transducer generates and detects longitudinal acoustic waves at the flat surface of a Z-cut sapphire rod. The acoustic signal generated in the rod is focused by an acoustic lens with a cylindrical concave surface at the other end of the rod. Experimental results reported in this chapter were primarily obtained with a Honda AMS-5000™ ultrasonic measurement system and a line-focus acoustic lens provided by Tohoku University. The cylindrical concave surface has a radius of 1 mm and an aperture half-angle of 60°. The operating frequency is around 225 MHz, and the focal length of the lens is 1.15 mm. For efficient transmission of acoustic waves through the lens–couplant interface, a chalcogenide glass film with a quarter-wavelength thickness is deposited on the cylindrical concave surface. The specimen is placed on a mechanical stage, translated in the vertical direction, and rotated around the

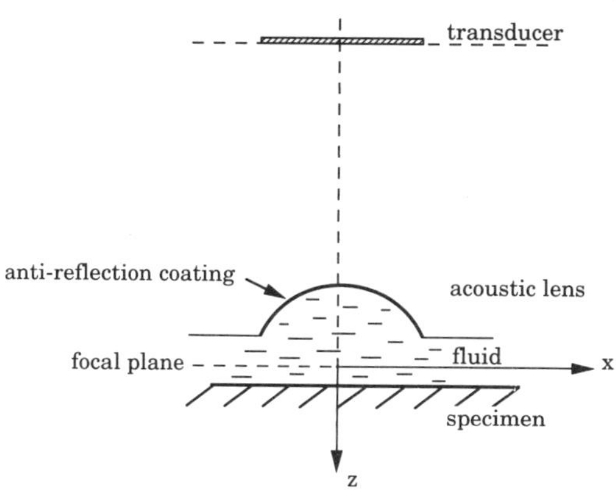

Figure 5.2. Geometry of a transducer, lens, and specimen.

axis of the rod. A drop of distilled water at room temperature is used as the coupling fluid.

A radiofrequency (r.f.) tone burst of 0.5-μsec width and repetition rate of 20 kHz excites the transducer to generate single-frequency plane acoustic waves. Plane waves propagate along the buffer rod; they are refracted at the concave surface of the lens and focused through the coupling fluid into the specimen. Reflection from the specimen is easiest discussed in terms of ray theory. For many materials, two kinds of rays return to the transducer, which now acts as a receiver. One kind that always occurs carries the specular reflection from normal incidence on the specimen. The other kind, which requires a combination of lens angle and material properties of the specimen to generate leaky surface waves, carries the radiation of the surface waves excited by critical angle incidence on the specimen. The voltage output of the transducer displays the interference of these two kinds of rays. When the specimen is translated toward the lens, the voltage output of the transducer is recorded to form a $V(z)$ curve.

5.2.4. Specimen Preparation

In this chapter, experimental results are presented for $V(z)$ curves of isotropic materials, such as aluminum and glass; anisotropic materials, such as cubic-crystalline silicon and magnesium oxide (MgO), and isotropic and anisotropic thin-films deposited on isotropic and anisotropic substrates. The following thin-film/substrate combinations have been considered by the present authors: gold on glass, diamond and carbon on silicon, TiN, NbN, and VN on MgO, and transition metal nitride superlattice films on MgO. Other authors have considered gold films on fused quartz,[4,5] gold film on fused silica,[6] gold film on silicon,[7] machining damaged layers on beryllium[8] and on silicon,[9] a-C:H coating on fused quartz and silicon and GaAs wafers implanted with silicon ions.[10]

The single-crystal transition metal nitride films (TiN, NbN, VN) used in experiments reported here were deposited by an ultrahigh vacuum-reactive magnetron-sputtering technique. The process was carried out in a 36-cm-diameter all-metal-sealed stainless-steel chamber evacuated by a 360 l/s turbomolecular pump to a base pressure of $\sim 5 \times 10^{-9}$ Torr. The system has a water-cooled, 5-cm-diameter DC magnetron-sputtering source with Ti (purity >99.7%), Nb (purity >99.9%), or V(purity >99.9%), aimed directly at the substrate position to provide metal fluxes for film deposition. Substrates were polished ($12 \times 12 \times 0.5$ mm) MgO (001) wafers ultrasonically cleaned in trichloroethylene, acetone, and methanol, then blown dry with dry N_2 before insertion in the chamber. The substrates were resistively heated to 700°C for 1 hour prior to deposition to obtain a clean, defect-free surface, and they were maintained at that temperature during deposition. The deposition system and technique are described in detail by Mirkarimi *et al.*[11] Low-energy electron diffraction (LEED) measurements were used to verify that the grown films are single crystals of the same orientation as the substrate.

Transition metal nitride superlattice films were also deposited on cubic crystal MgO substrates using an ultrahigh vacuum-reactive magnetron-sputtering system. The TiN/NbN superlattice films were grown on MgO(001) wafers from two water-cooled 5-cm diameter DC magnetron sputter sources with Ti and Nb targets. The sputtering gas was an Ar-N_2 mixture (gas purity >99.999%) with a total pressure of 13 mTorr and N_2 partial pressure of 4.2 mTorr. The Ti source voltage was 460 V, and the current was 0.6 A, resulting in a TiN deposition rate of 0.25 μm/h. The Nb source voltage was 520 V, and the current was 0.6 A, yielding a NbN deposition rate of 0.74 μm/h. The thicknesses of the superlattice films deposited in this way were in the range of 0.4–3.8 μm. A computer-controlled shutter modulated the sputtered fluxes for superlattice deposition. The ratio of the TiN and NbN layer deposition times was maintained at unity in all cases. While switching source fluxes, the shutter shadowed both targets for about 0.15 second. The Ti and Nb targets were sputter cleaned in pure Ar for 10–15 minutes. The substrate temperature during deposition was 700 ± 30°C, as measured by an optical pyrometer on TiN-coated MgO, using an emissivity of 0.3. Superlattice deposition was initiated with a TiN layer. Further details can be found in the paper by Shinn *et al.*[(12)]

In situ LEED and Auger electron spectroscopy (AES) measurements were carried out immediately after deposition using a retarding field analyzer. After deposition samples were cooled for 20 minutes before being moved to the analysis position to minimize outgassing from the LEED/AES. Spotty LEED patterns were observed, indicating that the films were single crystals. The AES spectra results showed that the TiN and NbN films were stoichiometric. The ratio of TiN layer thickness to the total superlattice period was 0.30 ± 0.05, as measured by energy-dispersive X-ray analysis and a calibration of the deposition rates. As-grown films were also examined using an X-ray diffractometer.

Film thicknesses h were measured in two ways: For the superlattices, the period Λ was extracted from the X-ray diffraction results, and the film thickness was then estimated by multiplying Λ by the number of periods deposited. Alternatively part of the substrate can be masked by Ta foil, and after deposition, h can be measured with a profilometer.

To avoid the effects of surface roughness, specimens should be locally flat with a roughness not exceeding a tenth of a wavelength. Thin-film specimens were used as processed, without polishing.

5.3. Measurement Model for $V(z)$ Curves

A clear understanding of the relevant ultrasonic interference phenomena that give rise to a $V(z)$ curve is a requirement for correctly using quantitative acoustic microscopy. For the point-focus acoustic microscopy, a calculation of $V(z)$ curves for layered isotropic materials has been presented by Kundu.[13] This section describes calculating $V(z)$ curves for layered anisotropic materials for the line-focus acoustic microscope by analytical (Sections 5.3.1–5.3.2 and numerical (Section 5.3.3) modeling.

An analytical approach to calculating line-focus $V(z)$ curves for anisotropic-layer/anisotropic-substrate configurations was presented by Lee *et al.*[14] The method follows the Fourier optical approach, in which the $V(z)$ curve is considered as a Fourier integral over the product of characteristic functions of the acoustic lens and the reflection coefficient of the fluid-loaded specimen. Thus when the characteristic functions are known, the $V(z)$ curve can be directly related to the reflection coefficient of the specimen through a Fourier-type integral. Accurate characteristic functions are essential for calculating $V(z)$ curves by this approach. Attenuation in the coupling fluid, angular dependence of the transmission by the antireflection coating on the lens surface, and the actual focal length of the lens must be carefully taken into consideration.

Figure 5.2 shows the general configuration of an acoustic probe and a specimen. The acoustic probe consists of a buffer rod with a transducer at one end and a cylindrical lens at the other. The lens, which is coated with a thin layer to reduce reflection, is coupled to the specimen through a fluid, usually water. Ultrasonic waves generated by the transducer propagate through the rod, and they are focused by the lens into a line-focused beam in the coupling fluid. The focused beam is reflected by the specimen and returned to the trans ducer to produce a voltage. The output voltage of the transducer can be written as

$$V(z) = \int_{-\infty}^{+\infty} \exp(2\, ik_z z) L_1(k_x) L_2(k_x) R(k_x)\, \mathrm{d}k_x \tag{1}$$

where $k_z = (k_w^2 - k_x^2)^{1/2}$ and k_w is the wave vector in the coupling fluid. $L_1(k_x)$ is the angular spectrum of the incident wave field at the focal plane generated by a plane wave propagating in the buffer rod toward the lens, and $L_2(k_x)$ is the voltage response of the transducer when a plane wave of unit amplitude and wave vector (k_x, k_z) is insonifying the lens. $R(k_x)$ is the reflection coefficient of the fluid-loaded specimen. Equation (1) is analogous to the formula given by Somekh *et al.*[15]; see also Briggs[3] except that these authors have combined the terms $L_1(k_x)$ and $L_2(k_x)$ into a pupil function.

5.3.1. Characteristic Functions of the Acoustic Lens

In Eq. (1), $L_1(k_x)$ and $L_2(k_x)$ are characteristic functions of the acoustic lens, which are to be determined. In the following analysis, it is assumed that a plane wave is generated by the transducer and propagates through the buffer rod.

5.3.1.1. Angular Spectrum Function $L_1(k_x)$

Calculating the angular spectrum function $L_1(k_x)$ follows the method presented by Li *et al.*[16] with some modifications. Based on diffraction theory and some assumptions and approximations discussed by Atalar,[17] the function $L_1(k_x)$ is expressed as

$$L_1(k_x) = i\,r \int_{-\theta m}^{\theta m} T_{sf}(\theta) \cos\theta' \exp\left[-i \frac{k_w}{2f_L}\left(r \sin\theta + f_L \frac{k_x}{k_w}\right)^2\right] d\theta \qquad (2)$$

where r and θ_m are the radius of curvature and the half-aperture angle of the lens, respectively; f_L is its focal length; $\cos\theta' \approx (1 - \bar{c} \cdot \sin^2\theta)^{1/2}$; and $\bar{c}$ is the ratio of the longitudinal wave speed v_l in the lens to the wave speed v_w in a coupling fluid. In Eq. (2), $T_{sf}(\theta)$ is the actual transmission coefficient of an incident wave from the lens to the fluid at incident angle θ through a coating layer. The attentuation effect of the coupling fluid (water) is included by taking the wave number as a complex number, i.e., $k_w = \omega/v_w + i\,\alpha_w$, where α_w is the attenuation factor in water at the circular frequency ω. Although the lens is usually made of a slightly anisotropic material, such as sapphire, it is assumed to be isotropic in this calculation. An approximation for $T_{sf}(\theta)$ has been given by Achenbach *et al.*[18] by modeling the coating layer as a spring mass system, but, Brekhovskikh[19] offers an exact solution for $T_{sf}(\theta)$.

5.3.1.2. Response Function $L_2(k_x)$

By definition the response function $L_2(k_x)$ is the voltage response of the transducer when a plane wave with wave vector (k_x, k_z) insonifies the lens. Atalar[17] gives a formula for the voltage output as an integration over the incident wave field $u_0^+(x_0)$ and the reflected wave field $u_0^-(x_0)$ at the plane of the transducer

$$L_2(k_x) = C \int_{-a}^{+a} u_0^+(x_0) u_0^-(x_0)\, dx_0 \qquad (3)$$

where a is the radius of the transducer and C is a calibration constant. It is assumed that the transducer excites a uniform plane wave field of unit amplitude, i.e.,

$$u_0^+(x_0) = 1 \qquad (4)$$

and that the reflected wave field insonifying the lens can be obtained by geometrical ray theory similar to the procedure in Kundu.[13] As shown in Fig. 5.3, a plane wave is incident on the lens at an angle θ_0. From geometrical ray theory, we have

$$\theta_1 = \sin^{-1}\left[\frac{v_l}{v_w} \sin(\theta_s - \theta_0)\right] \qquad (5)$$

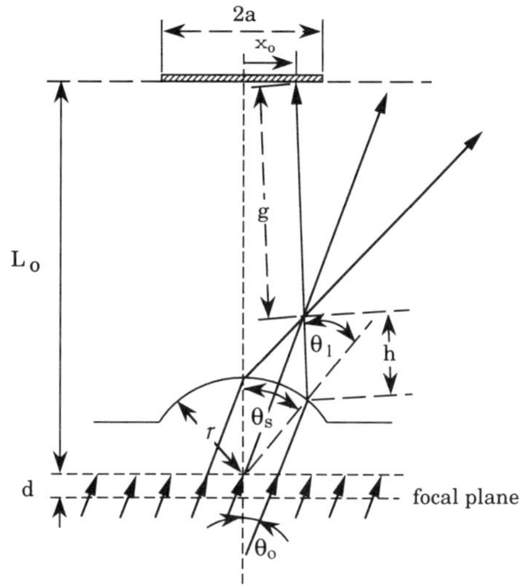

Figure 5.3. A ray diagram for the signal received by the transducer when a plane wave insonifies the lens at angle θ_o.

and

$$x_0 = r \sin \theta_s + (L_0 - r \cos \theta_s) \tan (\theta_s - \theta_1) \tag{6}$$

Equation (6) shows that there is a one-to-one relation between x_0 and θ_s. The distance parameters h and g in Fig. 5.3 are

$$h = r \frac{\sin (\theta_s - \theta_0)}{\sin [\theta_1 - (\theta_s - \theta_0)]} \tag{7}$$

and

$$g = [(L_0 - r \cos \theta_0)^2 + (x_0 - r \sin \theta_s)^2]^{1/2} - h \tag{8}$$

For an (x,z) system with its origin at the focal line of the lens, the displacement of the incident wave u^0_{in} on the lens aperture is

$$u^0_{in}(\theta_s) = \exp(i\, d\, k_z) \exp[i\, (k_x\, r \sin \theta_s + k_z\, r \cos \theta_s)] \tag{9}$$

where $d = f_L - r$ and $k_x = (\omega/v_w) \sin \theta_0$. It can be shown by ray theory, see, e.g., Kundu,[13] that the disturbance that reaches the transducer at position x_0 corresponds to an incident wave on the lens aperture of the form

$$u^-_0(x_0) = \frac{\sqrt{h}}{\sqrt{g}} \exp[i\, k_L(g + h)]\, T_{fs}(\theta_s - \theta_0)\, u^0_{in}(\theta_s) \tag{10}$$

where θ_s is related to x_0 by Eqs. (5) and (6) and $T_{fs}(\theta)$ is the transmission coefficient from the fluid through the coating layer to the lens at incident angle θ. Substituting Eqs. (4) and (10) into Eq. (3) yields the function $L_2(k_x)$.

5.3.1.3. Numerical Calculations of $L_1(k_x)$ and $L_2(k_x)$

For the line-focus acoustic microscope, the angular spectrum function and the response function have been calculated[14,20]. The geometrical parameters of the lens (see Fig. 5.3) used in the calculation are r = 1 mm, L_O = 13 mm, a = 0.865 mm, and $\theta_m = 60°$. For the coupling water, v_w = 1490 m/s, ρ_w = 998 kg/m^3, and α_w = 1100 Np/m. For the sapphire lens, v_l = 11100 m/s, v_t = 6600 m/s, and ρ = 3986 kg/m^3. For the antireflection coating layer of chalcogenide glass, the longitudinal wave velocity is 2250 m/s, the density is 4640 kg/m^3, and the thickness is 2.45 μm. The transverse wave velocity of the coating layer material is not available, but fortunately it has very little effect on the result, and it can be taken as half the longitudinal wave speed.

For the focal length f_L, an approximation based on the paraxial assumption gives $f_L = r/(1-\bar{c}) \approx 1.156$ mm. However due to the large lens angle (θ_m = 60°), the paraxial assumption is no longer valid. Based on geometrical ray theory, the actual focal length should fall in the range of 1.156–1.146 mm. In the present calculation, the focal length is taken as the median value, namely, f_L = 1.15 mm. It is later shown that this choice eliminates the mysterious mismatch noted in earlier work of the positions of the largest peaks in the calculated and measured $V(z)$ curves.

Calculated results for the coefficient of transmission from the lens to the fluid $T_{sf}(\theta)$ and from the fluid to the lens $T_{fs}(\theta)$ are shown in Figs. 5.4(a) and (b), respectively. Since the longitudinal wave speed of sapphire is much greater than that of water, the $T_{fs}(\theta)$ has a total reflection angle defined by $\theta_{cr} = \sin^{-1}(v_w/v_l)$ of about 7.7°. For an angle greater than θ_{cr}, the transmission coefficient $T_{fs}(\theta)$ is zero. Amplitudes and phases of the calculated values of $L_1(k_x)$ and $L_2(k_x)$ are shown in Figs. 5.5(a) and (b), respectively. These results are used in the sequel.

Note that the calculated values of $L_1(k_x)$ and $L_2(k_x)$ are not the same. $L_1(k_x)$ defined in Eq. (2) is the angular spectrum in the focal plane for a plane wave propagating through the buffer rod. On the other hand, $L_2(k_x)$ defined in Eq. (3) is the voltage response of the transducer when a plane wave of unit amplitude and wave vector (k_x, k_z) is insonifying the lens. Such a plane wave gives rise to a spectrum of plane waves in the buffer rod, not a single plane wave as assumed for the starting point of the calculation of $L_1(k_x)$. Since $L_2(k_x)$ results from a superposition of a number of plane waves that reach the transducer, a reciprocity relation does not apply for $L_1(k_x)$ and $L_2(k_x)$.

5.3.2. Reflection Coefficient for the Liquid/Layered–Solid Interface

The reflection coefficient $R(k_x)$ of an acoustic plane wave incident on a liquid–specimen interface is presented in this section. For a specimen of isotropic material, the reflection coefficient is known in closed form. For an anisotropic

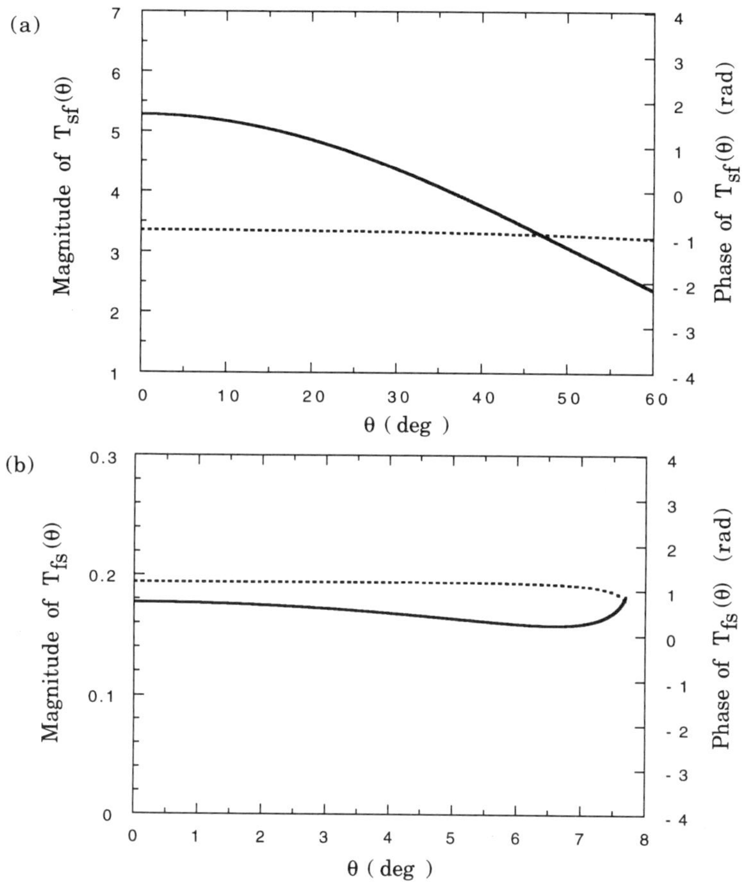

Figure 5.4. Transmission coefficients for transmission from (a) lens to coating to fluid and (b) fluid to coating to lens. *Solid line,* magnitude; *dashed line,* phase (Ref. 14).

solid half-space, a method for calculating the reflection coefficient was proposed by Nayfeh.[21] The method, which has been extended to layered anisotropic materials by Chimenti and Nayfeh[22] and Nayfeh,[23] is adopted here to calculate the reflection coefficient for a single anisotropic layer deposited on an anisotropic substrate.

5.3.2.1. Reflection by an Isotropic Solid

The reflection coefficient for the interface of a liquid and an isotropic solid half-space can be written in the following form[24]:

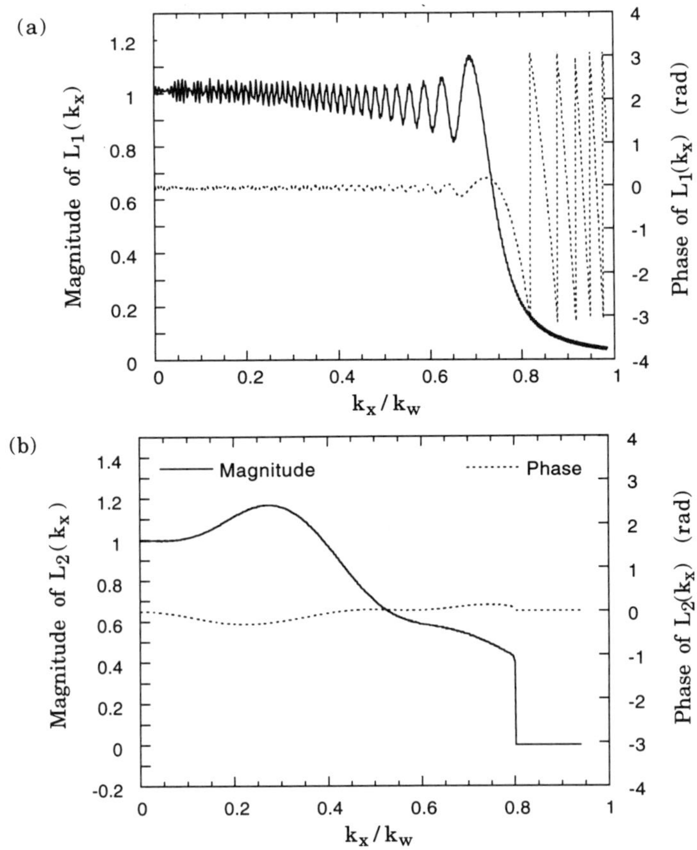

Figure 5.5. Characteristic functions of the acoustic lens: (a) the angular spectrum $L_1(k_x)$ and (b) the response function $L_2(k_x)$ (Ref. 14).

$$R(k_x) = \frac{(2k_x^2 - k_t^2)^2 - 4k_x^2\sqrt{(k_x^2 - k_l^2)(k_x^2 - k_t^2)} - i\,(\rho_w/\rho)\,\sqrt{(k_x^2 - k_l^2)/(k_w^2 - k_x^2)}}{(2k_x^2 - k_t^2)^2 - 4k_x^2\sqrt{(k_x^2 - k_l^2)(k_x^2 - k_t^2)} + i\,(\rho_w/\rho)\,\sqrt{(k_x^2 - k_l^2)/(k_w^2 - k_x^2)}} \tag{11}$$

where the transverse wave number $k_x = k_w \sin\theta$, ρ_w and ρ are the mass densities of the coupling liquid and the solid, $k_j = \omega/v_j$ (j = w,l,t), v_l and v_t are the longitudinal and transverse wave velocities of the solid, and v_w is the wave velocity in the liquid. The reflection coefficient for a solid composed of isotropic layers can be expressed as

$$R(k_x) = \frac{\Gamma_s - (\rho_w/\rho)\,\Gamma_w}{\Gamma_s + (\rho_w/\rho)\,\Gamma_w} \tag{12}$$

where Γ_s and Γ_w are 6×6 matrices (see Ref. 25).

5.3.2.2. Reflection by an Anisotropic Layer/Anisotropic Substrate

Consider the configuration of a fluid(water)-loaded single-layer/substrate specimen shown in Fig. 5.6. The materials of both the layer and the substrate are anisotropic, and they may possess as low as monoclinic symmetry. The principal crystalline axes of the layer are x_1', x_2', x_3'. The direction normal to the specimen surface coincides with a principal axis of the layer and substrate materials. A plane wave with wave vector $\tilde{k}_w$ is incident on the specimen under an angle of incidence θ with the x_3' axis. A new coordinate system (x_1, x_2, x_3) is defined such that $\tilde{k}_w$ is located in the x_1x_3 plane. The x_1 axis makes an angle ϕ with the x_1' axis.

Material constants c_{ijkl} relative to the new coordinates (x_1, x_2, x_3) are related to those relative to the natural crystalline axes c'_{ijkl} by

$$c_{ijkl} = \beta_{im}\,\beta_{jn}\,\beta_{ko}\,\beta_{lp}\,c'_{mnop} \tag{13}$$

where

$$\beta_{ij} = \begin{bmatrix} \cos\phi & \sin\phi & 0 \\ -\sin\phi & \cos\phi & 0 \\ 0 & 0 & 1 \end{bmatrix} \tag{14}$$

The constitutive equation is

$$\sigma_{ij} = c_{ijkl}\,u_{k,l} \tag{15}$$

and the equation of motion is

$$c_{ijkl}\,u_{k,lj} = \rho\ddot{u}_i \tag{16}$$

The elastic constants c_{ijkl} (i,j,k,l = 1,2,3,4) are related to c_{mn} (m,n = 1,2, . . . , 6) by the contracting subscript notation 11→1, 22→2, 33→3, 23→4, 13→5, 12→6.

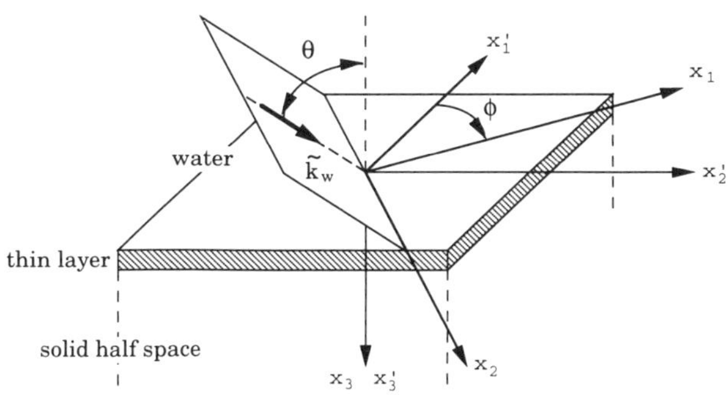

Figure 5.6. The incident wave and geometry of a water-loaded thin-film/substrate configuration.

Consider a homogeneous plane wave solution

$$u_i = v_i \exp[i\, k\, (x_1 + l\, x_3 - c\, t)] \tag{17}$$

where $k = k_w \sin\theta$, $c = c_w/\sin\theta$, and l is an unknown constant. Substituting Eq. (17) into Eq. (16) leads to the following system of equations:

$$K_{ij}\, v_j = 0 \qquad (i, j = 1, 2, 3) \tag{18}$$

with

$$K_{11} = c_{11} - \rho\, c^2 + c_{55} l^2, \quad K_{12} = K_{21} = c_{16} + c_{45} l^2, \quad K_{13} = K_{31} = (c_{13} + c_{55}) l$$
$$K_{22} = c_{66} - \rho\, c^2 + c_{44} l^2, \quad K_{23} = K_{32} = (c_{36} + c_{45}) l, \quad K_{33} = c_{55} - \rho\, c^2 + c_{33} l^2 \tag{19}$$

The condition that the determinant of K_{ij} must vanish yields a sixth-order polynomial equation for l

$$l^6 + A_1\, l^4 + A_2\, l^2 + A_3 = 0 \tag{20}$$

Coefficients A_1, A_2, and A_3, are listed in the appendix of the paper by Nayfeh.[21] For each root l_q of Eq. (20), there is a corresponding eigenvector $\tilde{v}^q$. The l's are arranged such that the Poynting vectors corresponding to l_1, l_3, and l_5 are directed into the solid.

Displacements in the layer and the substrate can then be written as

$$u_j^L = \sum_{q=1}^{6} U_q\, v_j^{q(L)} \exp[i\, k(x_1 + l_q^L\, x_3 - c\, t)] \tag{21a}$$

and

$$u_j^S = \sum_{q=1,3,5} W_q\, v_j^{q(S)} \exp\, [i\, k(x_1 + l_q^S\, x_3 - c\, t)] \tag{21b}$$

where U_q and W_q are unknown constants and superscripts L and S denote quantities in the layer and the substrate, respectively. On substituting Eq. (21a) into Eq. (15), the displacements and stress components σ_{i3} (i = 1,2,3) for the layer can be written as

$$\begin{Bmatrix} u_1^L \\ u_2^L \\ u_3^L \\ \sigma_{33}^L \\ \sigma_{13}^L \\ \sigma_{23}^L \end{Bmatrix} = \begin{bmatrix} T^L(x_3) \end{bmatrix} \cdot \begin{Bmatrix} U_1 \\ U_2 \\ U_3 \\ U_4 \\ U_5 \\ U_6 \end{Bmatrix} \tag{22}$$

For the substrate, substituting Eq. (21b) into Eq. (15) yields

$$\begin{Bmatrix} u_1^s \\ u_2^s \\ u_3^s \\ \sigma_{33}^s \\ \sigma_{13}^s \\ \sigma_{23}^s \end{Bmatrix} = \begin{bmatrix} T^s(x_3) \end{bmatrix} \cdot \begin{Bmatrix} W_1 \\ 0 \\ W_3 \\ 0 \\ W_5 \\ 0 \end{Bmatrix} \tag{23}$$

For $q = 1,2, \ldots, 6$, elements of the matrices $[T^L]$ and $[T^S]$ follow from the following expressions by substituting appropriate quantities for the layer and the substrate, respectively:

$$T_{iq}(x_3) = v_i^q \exp(i\, k\, l_q\, x_3) \qquad (i = 1,2,3)$$

$$T_{4q}(x_3) = i\, k\, (c_{13}v_1^q + c_{36}v_2^q + c_{33}l_q v_3^q) \exp(i\, k\, l_q\, x_3)$$

$$T_{5q}(x_3) = i\, k\, [c_{55}(l_q v_1^q + v_3^q) + c_{45}\, l_q v_2^q] \exp(i\, k\, l_q\, x_3)$$

and

$$T_{6q}(x_3) = i\, k\, [c_{45}(l_q v_1^q + v_3^q) + c_{44} l_q v_2^q] \exp(i\, k\, l_q\, x_3) \tag{24}$$

Beginning with Eq. (22), the term $\exp[i\, k(x_1 - c\, t)]$ has been omitted. Constants W_2, W_4, and W_6 are set equal to zero to keep the expressions in the substrate bounded as $x_3 \to \infty$.

At the interface of the film and the substrate, displacements and stresses σ_{i3} are continuous, i.e.,

$$\begin{bmatrix} T^L(x_3 = h) \end{bmatrix} \cdot \begin{Bmatrix} U_1 \\ U_2 \\ U_3 \\ U_4 \\ U_5 \\ U_6 \end{Bmatrix} = \begin{bmatrix} T^s(x_3 = h) \end{bmatrix} \cdot \begin{Bmatrix} W_1 \\ 0 \\ W_3 \\ 0 \\ W_5 \\ 0 \end{Bmatrix} \tag{25}$$

where h is the thickness of the film. It follows that at $x_3 = 0$, displacements and stress components in the layer can be written as

$$\begin{Bmatrix} u_1^L \\ u_2^L \\ u_3^L \\ \sigma_{33}^L \\ \sigma_{13}^L \\ \sigma_{23}^L \end{Bmatrix}_{x_3=0} = \begin{bmatrix} H \end{bmatrix} \cdot \begin{Bmatrix} W_1 \\ 0 \\ W_3 \\ 0 \\ W_5 \\ 0 \end{Bmatrix} \tag{26}$$

where the matrix [H] is defined as

$$[\mathrm{H}] = [\mathrm{T}^{\mathrm{L}}(0)] \cdot [\mathrm{T}^{\mathrm{L}}(h)]^{-1} \cdot [\mathrm{T}^{\mathrm{s}}(h)] \tag{27}$$

To determine the reflection coefficient for a plane wave incident from the liquid onto the liquid–solid interface, we need solutions for the liquid similar to those in Eq. (26). The sum of the incident and reflected waves in the liquid can be written as

$$\mathrm{u}^{\mathrm{w}} = \sum_{\mathrm{q}=1}^{2} \mathrm{Z_q}\, \mathrm{v_q^w} \exp\{i\, k[\mathrm{x}_1 + (-1)^{\mathrm{q}+1}\, l_{\mathrm{w}} \mathrm{x}_3 - \mathrm{ct}]\} \tag{28}$$

Since a liquid does not support shear deformation, displacements and stress components for the liquid can be written as

$$\begin{Bmatrix} \mathrm{u_1^w} \\ \mathrm{u_2^w} \\ \mathrm{u_3^w} \\ \sigma_{33}^{\mathrm{w}} \\ \sigma_{13}^{\mathrm{w}} \\ \sigma_{23}^{\mathrm{w}} \end{Bmatrix} = \begin{bmatrix} 1 & 1 \\ 0 & 0 \\ \mathrm{v_1^w} & \mathrm{v_2^w} \\ i\, k\, \rho_w\, c^2 & i\, k\, \rho_w\, c^2 \\ 0 & 0 \\ 0 & 0 \end{bmatrix} \begin{Bmatrix} \mathrm{Z_1} \\ \mathrm{Z_2} \end{Bmatrix} \tag{29}$$

where $\mathrm{v_1^w} = -\mathrm{v_2^w} = l_w = q_w/k$. Here $q_w = (k_w{}^2 - k_x{}^2)^{1/2}$. The continuity of normal displacement and normal stress at the interface $x_3 = 0$ provides

$$\begin{Bmatrix} \mathrm{u_3^L} \\ \sigma_{33}^{\mathrm{L}} \\ \sigma_{13}^{\mathrm{L}} \\ \sigma_{23}^{\mathrm{L}} \end{Bmatrix}_{\mathrm{x}_3=0} = \begin{Bmatrix} \mathrm{u_3^w} \\ \sigma_{33}^{\mathrm{w}} \\ 0 \\ 0 \end{Bmatrix}_{\mathrm{x}_3=0} \tag{30}$$

Substituting Eqs. (26) and (29) into Eq. (30) and rearranging matrix elements yields

$$\begin{bmatrix} \mathrm{H}_{31} & \mathrm{H}_{33} & \mathrm{H}_{35} & q_w/k \\ \mathrm{H}_{41} & \mathrm{H}_{43} & \mathrm{H}_{45} & -i\, k\, \rho_w\, \mathrm{c}^2 \\ \mathrm{H}_{51} & \mathrm{H}_{53} & \mathrm{H}_{55} & 0 \\ \mathrm{H}_{61} & \mathrm{H}_{63} & \mathrm{H}_{65} & 0 \end{bmatrix} \begin{Bmatrix} \mathrm{W}_1 \\ \mathrm{W}_3 \\ \mathrm{W}_5 \\ \mathrm{Z}_2 \end{Bmatrix} = \begin{Bmatrix} q_w/k \\ i\, k\, \rho_w\, \mathrm{c}^2 \\ 0 \\ 0 \end{Bmatrix} \mathrm{Z}_1 \tag{31}$$

By using Eq. (31), the reflection coefficient R, which is the ratio of the reflected wave amplitude Z_2 to the incident wave amplitude Z_1, is then obtained as a function of θ and ϕ

$$R = \frac{(\mathrm{A} - \beta\, \mathrm{B})}{(\mathrm{A} + \beta\, \mathrm{B})} \tag{32}$$

where

$$A = \begin{vmatrix} H_{41} & H_{43} & H_{45} \\ H_{51} & H_{53} & H_{55} \\ H_{61} & H_{63} & H_{65} \end{vmatrix} \qquad B = \begin{vmatrix} H_{31} & H_{33} & H_{35} \\ H_{51} & H_{53} & H_{55} \\ H_{61} & H_{63} & H_{65} \end{vmatrix}$$

and $\beta = i\, \rho_w\, \omega^2/q_w$.

An analogous approach has been used to determine the reflection coefficient for multilayered configurations by multiplying the transfer matrices of n layers as in Ref. 23

$$[\mathrm{H}] = [\mathrm{T}_1(0)]\cdot[\mathrm{T}_1(h_1)]^{-1}\cdot[\mathrm{T}_2(h_1)]\cdot[\mathrm{T}_2(h_2)]^{-1}\ldots[\mathrm{T}_n(h_n)]^{-1}\cdot[\mathrm{T}^{\mathrm{S}}(h_n)] \tag{33}$$

The reflection coefficient for a bare anisotropic half-space has the same form as Eq. (32), i.e.,

$$R = (\mathrm{A}_1 - \beta_1\, \mathrm{B}_1)/(\mathrm{A}_1 + \beta_1\, \mathrm{B}_1) \tag{34}$$

where

$$A_1 = \begin{vmatrix} D_{11} & D_{13} & D_{15} \\ D_{21} & D_{23} & D_{25} \\ D_{31} & D_{33} & D_{35} \end{vmatrix} \qquad B_2 = \begin{vmatrix} W_1 & W_3 & W_5 \\ D_{21} & D_{23} & D_{25} \\ D_{31} & D_{33} & D_{35} \end{vmatrix}$$

and $\beta_1 = \rho_w\, \omega^2/(k\, q_w)$. Here W_q and D_{iq} ($q = 1,3,5$; $i = 1,2,3$) are expressions in terms of the material constants as stated in Section 2 of the paper by Nayfeh.[21]

5.3.3. Numerical Approach

Numerical techniques have been employed as an alternative to calculating $V(z)$ curves for the line-focus acoustic microscope. The boundary element method (BEM) has been used to calculate $V(z)$ curves for isotropic specimens[18] and isotropic-layer/substrate configurations.[26] A technique combining the finite element method (FEM) and BEM has been used to calculated $V(z)$ curves for anisotropic-layer/isotropic-substrate configurations.[27]

For certain configurations, these numerical techniques have advantages over the Fourier optics approach discussed in Sections 5.3.1–5.3.2. For example it is quite complicated to determine the reflection coefficient for a specimen containing a crack. On the other hand, once the required programming has been done, the BEM approach can be used for specimens containing defects of quite arbitrary shape. Examples for surface-breaking and subsurface cracks of various orientations have been given by Ahn *et al.*[28] and Ahn and Achenbach.[29] Since the BEM technique may however suffer convergence problems for very thin layers, the combined FEM/BEM approach is recommended for film/substrate configurations, with the FEM applied in the layer and the BEM in the substrate; see Liu *et al.*[27] The FEM/BEM approach has the additional advantage that anisotropic and inhomogeneous material properties or the presence of a discrete inhomogeneity in the layer can be accommodated without difficulties once the FEM program is available.

An electromechanical reciprocity identity is used to relate the change in voltage to the mechanical wave fields at the surface of the specimen. Because this relation plays an essential role in the numerical approach, its derivation, which follows closely that given by Auld,[30] is briefly reviewed. The starting point is the following identity:

$$(-i\,\omega\, u_{k1}\, \sigma_{jk2} + i\,\omega\, u_{k2}\, \sigma_{jk1} + e_{jkm}\, E_{k1}\, H_{m2} - e_{jkm}\, E_{k2}\, H_{m1})_{,j} = 0 \tag{35}$$

which can be derived from equations governing the electrical and mechanical fields in a piezoelectric solid and which is valid in a region free of sources. In Eq. (35), subscripts 1 and 2 indicate two different solutions to the piezoelectric equations: u_i is a component of the particle displacement, σ_{ij} is a component of the stress, E_i is a component of the electric field, and H_i is a component of the magnetic field. The term e_{ijk} is the alternating tensor. Next this equation is integrated over a surface that surrounds the transducer, the buffer rod, and the lens. The electromagnetic field is assumed to be zero everywhere except where the surface of integration cuts the coaxial cable coming from the terminals of the transducer. The elastodynamic field is assumed to be zero everywhere except over a surface at or below the lens surface. Field 1 is excited if the specimen is not present, and Field 2 is excited in the presence of the specimen. Following Auld,[30] integration over the coaxial cable cross section is found to be proportional to the change in voltage caused by the specimen. The change in voltage δV is given by

$$\delta V = C \int_{S_l} (u_i^{in}\, t_i - u_i\, t_i^{in})\, dS \tag{36}$$

Here u_i^{in} and t_i^{in} are components of the incident particle displacement and traction, u_i and t_i are those of the total particle displacement and traction, and C is a normalizing constant. The contour integration S_l is chosen to be the surface of the specimen.

To calculate the incident wave field at the specimen's surface, a model of the propagation and subsequent focusing of the incident wave through the buffer rod, lens, and coupling fluid has been developed. It is described in Section 2 of the paper by Achenbach *et al.* [18] The initial assumption is that the transducer radiates a beam approximately Gaussian in profile and harmonic time-dependence at the aperture of the lens. This assumption can be checked only by comparing the model's predictions with measurements (however the model can readily be changed to accommodate alternative beam profiles). The wave field transmitted by the lens, which is the field incident on the surface of the specimen, is calculated by formulating the problem as a system of integral equations for the lens immersed in water and solving them using the BEM. From this calculation we determine the wave fields incident on the specimen u_i^{in} and t_i^{in}. The model produces a sharply focused acoustic beam in a fluid. It remains to calculate wave fields reflected by the specimen. This can be done by the BEM or the combined FEM/BEM technique, as shown in the previously referenced papers.

Once displacement and traction components on the surface of the specimen have been calculated, the corresponding transducer voltage can be obtained from Eq. (36). The $V(z)$ curve is subsequently obtained by calculating the voltage for various z. Examples of $V(z)$ curves calculated by these numerical approaches are discussed by Ahn *et al.* [(26)] and Liu *et al.* [(27)]

5.3.4. Theoretical Results

Using the characteristic functions of the acoustic lens, $L_1(k_x)$ and $L_2(k_x)$, discussed in Section 5.3.1 and the reflection coefficient of the specimen $R(k_x)$, discussed in Section 5.3.2, $V(z)$ curves for the line-focus acoustic microscope were calculated for various isotropic and anisotropic materials and various anisotropic-film/substrate configurations.[(14,20)] For anisotropic configurations, $V(z)$ curves for specific directions of wave propagation were calculated for various values of the normalized film thickness, while for a specific normalized film thickness, V(z) curves were calculated as functions of the propagation direction. The calculated results were compared with experimental results.

Specimens considered in these calculations can be divided into four groups: (1) isotropic materials, such as aluminum and glass; (2) anisotropic materials, such as cubic crystal silicon and MgO; (3) anisotropic single layers deposited on anisotropic substrates, such as TiN films and NbN films deposited on MgO substrates; (4) alternating layers of anisotropic materials deposited on an anisotropic substrate, such as TiN/NbN superlattice films deposited on MgO substrates. Details of the experimental procedures are described in Section 5.4.

The $V(z)$ curves for isotropic aluminum and glass are shown in Figs. 5.7(a) and (b), respectively. The calculated $V(z)$ curves show good agreement with experimental results. The good agreement has been achieved because the effects of the anti-reflection coating on the lens surface, the wave attenuation in the coupling fluid, and a more accurate estimate of the focal length have been taken into account. In Figs. 5.7(a) and (b) oscillations in the range of $-200\ \mu\text{m} < z < -30\ \mu\text{m}$ are dominantly related to the leaky Rayleigh wave, while oscillations in the range of $z < -200\ \mu\text{m}$ are related to the leaky surface-skimming longitudinal wave.

Figures 5.8(a) and (b) show $V(z)$ curves for propagation along the [100] direction on the (001) plane of cubic-crystalline silicon and MgO, respectively.

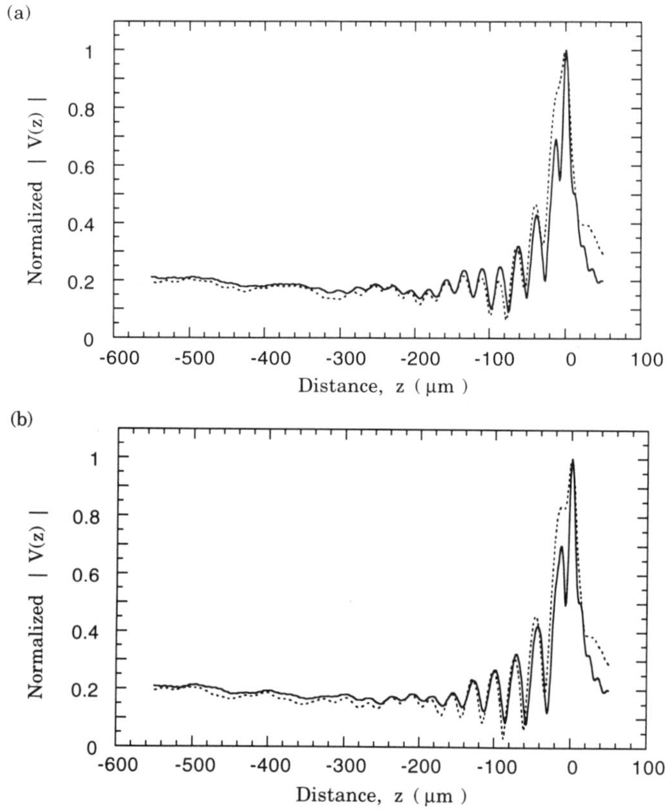

Figure 5.7. A comparison of calculated (*solid line*) and measured (*dashed line*) $V(z)$ curves for (a) aluminum and (b) glass (Ref. 14).

The amplitude of the oscillating $V(z)$ curve for MgO shown in Fig. 5.8(b) increases as the specimen is raised because the reduction of attenuation in the water is greater than the increase of attenuation in the MgO specimen. Figures 5.9(a) and (b) show results for a 2.8-μm TiN film and a 1.1-μm NbN film, both deposited on an MgO substrate. The calculated $V(z)$ curves agree very well with the measured ones, not only in the spacing of peaks and dips but also in the oscillation amplitudes.

Figure 5.10 shows the result for a superlattice film composed of 65 pairs of TiN/NbN layers, deposited on MgO. Each single layer consists of two single-crystal layers of TiN and NbN. The film has a total thickness of 2.36 μm. The TiN fraction, defined as $d_T/(d_T + d_N)$, where d_T and d_N are the thicknesses of

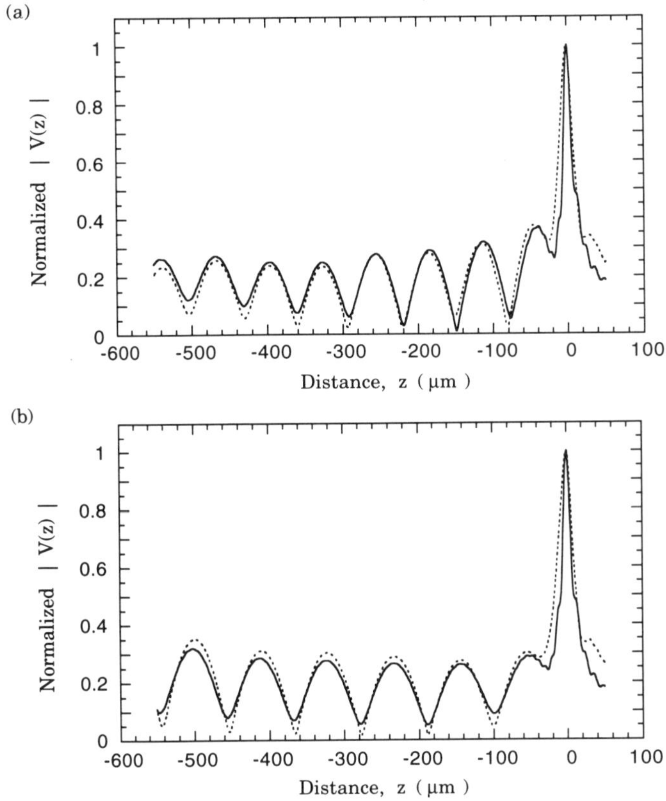

Figure 5.8. A comparison of calculated and measured $V(z)$ curves for propagation along the [100] direction on the (001) plane of (a) bare silicon and (b) bare MgO (Ref. 14).

individual TiN and NbN layers, is 0.3. The reflection coefficient of this 130-layer/substrate configuration is calculated from Eqs. (32) and (33). Each layer is now represented by a matrix, and the whole film is represented by the multiplication of 130 matrices. Again the curves show excellent agreement between theoretical and experimental results.

It should be mentioned that SAW velocities of the specimens range from 3000–6000 m/s, which is approximately the range for which the acoustic microscope with an acoustic lens of half-aperture angle 60° can produce a $V(z)$ curve with sufficient peaks and dips to analyze SAW velocity and attenuation. Theoretical $V(z)$ curves agree very well with experimental curves up to $z = -550$ μm, which is almost the maximum defocus distance of the acoustic microscope.

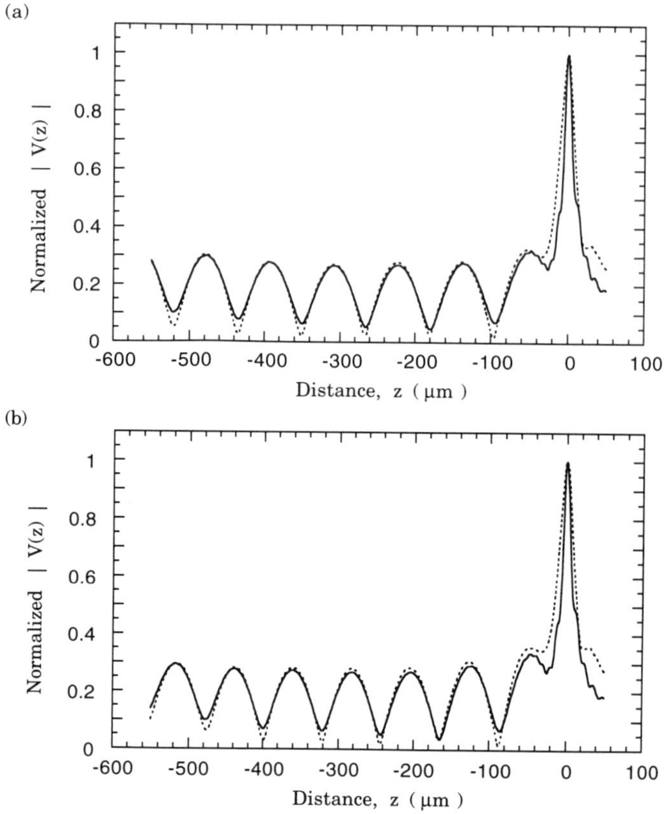

Figure 5.9. A comparison of calculated (*solid line*) and measured (*dashed line*) $V(z)$ curves for propagation along the [100] direction on the (001) plane of (a) MgO coated with 2.8-μm TiN and (b) MgO coated with 1.1-μm NbN (Ref. 14).

5.4. Measuring and Calculating SAW Velocities

Velocities of leaky modes of elastic wave propagation can be obtained from measured $V(z)$ curves. The SAW velocity v has a functional relationship with the period Δz of the oscillation in the $V(z)$ curve as[31,32]

$$v = v_w \left[1 - \left(1 - \frac{v_w}{2f\Delta z} \right)^2 \right]^{-1/2} \tag{37a}$$

$$= (v_w f \Delta z)^{1/2} \left(1 - \frac{v_w}{4f\Delta z} \right)^{-1/2} \tag{37b}$$

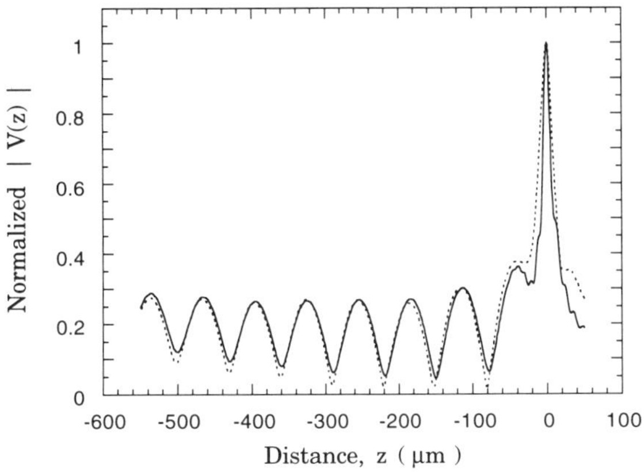

Figure 5.10. A comparison of calculated (*solid line*) and measured (*dashed line*) $V(z)$ curves for propagation along the [100] direction of a 2.36-μm TiN/NbN superlattice film deposited on the (001) plane of MgO (Ref. 14).

where v_w is the wave velocity in the coupling water (1490 m/s at 23° C) and f is the wave frequency. Equation (37b) implies that for $v_w/(4f\Delta z) << 1$, the surface wave velocity v is approximately proportional to $(\Delta z)^{1/2}$. The period Δz, and thus the SAW velocity of the specimen, are obtained by processing the $V(z)$ data. The processing procedure consists of three main steps. The first step is subtracting the geometric effect of the acoustic lens, which is often denoted by $V_L(z)$, from the $V(z)$ data. The $V_L(z)$ data are approximately measured as the $V(z)$ data of a material, such as Teflon, that does not generate an SAW for a suitable range of incidence angles. By subtracting the $V_L(z)$ data from the $V(z)$, we obtain the $V_I'(z)$ curve, which shows periodic oscillations and decaying amplitudes for increasing $|z|$. The second step is a spectral analysis, such as the fast Fourier transform, of the $V_I'(z)$ data to obtain the velocity and attenuation of each mode from multiple mode data. A digital filtering with the acoustic properties of a desirable mode reduces the effects of other modes and yields $V_I(z)$ data. Finally a fast Fourier transform analysis of the $V_I(z)$ data determines the period Δz and the exponential amplitude decay per unit z distance of a mode of leaky SAWs. For further details of this procedure see Ref. 32.

5.4.1. Characteristic SAW Equations

In a theoretical analysis, the velocities of leaky SAWs on a bare or film-coated solid can be calculated from the characteristic equation relating phase velocity to the frequency, thin-film parameters, and the wave propagation direction. The characteristic equation of leaky SAWs can be obtained by setting to zero the denominator of the reflection coefficient for a plane acoustic wave incident on the interface between the water and the specimen.

The location of the roots of the characteristic equation in the complex k_x-plane determines the nature of the corresponding wave motion. For the characteristic equation of a fluid–isotropic-solid interface, the relation between root location and wave type has been investigated by Tsukahara *et al.*[(33)] For fluid-loaded anisotropic solids and isotropic- and anisotropic-film/substrate configurations, this equation has been discussed in some detail by Sklar.[10]

For an isotropic solid, the characteristic equation is obtained from the denominator of the reflection coefficient given by Eq. (11)

$$(2k_x^2 - k_t^2)^2 - 4k_x^2\sqrt{(k_x^2 - k_l^2)(k_x^2 - k_t^2)} + i\left(\frac{\rho_w}{\rho}\right)\sqrt{\frac{(k_x^2 - k_l^2)}{(k_w^2 - k_x^2)}} = 0 \qquad (38)$$

Equation (38) is the same as the characteristic equation for leaky Rayleigh waves derived by Viktorov.[(34)] The characteristic equation for a solid composed of isotropic layers can be obtained from the reflection coefficient given by Eq. (12)

$$\Gamma_s + \left(\frac{\rho_w}{\rho}\right)\Gamma_w = 0 \qquad (39)$$

For an anisotropic solid, the characteristic equation is obtained from Eq. (34)

$$A_1 + \beta_1 B_1 = 0 \qquad (40)$$

For an anisotropic layered solid and wave propagation under an angle ϕ with the crystalline axis x_1', the characteristic equation is obtained from Eq. (32)

$$A + \beta B = 0 \qquad (41)$$

For wave propagation in the direction of crystalline symmetry, i.e., in the [100] or the [110] directions on the (001) plane of a thin-film/substrate configuration, the in-plane motion uncouples, and Eq. (37) is simplified to

$$\overline{A} + \left(\frac{i\,\rho_w\,\omega^2}{q_w}\right)\overline{B} = 0 \qquad (42)$$

where

$$\overline{A} = \begin{vmatrix} \overline{H}_{31} & \overline{H}_{33} \\ \overline{H}_{41} & \overline{H}_{43} \end{vmatrix} \qquad \overline{B} = \begin{vmatrix} \overline{H}_{21} & \overline{H}_{23} \\ \overline{H}_{41} & \overline{H}_{43} \end{vmatrix}$$

The terms $\overline{H}_{iq}$ (q = 1,3; i = 2,3,4) are elements of the matrix defined by

$$[\overline{H}] = [\overline{T}^L(0)] \cdot [\overline{T}^L(h)]^{-1} \cdot [\overline{T}^S(h)] \qquad (43)$$

where h is the thickness of the film and the superscripts L and S represent quantities in the layer and the substrate, respectively. Elements of the matrices

$$[\overline{T}^{L}] \text{ and } [\overline{T}^{S}]$$

can be obtained from the following expressions by substituting appropriate quantities for the layer and the substrate:

$$\overline{T}_{1q}(x_3) = \exp(i\,k\,l_q\,x_3) \qquad \overline{T}_{2q}(x_3) = v^q \exp(i\,k\,l_q\,x_3)$$
$$\overline{T}_{3q}(x_3) = i\,k\,(c_{12} + c_{11}\,l_q\,v^q)\exp(i\,k\,l_q\,x_3)$$
$$\overline{T}_{4q}(x_3) = i\,k\,c_{44}\,(l_q + v^q)\exp(i\,k\,l_q\,x_3)$$

and

$$v^q = -\frac{K_{13}(l_q)}{K_{33}(l_q)} \tag{44}$$

where q = 1, 2, 3, 4. In Eq. (44), l_q^L and l_q^S (q = 1,2,3,4) are the roots of the following Christoffel equation into which again appropriate quantities for the layer and the substrate must be substituted

$$K_{11}(l)\,K_{33}(l) - [K_{33}(l)]^2 = 0 \tag{45}$$

where

$$K_{11}(l) = c_x + c_{44}\,l^2 - \rho\left(\frac{\omega}{k}\right)^2, \qquad K_{13}(l) = (c_{12} + c_{44})\,l$$

and

$$K_{33}(l) = c_{44} + c_{11}\,l^2 - \rho\,(\omega/k)^2$$

Here $c_x = c_{11}$ for the [100] direction, and $c_x = (c_{11} + c_{12})/2 + c_{44}$ for the [110] direction. The roots l_q^S for the substrate are arranged so that the imaginary parts of kl_1^S and kl_3^S are positive, i.e., the Poynting vectors corresponding to l_1^S and l_3^S are directed into the solid.

For a bare anisotropic solid, the characteristic equation for wave propagation in the [100] and [110] directions on the (001) plane follows from Eq. (40)

$$\overline{A}_1 + \left[\frac{\rho_w\,\omega^2}{k\,q_w}\right]\overline{B}_1 = 0 \tag{46}$$

where

$$\overline{A}_1 = \begin{vmatrix} \overline{D}_{11} & \overline{D}_{13} \\ \overline{D}_{21} & \overline{D}_{23} \end{vmatrix}, \qquad \overline{B}_1 = \begin{vmatrix} \overline{W}_1 & \overline{W}_3 \\ \overline{D}_{21} & \overline{D}_{23} \end{vmatrix}$$

The terms $\overline{W}_q$ and $\overline{D}_{iq}$ where (q = 1,3; i = 1,2) can now be simplified to

$$\overline{W}_q = -\frac{K_{13}(l_q)}{K_{33}(l_q)}, \qquad \overline{D}_{1q} = c_{12} + c_{11}\,l_q\,\overline{W}_q, \qquad \overline{D}_{2q} = c_{44}\,(l_q + \overline{W}_q) \tag{47}$$

and l_q are the roots of Eq. (45) (the Christoffel equation). The l roots are arranged so that the imaginary parts of kl_1 and kl_3 are positive, i.e., the Poynting vectors corresponding to l_1 and l_3 are directed into the solid.

The characteristic Eqs. (38)–(42) and (46) take account of the loading of the fluid on the solid, and they govern wave mode velocities in the actual configuration of a film-covered specimen coupled with water in an acoustic microscope. Calculating the roots of equations that include the fluid-loading effect is however somewhat complicated because the wave number k has a complex value, with the imaginary part of the roots related to attenuation due to leakage of the wave motion into the fluid. Fortunately the effect of fluid-loading on wave mode velocity is small and usually negligible.[34] Therefore especially for the iterative calculation employed in the inverse method discussed in Section 5.5, it is useful to calculate the velocity of SAWs without the effect of fluid-loading. The characteristic equation for wave mode velocities on an anisotropic-film-covered solid without fluid-loading can be simply obtained from the characteristic equation with fluid coupling by setting $\rho_w = 0$. For wave propagation in the crystalline symmetry directions, we obtain from Eq. (42)

$$F_1(\bar{h}, v) \equiv \begin{vmatrix} \overline{\mathrm{H}}_{31} & \overline{\mathrm{H}}_{33} \\ \overline{\mathrm{H}}_{41} & \overline{\mathrm{H}}_{43} \end{vmatrix} = 0 \tag{48}$$

where $\bar{h}$ $(= h/\lambda_s)$ is the normalized film thickness and λ_s is the wavelength of transverse waves in the substrate.

A characteristic equation similar to Eq. (48) was derived differently by Kim and Achenbach[35] using the general equations presented by Farnell and Adler.[36] From the traction-free boundary condition at the free surface of a film of thickness h and from the continuity of tractions and displacements at the interface between the film and the substrate, the following equation was obtained:

$$F_2(\bar{h}, v) \equiv |a_{ij}| = 0 \qquad (i, j = 1, 2, \ldots, 6) \tag{49}$$

For $j = 1$ and 2, the elements of the determinant in Eq. (49) are

$$a_{1\mathrm{j}} = 1 \qquad a_{2\mathrm{j}} = \mathrm{v}_\mathrm{m}^\mathrm{S} \qquad a_{3\mathrm{j}} = \mathrm{c}_{12}^\mathrm{S} + \mathrm{c}_{11}^\mathrm{S}\, l_\mathrm{m}^\mathrm{S}\, \mathrm{v}_\mathrm{m}^\mathrm{S}$$

$$a_{4\mathrm{j}} = \mathrm{c}_{44}^\mathrm{S}\,(l_\mathrm{m}^\mathrm{S} + \mathrm{v}_\mathrm{m}^\mathrm{S}) \qquad a_{5\mathrm{j}} = a_{6\mathrm{j}} = 0$$

For $j = 3$, 4, 5, and 6, the elements are

$$a_{1\mathrm{j}} = -1 \qquad a_{2\mathrm{j}} = -\mathrm{v}_\mathrm{n}^\mathrm{L} \qquad a_{3\mathrm{j}} = -\mathrm{c}_{12}^\mathrm{L} - \mathrm{c}_{11}^\mathrm{L}\, l_\mathrm{n}^\mathrm{L}\, \mathrm{v}_\mathrm{n}^\mathrm{L} \qquad a_{4\mathrm{j}} = -\mathrm{c}_{44}^\mathrm{L}(l_\mathrm{n}^\mathrm{L} + \mathrm{v}_\mathrm{n}^\mathrm{L})$$

$$a_{5\mathrm{j}} = (c_{12}^\mathrm{L} + \mathrm{c}_{11}^\mathrm{L}\, l_\mathrm{n}^\mathrm{L}\, \mathrm{v}_\mathrm{n}^\mathrm{L}) \exp(-i\, l_\mathrm{n}^\mathrm{L} k\, h) \qquad a_{6\mathrm{j}} = (l_\mathrm{n}^\mathrm{L} + \mathrm{v}_\mathrm{n}^\mathrm{L}) \exp(-i\, l_\mathrm{n}^\mathrm{L} k\, h)$$

Here $\mathrm{v}_\mathrm{m}^\mathrm{S}$ and $\mathrm{v}_\mathrm{n}^\mathrm{L}$ are normalized eigenvectors defined by Eq. (44). The superscripts m $(= 2j-1)$ and n $(=j - 2)$ denote the mth and nth solutions of the eigenvalue problems defined by Eq. (45) for the layer and the substrate, respectively. When the elastic constants c_{11}, c_{12}, and c_{44} for the layer and the substrate are known, Eq. (49) is the dispersion equation that yields the phase velocity v as a function of the normalized thickness $\bar{h}$. On the other hand, when v is known as a function of $\bar{h}$, Eq. (49) can be used to determine the elastic constants c_{11}, c_{12}, and c_{44} of the film.

For known elastic constants, computing the SAW velocity proceeds as follows:

1. A value for the SAW velocity v is assumed and substituted into Eq. (45).
2. The eigenvalue problem defined by Eq. (45) is solved for (l^S_m, v^S_m) and (l^L_n, v^L_n).
3. The boundary condition determinant given by Eq. (49) is evaluated.
4. The phase velocity v is searched iteratively until this determinant becomes close to zero within a desired accuracy.

5.4.2. SAW Dispersion Curves

The dependence of wave mode velocities on film thickness and frequency, which produces the dispersion effect, was studied theoretically by Ewing *et al.*[37] for isotropic materials and by Farnell and Adler[36] for anisotropic as well as isotropic materials. Using Eq. (49) based on the Farnell and Adler approach,[36] velocities for a thin film and a substrate of specific anisotropies were calculated by Kim and Achenbach.[35] Both the film and the substrate were taken to display cubic symmetry, and the crystalline axes of the film, which are parallel and normal to its free surface, coincide with those of the substrate.

Wave mode velocity varies with film thickness and wavelength. Velocity curves of a wave mode versus h/λ_s, where h is the film thickness and λ_s is the wavelength of tranverse waves in the substrate, are called dispersion curves. When the thickness of the film is smaller than the wavelength, the wave motion penetrates through the film into the substrate. Depending on the acoustic properties of the film relative to those of the substrate, the wave velocity increases or decreases with the ratio of film thickness and wavelength. When wave velocity in the film material is larger than the corresponding wave velocity in the substrate material, there exists a single propagating mode, the Rayleigh wave mode, whose phase velocity increases with h/λ_s. Many modes exist when wave velocity in the film material is smaller than the corresponding wave velocity in the substrate material. Two of these modes are the most relevant ones, as discussed by Kushibiki *et al.*[5] The lowest mode is usually called the Rayleigh mode. At $h/\lambda_s = 0$ its velocity equals the velocity of leaky Rayleigh waves on the substrate material, and as h/λ_s increases, the velocity decreases to asymptotically approach the velocity of leaky Rayleigh waves of the layer material. The second mode can be considered to consist of two parts: the leaky pseudo-Sezawa mode and the leaky Sezawa mode, named after the seismologist who first investigated these modes.[38] The leaky pseudo-Sezawa mode exists at small h/λ_s; at $h/\lambda_s = 0$ its velocity equals the velocity of leaky surface-skimming longitudinal waves in the substrate material. The velocity of the pseudo-Sezawa mode decreases to that of transverse waves in the substrate at a specific cutoff value of h/λ_s. With a possible jump in the phase velocity, the leaky Sezawa mode starts at that cutoff value of h/λ_s, and its phase velocity asymptotically decreases to the transverse wave velocity in the layer as h/λ_s further increases. For isotropic films on isotropic or anisotropic substrates, SAW velocity measurements were used to determine film thickness[2,4,7,8,39,40] and the film's elastic properties.[5,6,41]

For the two lowest modes, Fig. 5.11 shows dispersion curves for gold films deposited by sputtering on glass slides.[35] Velocities of the Rayleigh mode and the pseuo-Sezawa mode were obtained for gold film thicknesses of 0.03, 0.11, 0.145, 0.22, 0.33, and 0.59 μm at 225 MHz. Measured velocities, represented by squares, are compared with calculated values, represented by lines. The dispersion curves in Fig. 5.11 show typical negative dispersion.

For anisotropic-film/substrate configurations, wave velocities vary with the propagation direction. For propagation in the [100] and [110] directions, Fig. 5.12 shows SAW velocities measured at various frequencies for TiN films of various thicknesses deposited on MgO substrates as functions of normalized thickness h/λ_s.[42] Measured velocities, represented by squares, were obtained for 0.1-μm film at 225 MHz, for 0.44-μm film at 225 MHz and 255 MHz, and for 1.2- and 2.8-μm film at 195 MHz, 225 MHz, and 255 MHz, respectively, to yield data for nine data points. An additional data point was obtained for $h/\lambda_s = 0$, where the phase velocity equals the one for a bare substrate. The two curves

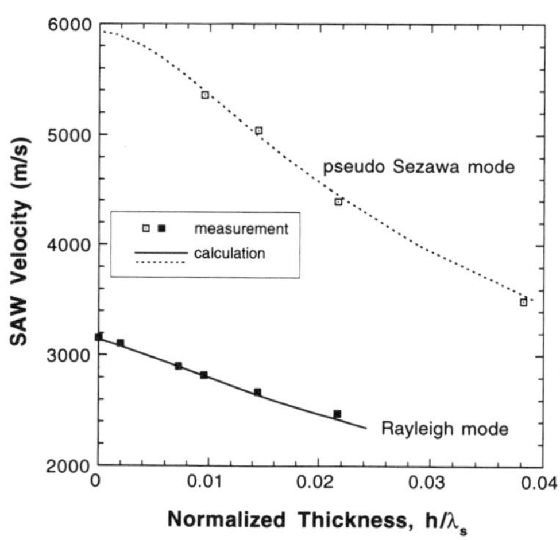

Figure 5.11. Dispersion curves of surface acoustic waves on glass coated with gold films (Ref. 35).

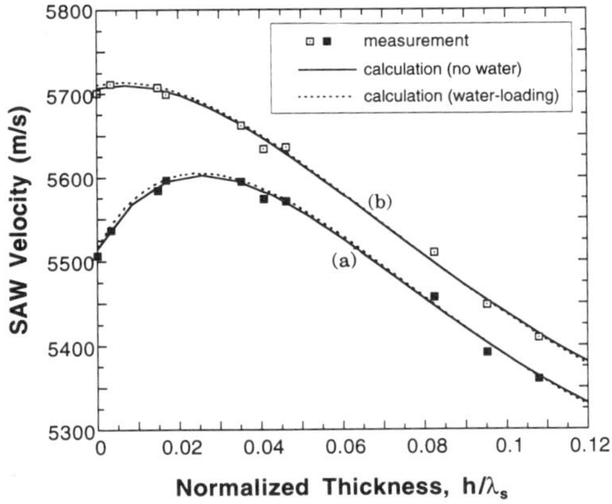

Figure 5.12. Dispersion curves of SAW on the (001) plane of TiN films epitaxially deposited on MgO substrates; (a) [100] direction, (b) [110] direction.

in Fig. 5.12 represent the regular SAW in the [100] direction and the pseudo-SAW in the [110] direction on the (001) plane. Solid lines indicate dispersion curves calculated from Eq. (49) and the elastic constants of TiN (c_{11} = 625 GPa, c_{12} = 165 GPa, c_{44} = 163 GPa, ρ = 5.39 g/cm^3). Dashed lines indicate dispersion curves calculated from Eq. (42) for the water-loaded specimen. The solid lines and the dashed lines are very close to each other, which means that the effect of the water-loading on the SAW velocity is very small. Curves in Fig. 5.12 show negative dispersion over most of their ranges. At small values of h/λ_s, the dispersion is however positive because the SAW velocity of the film material is close to that of the substrate material.[36]

For a silicon substrate coated with a diamondlike film, the dependence of the SAW velocities on the normalized thickness shows positive dispersion over most of the range of h/λ_s (see Ref. 43).

5.4.3. Anisotropic SAW Velocities

The SAW velocities measured for specific propagation directions on certain anisotropic materials may represent the regular SAW or the pseudo-SAW. The latter is called a pseudo-SAW because its velocity is larger than that of a bulk wave, namely, the out-of-plane transverse wave. Figures 5.13(a) and (b) show the theoretical velocities of surface acoustic waves on the (001) plane of silicon and MgO, respectively. The thick solid line and the thick dashed line represent the regular SAW and pseudo-SAW, respectively, without water-loading. These curves, from the paper by Kim and Achenbach,[35] were calculated using the equations of Farnell.[44] The same results are obtained using Eq. (40) with β_1 = 0. The thin solid line and the thin dashed line in Figs. 5.13(a) and (b) are the

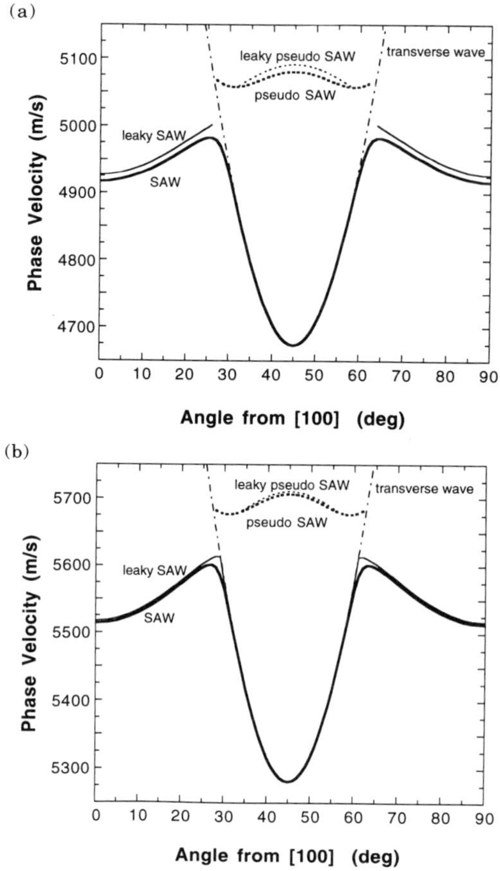

Figure 5.13. A comparison of theoretical velocities of SAWs and corresponding leaky SAWs for (001) cubic crystals; (a) silicon, (b) MgO.

velocities of the corresponding leaky SAWs. These were calculated using Eq. (40). The effect of water-loading, which slightly increases the SAW velocities, is negligibly small, as was earlier noted for isotropic materials by Viktorov.[34]

Velocities measured by LFAM are represented by open circles in Fig. 5.14(a) for silicon and in Fig. 5.14(b) for MgO. Figure 5.14(b) includes results for a 1.1-μm NbN film epitaxially deposited on an MgO substrate. Measured velocities are compared with velocities obtained from the measurement model, i.e., calculated $V(z)$ curves,[14] and they show good agreement. The anisotropic SAW velocity curves in Figs. 5.14(a) and (b) show cubic symmetry with respect to the crystalline directions (0° or 90°) and with respect to the bisectrix of the crystalline directions.

As shown in Figs. 5.14(a) and (b), for propagation on the (001) plane of cubic crystals, acoustic microscopy measurements yield the phase velocity of

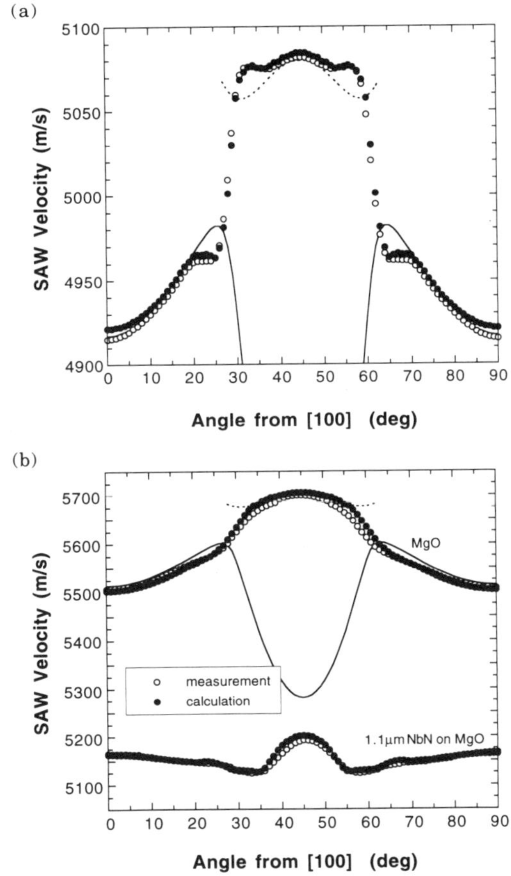

Figure 5.14. A comparison of directional variation of SAW velocities measured by acoustic microscopy and calculated from a $V(z)$ model; (a) silicon, (b) MgO and 1.1-μm NbN on MgO.

the regular SAW in one range of propagation directions and that of pseudo-SAW in the complementary range. The reason for this dichotomy is explained by studying displacement components of surface waves presented by Lim and Farnell.[45] The displacement of the regular SAW near the [100] direction includes longitudinal and vertically polarized transverse components. In the range of [100]–[110], there is an additional out-of-plane transverse component. As the propagation direction approaches [110], the out-of-plane transverse component becomes dominant, and hence the acoustic microscope cannot detect the regular SAW near [110]. Displacement components of the pseudo-SAW near [110] are longitudinal and vertical transverse, similar to the regular SAW near [100]. Therefore the acoustic microscope detects pseudo-SAW near [110]. In the range of 25°–30° from [100], the measured velocity is neither that of the regular SAW nor that of the pseudo-SAW but rather a superposition of these waves, since near this direction, both regular and pseudo-SAWs contribute to the phase shift of the measured voltage output.

5.5. Inverse Method

When the mass density and the thickness of the film and the elastic constants and the mass density of the substrate are known, elastic constants of the film can be determined from measured SAW velocities. Determining thin-film elastic constants can be based on data for the variation of SAW velocity with the film thickness to wavelength ratio for propagation in a specific direction, usually along a crystalline axis. To obtain sufficient data for a single thin-film specimen, the operating frequency of the acoustic microscope must vary over a wide range without loss of measuring accuracy, which may not be possible. Alternatively measurements can be carried out at a single frequency or at a few frequencies on several specimens with different film thicknesses. The disadvantage of this approach is that more specimens have to be prepared and it must be assumed that the elastic constants do not vary significantly for different film thicknesses. For anisotropic specimens, determining the elastic constants can be based on data for the variation of SAW velocity with propagation direction. The advantage of using the directional variation is that only one specimen is needed and the microscope can be operated at its optimal frequency. Each specimen with a different film thickness can be characterized individually.

5.5.1. Least-Square Curve Fit

First we consider the case when the elastic constants are obtained from measured SAW velocities in two specific directions by solving the dispersion equation given by Eq. (49). For a cubic crystal film, this equation is rewritten as

$$F_{\mathrm{Aq}} \equiv F(\bar{h}_{\mathrm{q}}, v_{\mathrm{q}}^{\mathrm{A}}, \mathrm{c}_{11}, \mathrm{c}_{12}, \mathrm{c}_{44}) \qquad F_{\mathrm{Bq}} \equiv F(\bar{h}_{\mathrm{q}}, v_{\mathrm{q}}^{\mathrm{B}}, \mathrm{c}_{11}, \mathrm{c}_{12}, \mathrm{c}_{44}) \tag{50}$$

where $(\bar{h}_{\mathrm{q}}, v_{\mathrm{q}}^{\mathrm{A}})$ and $(\bar{h}_{\mathrm{q}}, v_{\mathrm{q}}^{\mathrm{B}})$ denote sets of data ($\mathrm{q} = 1, 2, \ldots, N$) for the normalized film thickness $\bar{h}$ and the measured value of the SAW velocity v. The super- or subscripts A and B refer to the [100] and [110] directions, respectively. The inversion procedure determines a set of constants $(\mathrm{c}_{11}, \mathrm{c}_{12}, \mathrm{c}_{44})$ corresponding to the minimum of the following function defined by N data points for each direction[46]:

$$y = \frac{1}{2N}\sum_{\mathrm{q}=1}^{N} (w_{\mathrm{Aq}}F^{*}_{\mathrm{Aq}}F_{\mathrm{Aq}} + w_{\mathrm{Bq}}F^{*}_{\mathrm{Bq}}F_{\mathrm{Bq}}) \tag{51}$$

Here F^{*}_{Aq} and F^{*}_{Bq} are the complex conjugates of F_{Aq} and F_{Bq}, and w_{Aq} and w_{Bq}

are weighting coefficients assigned to each set of data; N is the number of data points for each direction. The weighting coefficients are initially assumed to be unity. They are adjusted such that for a calculation of Eq. (49) with an initial solution of the elastic constants and for a unit change of velocity, the change of terms in Eq. (51) is the same for each set of data.

To avoid weighting coefficients, the function y can alternatively be defined in terms of the measured and calculated velocities as

$$y = \frac{1}{2N} \sum_{q=1}^{N} [(V_{m,q}^{A} - V_{c,q}^{A})^2 + (V_{m,q}^{B} - V_{c,q}^{B})^2] \tag{52}$$

where $V_{c,q} = V(\bar{h}_q, c_{11}, c_{12}, c_{44})$ and $\bar{h}$ is the normalized film thickness h/λ_s. Subscripts m and c denote measured and calculated results.

Next we consider an anisotropic specimen, where the variation of the measured SAW velocity as a function of wave propagation direction is used to determine elastic constants. For this case, the function y is defined in terms of the measured and calculated velocities in N directions as

$$y = \frac{1}{N} \sum_{q=1}^{N} (V_{m}^{q} - V_{c}^{q})^2 \tag{53}$$

where $V_c^q = V(\phi_q, c_{11}, c_{12}, c_{44})$ and ϕ_q defines the direction of wave propagation relative to a particular crystalline axis, such as the [100] direction in a cubic crystal.

A systematic function minimization procedure known as the simplex method, initially suggested by Spendley *et al.*[47] and well established by Nelder and Mead,[48] proved to be useful for fitting a function of more than one variable to data.[49] In this work, the simplex method was used to determine elastic constants from experimental SAW data.

5.5.2. Simplex Method

This procedure is summarized in Fig. 5.15. A corresponding subroutine program is available.[50] Consider the minimization of function y of n variables (n is 3 for a cubic crystal). Initially select $(n+1)$ sets $P_0, P_1, \ldots, P_n$ of n variables. Hence each P_j ($j=0, 1, \ldots, n$) defines a point in n-dimensional space, and the $(n+1)$ points define the current simplex. The function value corresponding to P_j is denoted by y_j. The subscript j is assigned so that the y_j's are arranged in increasing order of magnitude, i.e., $y_0 < y_1 < \ldots < y_{n-1} < y_n$. The centroid M of $P_0, P_1 \ldots, P_{n-1}$ is calculated for later use. To eliminate the worst estimate y_n, P_n is replaced by a new point according to four types of operations: reflection, expansion, contraction, and shrinkage.

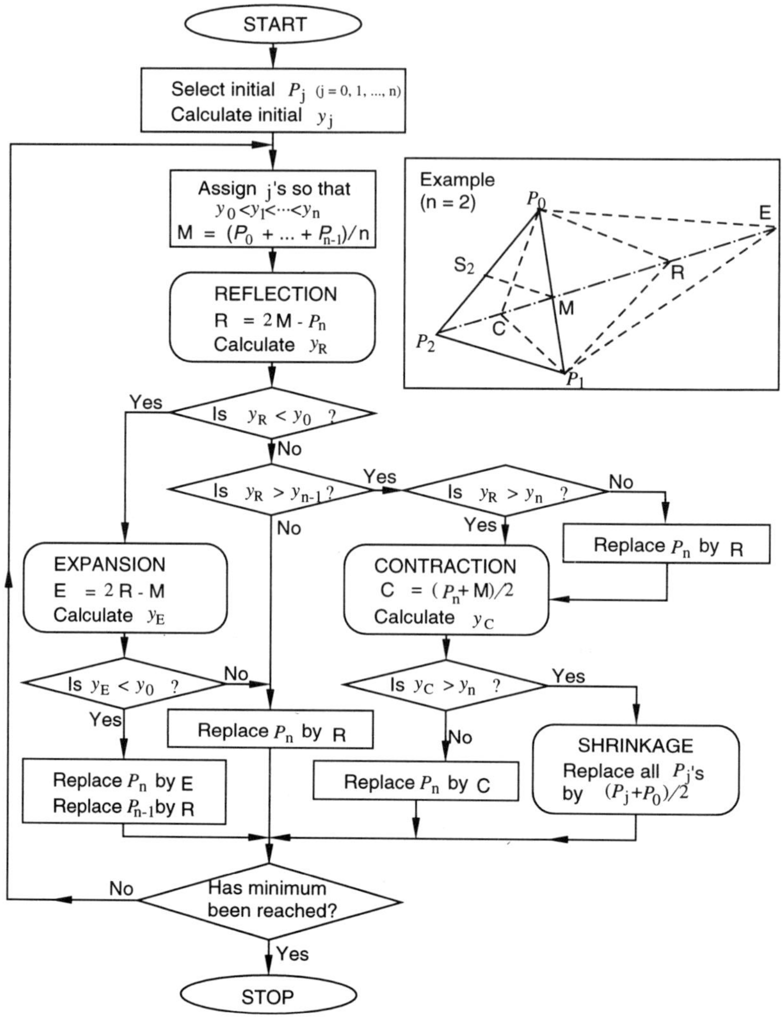

Figure 5.15. A flow chart for the algorithm of a simplex method to determine a set P minimizing a function y.

5.5.2.1. Reflection

The reflection of P_n is denoted by R, and its coordinates are defined by the relation

$$R - M = M - P_n \tag{54}$$

Thus R is a point on the line joining P_n and M on the far side of M from P_n. The corresponding value of the function is y_R. If y_R lies between y_0 and y_{n-1}, then P_n is replaced by R. With the new simplex, the process is restarted.

5.5.2.2. Expansion

If y_R is less than y_0, then R is expanded to E by the relation

$$E - R = R - M \tag{55}$$

The function value corresponding to E is y_E. If y_E is less than y_0, P_n and P_{n-1} are replaced by E and R, and the process is restarted. If y_E is not less than y_0, then the expansion has failed, and only P_n is replaced by R before restarting.

5.5.2.3. Contraction

If y_R is greater than y_{n-1}, then P_n is contracted to C by the relation

$$C - M = \frac{(P_n - M)}{2} \tag{56}$$

The function value corresponding to C is y_C. The P_n is replaced by C unless y_C is greater than y_n.

5.5.2.4. Shrinkage

If all three operations—reflection, expansion, and contraction—fail, then all P_js are replaced by $(P_j + P_0)/2$, and the process is restarted with the new sets of P_js.

The overall procedure is repeated until the function value reaches a minimum within a desired accuracy. The elastic constants c_{11}, c_{12}, and c_{44} corresponding to the minimum are the solution to the inverse problem.

5.5.3. Starting Points of Iteration

Selecting starting values for the simplex method is generally a process of trial and error. For thin-film specimens, elastic constants of the film material in bulk form can serve as good starting points. For bare solids of cubic symmetry, elastic constants may be estimated from measured SAW velocities V_0 in the [100] direction and V_{45} in the [110] direction by using the empirical relationship between these wave velocities and elastic constants of existing cubic crystals. To exploit this relationship, three nondimensional parameters are defined: the anisotropy factor

$$\eta = \frac{2\,c_{44}}{c_{11} - c_{12}}, \tag{57}$$

the normalized difference of SAW velocities V_0 and V_{45}

$$\gamma = \frac{V_{45} - V_0}{(V_{45} + V_0)/2}, \tag{58}$$

and the ratio of the SAW velocity V_0 to the transverse wave velocity V_t $(=\sqrt{c_{44}/\rho})$

$$\beta = \frac{V_0}{V_t} \tag{59}$$

The correlation between η and γ obtained from elastic constants and the mass density for 37 cubic crystals reported by Anderson[51] is displayed in Fig. 5.16(a), and the correlation between β and η is shown in Fig. 5.16(b).[20] Note that relationships between η and γ for $V_0 \leq V_{45}$ and between β and η are virtually linear. Therefore from the parameter γ obtained from the measured V_0 and V_{45}, η can be obtained from Fig. 5.16(a) and β from Fig. 5.16(b). The elastic constant c_{44} can then be estimated from Eq. (59), while the parameter η imposes a constraint given by Eq. (57) on c_{11} and c_{12}.

For MgO the measured V_0 and V_{45} are 5507 and 5701 m/s, respectively, which gives $\gamma = 0.035$. From Fig. 5.16, it then follows that $\eta = 1.6$ and $\beta = 0.85$. With the density known to be 3598 kg/m^3, the estimated c_{44} is 151 GPa. For Si the measured V_0 and V_{45} are 4916 and 5082 m/s, respectively, which gives $\gamma = 0.033$. From Fig. 5.16, $\eta = 1.6$ and $\beta = 0.85$. With the density known to be 2331 kg/m^3, the estimated c_{44} is 78 GPa. The starting values of c_{11} and c_{12} must still be estimated by trial and error, but the constraint given by η reduces the search from two dimensions to one dimension.

5.5.4. *Domain of Convergence*

Karim *et al.*[52] and Behrend *et al.*[6] have attempted to define a convergence domain for the simplex method by considering a two-dimensional domain of two variables for fixed values of other variables. However the concept of a convergence domain is not clear. For example, let A, B, C, D, and E be points in the two-dimensional space. It is quite possible that the method converges for the selection of (A, B, C) as initial points while it may diverge for (C, D, E). In that case, it is difficult to decide whether C is in the convergence domain or not. The triangle (A, B, C) yields a converging result while the triangle (C, D, E) yields a diverging one. Furthermore different values for the fixed variables could lead to different convergence domains, which renders the concept of a convergence domain still more ambiguous.

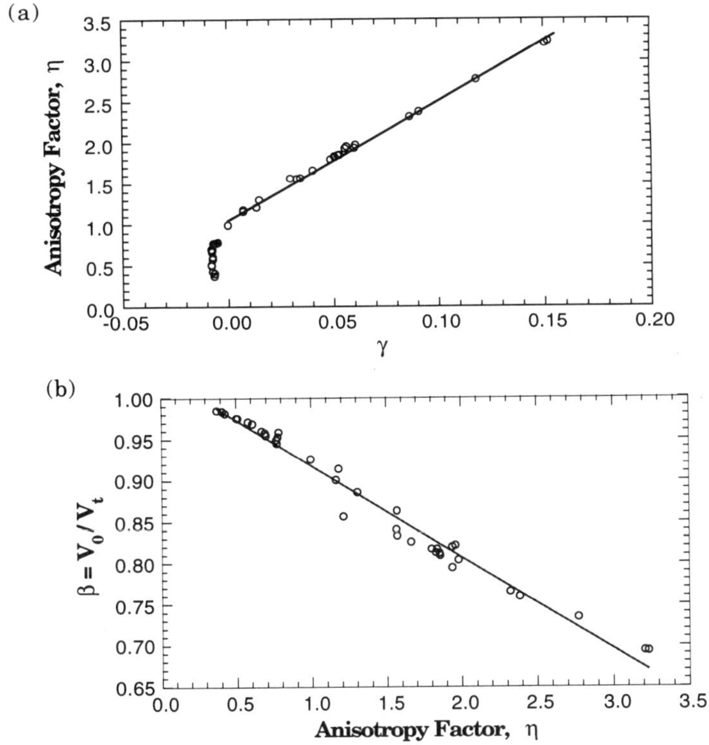

Figure 5.16. A correlation between nondimensional parameters of cubic crystals. (a) The anisotropy factor η and the SAW velocity parameter γ, $\gamma = 2(V_{45} - V_0)/(V_{45} + V_0)$, and (b) the ration β of the SAW velocity to the transverse wave velocity and the anisotropy factor η. $V_t = (c_{44}/\rho)^{1/2}$ (Ref. 60).

5.6. Elastic Constants of Single-Layer Films

The procedure of Section 5.5 was applied to determine elastic constants of thin films from SAW velocity data obtained from V(z) curves measured by acoustic microscopy.

5.6.1. Isotropic Films

The elastic properties of a gold film deposited on a fused silica substrate were determined by Behrend *et al.*[6] from measured dispersion data of the Sezawa wave mode. Three parameters, namely, the Young's modulus, the mass density, and the thickness, were determined. Figure 5.17 shows the dispersion curve of the Sezawa mode for gold film on fused silica.

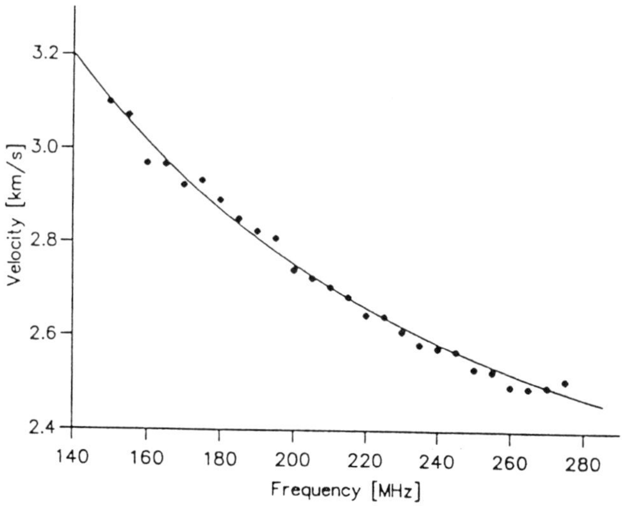

Figure 5.17. Dispersion curve of the Sezawa mode for gold film on fused silica substrate. Dots indicate measured data, and the solid line is the theoretical curve obtained using fitted parameters (Ref. 6).

The Young's modulus, the shear modulus, and Poisson's ratio of amorphous carbon (a-C) films deposited on (001)-oriented silicon substrates were determined by Kim and Achenbach.[53] The amorphous films are isotropic, but the specimens are anisotropic because the substrate is a cubic crystal. In Fig. 5.18, SAW velocities measured in the [100] and [110] directions are displayed by squares as functions of the normalized thickness h/λ_s. Measurements were carried out for 0.1- and 0.22-μm films at 225 MHz, for a 0.56-μm film at 225 and 255 MHz, and for 0.9- and 2.0-μm films at 195, 225, and 255 MHz, to yield data for 11 data points, including data for $h/\lambda_s = 0$.

For a thin-film mass density of $\rho = 2.2 \pm 0.1$ g/cm^3 obtained from measurements of volume and mass increase during film deposition, applying the method in Section 5.5 to the SAW dispersion data yields the elastic constants of the a-C films as $c_{11} = 72.5$ GPa and $c_{44} = 32.0$ GPa. The lines in Fig. 5.18 are dispersion curves calculated by using these elastic constants. Once the elastic constants c_{11} and c_{44} are determined, the corresponding Young's modulus E, shear modulus G, and Poisson's ratio ν can be obtained from the following relations:

$$E = \frac{c_{44}(3c_{11} - 4c_{44})}{c_{11} - c_{44}} \qquad G = c_{44} \qquad \nu = \frac{E}{2G} - 1 \tag{60}$$

In this manner, elastic constants of a-C films are obtained as $E = 70.7$ GPa, $G = 32.0$ GPa, and $\nu = 0.104$.

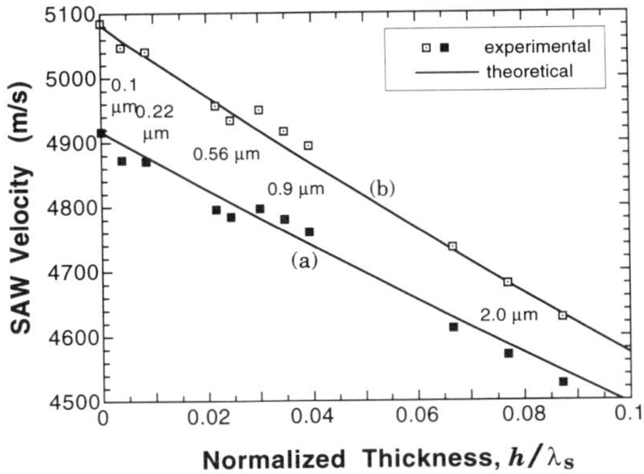

Figure 5.18. SAW dispersion curves for amorphous carbon films deposited on the (001) plane of silicon substrate; (a) along the [100] direction and (b) along the [110] direction (Ref. 53).

5.6.2. Cubic-Crystal Films

Elastic constants of cubic-crystal transition metal nitride films were determined by Kim *et al.*[42] from SAW dispersion data. The SAW velocities were measured on the (001) plane of TiN, VN, and NbN films epitaxially grown on MgO substrates. In Fig. 5.19, SAW velocities measured in the [100] and [110] directions for VN films on MgO substrates are displayed by squares as functions of the normalized film thickness h/λ_s. Measurements were carried out for 1.2- and 2.6-μm film at 195, 225, and 255 MHz. The mass density of VN is taken as 6.11 g/cm^3.[54] The elastic constants of VN determined from the inversion of

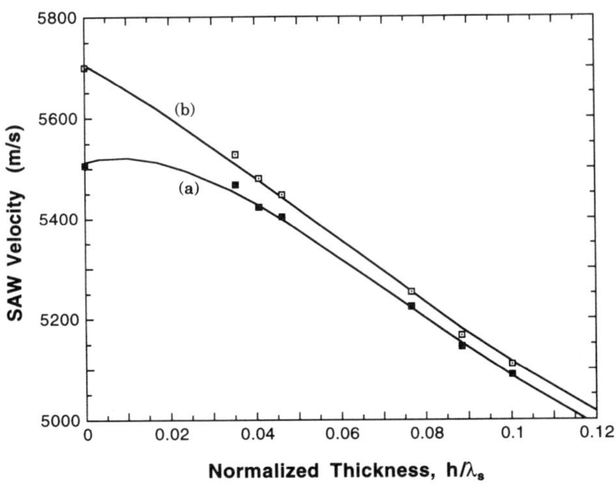

Figure 5.19. Dispersion curves of SAWs on the (001) plane of VN films epitaxially deposited on MgO substrates; (a) [100] direction, (b) [110] direction (Ref. 42).

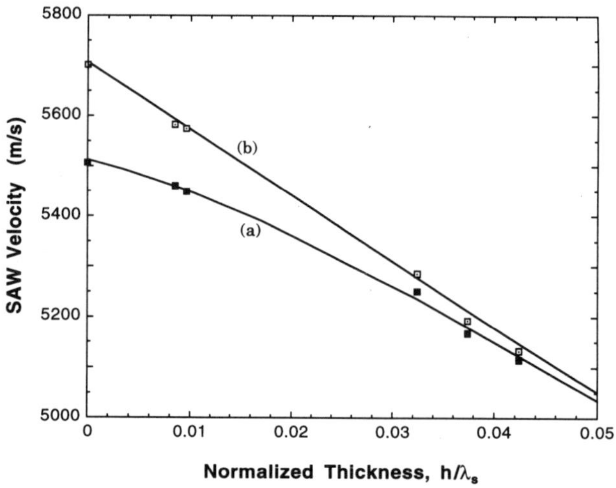

Figure 5.20. Dispersion curves of SAWs on the (001) plane of NbN films epitaxially deposited on MgO substrates; (a) [100] direction, (b) [110] (Ref. 42).

SAW dispersion data are $c_{11} = 533$, $c_{12} = 135$, and $c_{44} = 133$ GPa. The lines in Fig. 5.19 represent the dispersion curves calculated from these elastic constants. Similarly the elastic constants of TiN (mass density 5.39 g/cm^3) and NbN (mass density 8.43 g/cm^3) films were determined from measured SAW dispersion data shown in Fig. 5.12 and in Fig. 5.20, respectively. The elastic constants are listed in Table 5.1.

Once the elastic constants c_{11}, c_{12}, and c_{44} are determined, the corresponding compliance constants s_{11}, s_{12}, and s_{44} can be obtained by inverting the matrix of elastic constants. For cubic symmetry, we find

$$s_{11} = \frac{c_{11} + c_{12}}{(c_{11} - c_{12})(c_{11} + 2\,c_{12})} \qquad s_{12} = \frac{-c_{12}}{(c_{11} - c_{12})(c_{11} + 2\,c_{12})} \qquad s_{44} = \frac{1}{c_{44}} \tag{61}$$

Table 5.1. Elastic Constants of Single-Crystal TiN, NbN, and VN

Nitride material	Mass density (g/cm^3)	Stiffness (GPa)			Compliance (10^{-3} GPa^{-1})			Modulus (GPa)		
		c_{11}	c_{12}	c_{44}	s_{11}	s_{12}	s_{44}	E_{100}	E_{110}	E_{111}
TiN	5.39	625	165	163	1.80	−0.38	6.13	556	446	418
NbN	8.43	556	152	125	2.04	−0.44	8.00	490	357	328
VN	6.11	533	135	133	2.09	−0.42	7.52	478	368	342

Source: Ref. 42. Mass density cited in Ref. 54.

The anisotropy factor is defined from either stiffness constants or compliance constants as $\eta \equiv 2\, c_{44}/(c_{11} - c_{12}) = 2(s_{11} - s_{12})/s_{44}$. For an isotropic material, for example an amorphous carbon film, the anisotropy factor η is 1. The Young's moduli in the [100], [110], and [111] directions can be written as[55,56]

$$E_{100} = \frac{1}{s_{11}} \qquad E_{110} = \frac{1}{s_{11} - (1/2)\,[(s_{11} - s_{12}) - (1/2)\, s_{44}]} = \frac{1}{[s_{11} - (\eta/4 - 1/4)\, s_{44}]}$$

$$E_{111} = \frac{1}{s_{11} - (2/3)\,[(s_{11} - s_{12}) - (1/2)\, s_{44}]} = \frac{1}{[s_{11} - (\eta/3 - 1/3)\, s_{44}]} \tag{62}$$

The shear moduli in the [100], [110], and [111] directions can be written as[55]

$$G_{100} = \frac{1}{s_{44}} = c_{44} \qquad G_{110} = \frac{2}{s_{44} + 2\,(s_{11} - s_{12})} = \frac{c_{44}}{(1 + \eta)/2}$$

$$G_{111} = \frac{3}{s_{44} + 4\,(s_{11} - s_{12})} = \frac{c_{44}}{(1 + 2\eta)/3} \tag{63}$$

Compliance constants and elastic moduli corresponding to the determined constants are also listed in Table 5.1.

The elastic constants of cubic-crystalline films were also determined from the directional variation of measured SAW velocities. The advantage of using the directional variation of the SAW velocity instead of the SAW dispersion data is that sufficient data can be obtained from a single thin-film specimen. The directional variation of the SAW velocity has been used by Mendik and others[57] to determine the elastic constants of a cubic-crystalline solid without a layer. The method suggested by Mendik and others[57] can be used for the (111)- and (110)-oriented crystalline materials. For the (001)-oriented cubic-crystalline materials however, the transition of the measured SAW mode from regular to pseudo-SAWs should be taken into account.[58] In a range of propagation directions, the SAW velocity obtained from V(z) measurements is neither that of the regular SAW nor that of the pseudo-SAW but rather the one for a superposition of these waves,[35] as discussed in Section 5.4.3. The V(z) measurement model shows this same behavior, and it is therefore more consistent to compare the measured velocity with the one calculated from the V(z) measurement model rather than with the theoretical SAW velocity obtained from the relevant characteristic equation. Measurements carried out by using a dual-probe laser interferometer on a silicon disk, as reported by Huang and Achenbach,[59] show signals of both regular and pseudo-SAWs in the transition range and hence allow separate velocities to be determined. Outside the transition range, velocities obtained with the dual-probe laser interferometer agree with those obtained by LFAM.

Open circles in Figs. 5.21(a) and (b) are measured SAW velocities as functions of the direction of wave propagation on the (001) planes of TiN and VN films deposited on MgO substrates. Using the directional variation of the measured SAW velocities, elastic constants of these films were determined[60] by the inverse method described in Section 5.5. The elastic constants of TiN films are listed in Table 5.2, and results obtained for films of different thicknesses are compared with results determined from SAW dispersion data. Results for VN films are listed in Table 5.3 and compared in the same way. Solid lines in Figs.

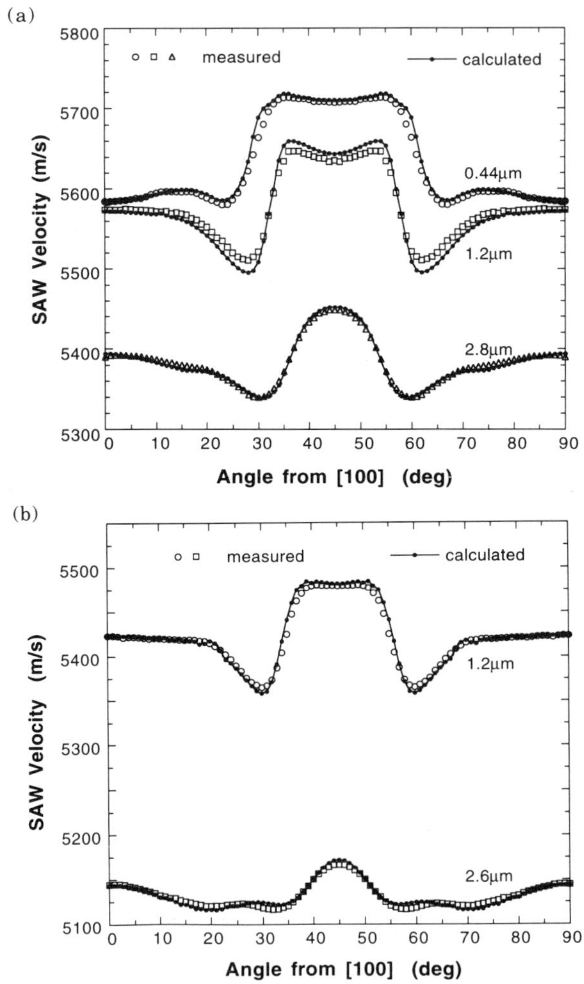

Figure 5.21. Directional variation of SAW velocities for films epitaxially deposited on the (001) plane of cubic-crystal MgO substrates; (a) TiN films, (b) VN films (Ref. 60).

Table 5.2. Elastic Constants of Single-Crystal TiN Films[a]

Film thickness (μm)	Elastic constants (GPa)		
	c_{11}	c_{12}	c_{44}
	625	165	163
0.44	635 (+2%)	145 (−12%)	168 (+3%)
1.2	608 (−3%)	153 (−7%)	171 (+5%)
2.8	627 (+0.3%)	163 (+2%)	162 (+1%)

[a]Elastic constants determined from the directional variation of SAW velocity are compared with elastic constants determined from SAW dispersion data (Ref. 60).

5.21(a) and (b) are velocities obtained from calculated $V(z)$ curves using these elastic constants. They show good agreement with measured velocities over the whole range of wave directions, which indicates that the inverse method was successfully applied.

5.6.3. Accuracy

If SAW velocities are less sensitive to variations in certain elastic constants, then conversely it is difficult to obtain these constants with sufficient accuracy from the velocities. It is therefore of interest to investigate the influence of variation of elastic constants on SAW dispersion curves. Some results are shown in Fig. 5.22 for a VN film on an MgO substrate. Figures 5.22(a) and (b), which correspond to wave propagation in the [100] and [110] directions, respectively, show the dispersion curves when one elastic constant is reduced by 10%. It is observed that all elastic constants contribute to changes in dispersion curves. This is contrary to a case discussed by Mal *et al.*,[61] which was concerned with determining five independent elastic constants of a composite laminate from plate wave dispersion data. Note from Figs. 5.22(a) and (b) that for a cubic-crystal film, the dispersion curve is less sensitive to changes in c_{12} than to changes in c_{11} and c_{44}. The same trend is observed in Fig. 5.23 for the directional variation

Table 5.3. Elastic Constants of Single-Crystal VN Films[a]

Film thickness (μm)	Elastic constants (GPa)		
	c_{11}	c_{12}	c_{44}
	533	135	133
1.2	530 (−1%)	121 (−10%)	138 (+4%)
2.6	537 (+1%)	142 (+5%)	132 (−1%)

[a]Elastic constants determined from the directional variation of SAW velocity are compared with the elastic constants determined from the SAW dispersion data (Ref. 60).

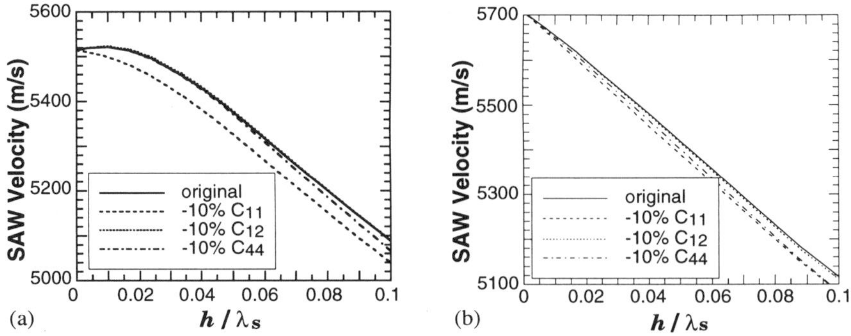

Figure 5.22. The effect of changes in c_{11}, c_{12}, and c_{44} of VN on SAW dispersion curves of a VN film/MgO substrate system; (a) along the [100] direction, (b) along the [110] direction (Ref. 53).

of SAW velocities for a 2.8-μm TiN film on an MgO substrate. Figure 5.23 shows the sensitivity of SAW velocities to changes of each elastic constant in the TiN film.

The nonlinear nature of the inversion process may result in the nonuniqueness of determined constants. Experience with the procedure used in Sections 5.6.1 and 5.6.2 indicates that the inversion process may reach more than one local minimum, but eventually it appears to find the absolute minimum. It should always be verified that the same results are obtained when the iteration procedure is started with different values of the elastic constants. With that precaution, we can assume that elastic constants determined by this method are correct.

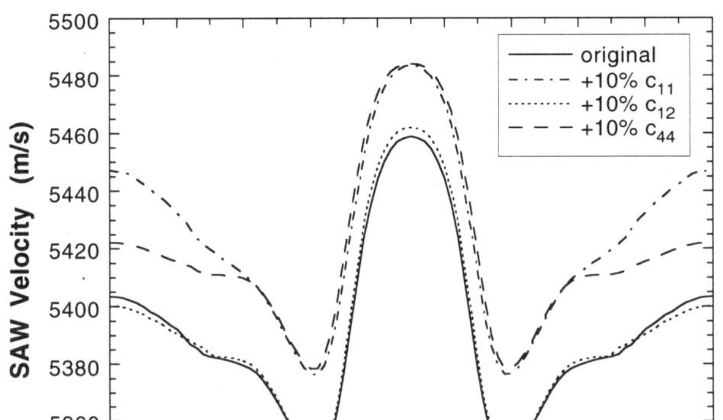

Figure 5.23. The effect of changes in c_{11}, c_{12}, and c_{44} of TiN on the directional variation of SAW velocities for a TiN film/MgO substrate system (Ref. 60).

Interpretation of SAW data does however require accurate information about film thickness and mass density, which is one potential source of uncertainty.

5.7. Effective Elastic Constants of Superlattices

A superlattice film consists of a large number of alternating thin layers of two materials with Period Λ, as shown in Fig. 5.24. It was shown that transition metal nitride superlattice films, such as TiN/NbN,[12] exhibit much higher hardness than homogeneous nitride films, and that they show hardness peaks at particular values of the period Λ.[62] Figure 5.25 is a cross-sectional TEM image of a TiN/NbN superlattice film.

Theoretical velocities of wave propagation modes in a superlattice deposited on a substrate can be obtained from the $V(z)$ measurement model in Section 5.2 by calculating the reflection coefficient for a multilayered film according to the method in Section 5.2.2. This was done for the lowest mode of a TiN/NbN superlattice composed of 65 pairs of TiN/NbN layers; results are shown in Fig. 5.26. The constituent layers in a superlattice are however very thin, of the order of nanometers, while actual wavelengths at 225 MHz are of the order of tens of micrometers. This substantial difference in characteristic length scales of the microstructure and the excitation suggests that the discrete nature of the superlattice can be ignored and the layered material can be modeled as a homogeneous but anisotropic solid, with an effective mass density and effective elastic constants that depend on the geometrical and material parameters of the layers. The effective mass density may be computed using the rule of mixtures, but the effective elastic constants require more complicated formulas for an accurate representation.

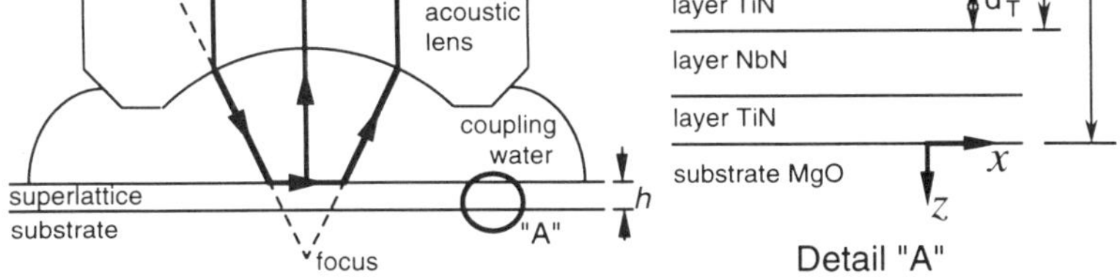

Figure 5.24. A schematic diagram of the acoustic probe and a TiN/NbN superlattice specimen.

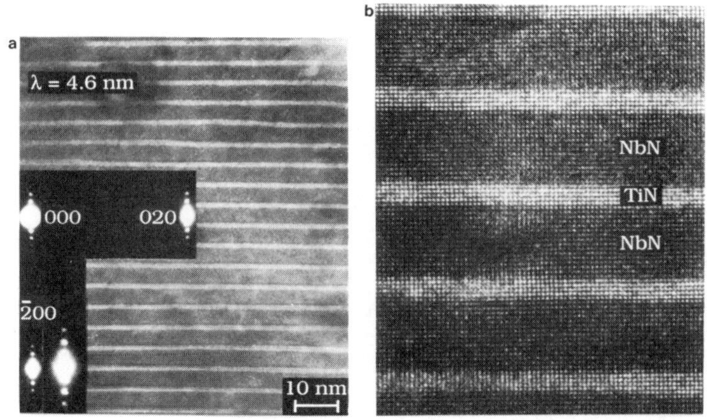

Figure 5.25. A cross-sectional transmission electron micrograph of a TiN/NbN (001) superlattice film with period $\Lambda = 4.6$ nm. (a) Selected area diffraction patterns, (b) a high-resolution lattice fringe image (Ref. 12).

If the elastic constants and thickness ratios of the constituent layers are known, the effective elastic constants of the superlattice can be calculated by using the analytical expressions given in Section 5.7.1. The elastic constants of the constituent layers must be obtained separately on the basis of acoustic microscopy measurements of single-layer film specimens. By calculating the elastic constants in this manner, it is assumed that the elastic constants of single-layer films are

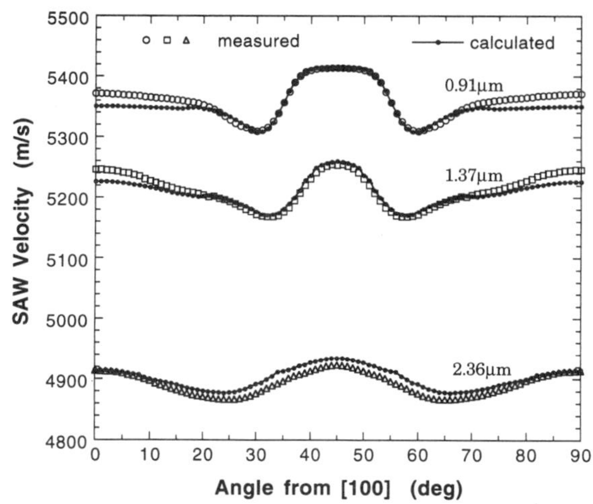

Figure 5.26. Directional variation of SAW velocities on TiN/NbN (001) superlattice films epitaxially deposited on MgO substrates (Ref. 14).

the same as the elastic constants of the same layer material in a superlattice. Another assumption implied by using the formulas in Section 5.7.1 is that interfaces between constituent layers are sharply defined as planes and contact between neighboring layers is perfect, i.e., both displacements and tractions are continuous. Finally it is assumed that stresses in superlattices generated by the processing procedure are sufficiently small, so that they do not affect the effective elastic constants of the superlattice. The appearance of residual stresses and their effects on the mechanical behavior has been discussed in considerable detail in the recent dissertation by Sklar.[10]

Naturally the effective elastic constants of superlattices can also be obtained experimentally from $V(z)$ measurements. A comparison of calculated and measured elastic constants indicates the validity of analytical expressions for effective elastic constants and of assumptions made in deriving these expressions.

5.7.1. Calculating Effective Elastic Constants

A derivation of the effective elastic constants of a superlattice composed of isotropic layers is found in Ref. 63. For the more general case, i.e., a superlattice composed of single-crystal layers, the effective elastic constants are derived by Grimsditch,[64] Grimsditch and Nizzoli,[65] Akcakaya and Farnell,[66] and Akcakaya *et al.*[67] This section presents expressions for the effective elastic constants of (001)-oriented superlattices.[68] Similar expressions for (111)-oriented superlattices can be found elsewhere.[69]

5.7.1.1. Expressions for Effective Elastic Constants

Consider a superlattice that consists of stacked cubic-crystal layers with the same orientation. The superlattice is considered to be tetragonal-symmetric because properties normal to the layers may differ from those in the planes of the layers. For an anisotropic material of tetragonal symmetry, the stress–strain relations are

$$\begin{Bmatrix} \sigma_x \\ \sigma_y \\ \sigma_z \\ \tau_{yz} \\ \tau_{zx} \\ \tau_{xy} \end{Bmatrix} = \begin{bmatrix} c_{11} & c_{12} & c_{13} & 0 & 0 & 0 \\ c_{12} & c_{11} & c_{13} & 0 & 0 & 0 \\ c_{13} & c_{13} & c_{33} & 0 & 0 & 0 \\ 0 & 0 & 0 & c_{44} & 0 & 0 \\ 0 & 0 & 0 & 0 & c_{44} & 0 \\ 0 & 0 & 0 & 0 & 0 & c_{66} \end{bmatrix} \begin{Bmatrix} \varepsilon_x \\ \varepsilon_y \\ \varepsilon_z \\ \gamma_{yz} \\ \gamma_{zx} \\ \gamma_{xy} \end{Bmatrix} \tag{64}$$

As shown in Fig. 5.24, the thicknesses of the TiN and NbN layers are d_T and d_N, respectively, and the fractions of TiN and NbN are

$$\bar{d}_T \equiv d_T/(d_T + d_N) = d_T/\Lambda \qquad \bar{d}_N \equiv d_N/(d_T + d_N) = d_N/\Lambda \tag{65}$$

Formulas for effective elastic constants of the tetragonal-symmetric superlattice can be obtained from general results of Grimsditch[64] as

$$c_{11} = \frac{1}{(\bar{d}_T/c^T_{11}) + (\bar{d}_N/c^N_{11})} + \frac{(c^T_{11} - c^N_{11})^2 - (c^T_{12} - c^N_{12})^2}{(c^T_{11}/\bar{d}_T) + (c^N_{11}/\bar{d}_N)} \qquad c_{33} = \frac{1}{(\bar{d}_T/c^T_{11}) + (\bar{d}_N/c^N_{11})}$$

$$c_{12} = \frac{(c^T_{12}/c^T_{11})\,(c^N_{12}/c^N_{11})}{(\bar{d}_T/c^T_{11}) + (\bar{d}_N/c^N_{11})} + \frac{(\bar{d}_T\, c^T_{12} + \bar{d}_N\, c^N_{12})\,[(c^T_{11} - c^T_{12})/\bar{d}_T + (c^N_{11} - c^N_{12})/\bar{d}_N]}{(c^T_{11}/\bar{d}_T) + (c^N_{11}/\bar{d}_N)}$$

$$c_{13} = \frac{\bar{d}_T\,(c^T_{12}/c^T_{11}) + \bar{d}_N\,(c^N_{12}/c^N_{11})}{(\bar{d}_T/c^T_{11}) + (\bar{d}_N/c^N_{11})} \qquad c_{44} = \frac{1}{(\bar{d}_T/c^T_{44}) + (\bar{d}_N/c^N_{44})}$$

and

$$c_{66} = \bar{d}_T\, c^T_{44} + \bar{d}_N\, c^N_{44} \tag{66}$$

where $(c^T_{11}, c^T_{12}, c^T_{44})$ and $(c^N_{11}, c^N_{12}, c^N_{44})$ are the elastic constants of the constituent layers of cubic symmetry. The effective mass density of the superlattice follows from the rule of mixtures

$$\rho = \bar{d}_T\, \rho_T + \bar{d}_N\, \rho_N \tag{67}$$

Equation (66) predicts that effective elastic constants do not depend on the superlattice period, but they do of course depend on the fractions of the two constituent layers.

The elastic constants of cubic-crystal TiN and NbN films have been determined from measured SAW dispersion data as $c^T_{11} = 625$, $c^T_{12} = 165$, and $c^T_{44} = 163$ GPa for TiN and $c^N_{11} = 556$, $c^N_{12} = 152$, and $c^N_{44} = 125$ GPa for NbN (see Section 5.6.2). The corresponding calculated effective elastic constants of TiN/NbN superlattices are shown in Fig. 5.27 as functions of the TiN fraction d_T/Λ. For TiN/NbN superlattices with a TiN fraction $d_T/\Lambda = 0.3$, the effective elastic constants calculated using Eq. (66) are $c_{11} = 577$, $c_{33} = 575$, $c_{12} = 156$, $c_{13} = 156$, $c_{44} = 134$, and $c_{66} = 136$ GPa; the effective mass density obtained from Eq. (67) is 7.52 g/cm^3.

5.7.1.2. *Verifying Calculated Constants*

As discussed in Section 5.4.3, theoretical SAW velocities can most easily be compared with results from measurements along the [100] and [110] directions. The velocities of SAWs propagating along a symmetry axis in either the [100] or the [110] direction on the (001) plane of a tetragonal-symmetric film deposited on a cubic-symmetric substrate were calculated using equations of Kim *et al.*[68] Measurements were carried out for specimens of various thicknesses (see Table 5.4) and at 195, 225, and 255 MHz. The SAW velocities from calculations and measurements for various film thicknesses and frequencies are given in Fig. 5.28

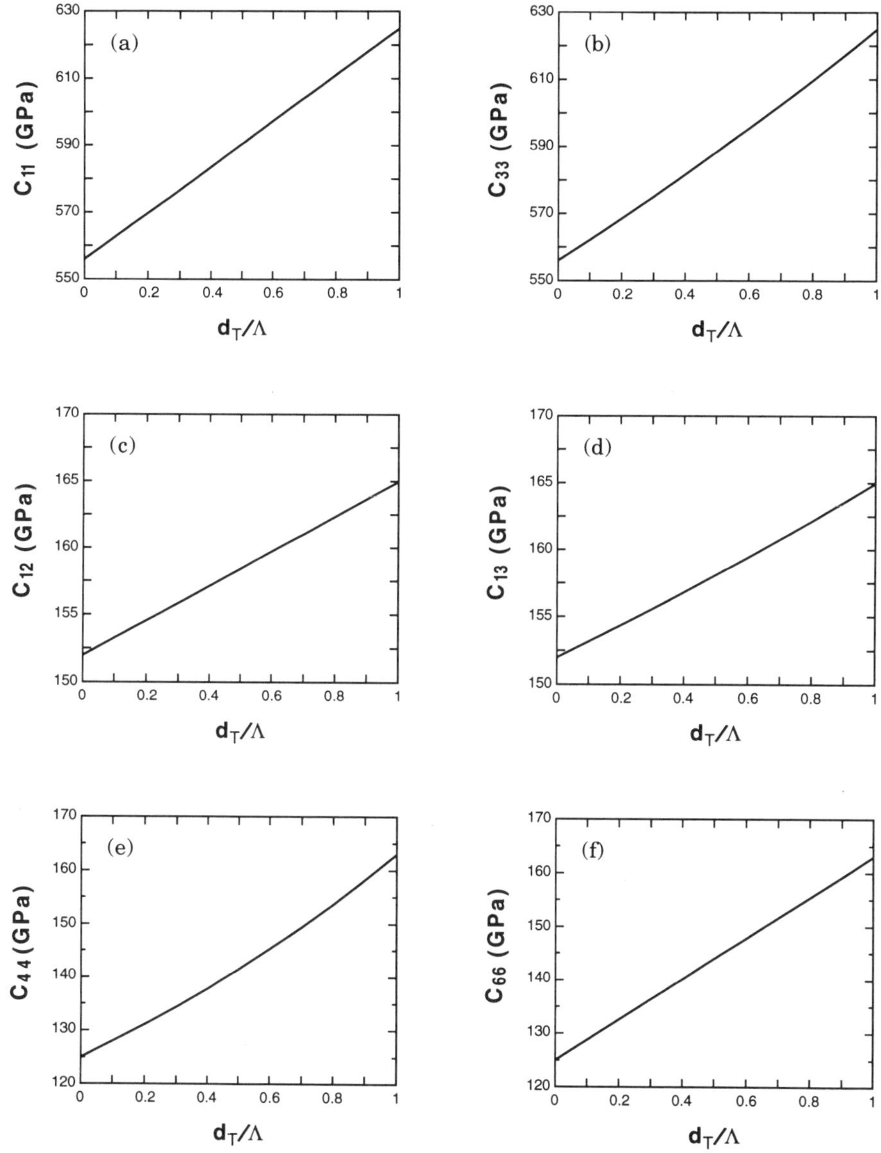

Figure 5.27. Effective elastic constants of TiN/NbN superlattices calculated as functions of the TiN fraction d_T/Λ, computed from formulas with elastic constants of constituent TiN and NbN layers (Ref. 68).

as functions of the normalized film thickness h/λ_s. Dashed lines in Fig. 5.28 represent the dispersion curves for a TiN/NbN superlattice film deposited on an MgO substrate, calculated from the elastic constants and the mass density of the superlattice film determined in this section and known elastic properties of the MgO substrate. The symbols are the measured SAW velocities.

Calculated results agree with experimental results for TiN/NbN superlattice specimens considered in this work. Other results however show that calculated results do not agree with experimental results if interfaces between superlattice layers are not sufficiently flat and sharp.[70] It appears that formulas given by Eq. (58) are valid only for superlattices with perfect interfaces.

5.7.2. Effective Elastic Constants Obtained from Measurements

From the experimental results displayed in Fig. 5.28, the effective elastic constants of TiN/NbN superlattices can be determined by the procedure in Section 5.5. However it is not necessary to determine six independent elastic constants from the given data. Based on Eq. (66), relationships between c_{11} and c_{33}, c_{12} and c_{13}, and c_{44} and c_{66} can be written as[71]

$$\frac{c_{11}-c_{33}}{c_{11}^{o}} = \left(\frac{\Delta_{11}}{c_{11}^{o}}\right)^2 \left(1-\frac{\Delta_{12}^{2}}{\Delta_{11}^{2}}\right) \frac{4\,\bar{d}_T\bar{d}_N}{1+(\bar{d}_N-\bar{d}_T)\,(\Delta_{11}/c_{11}^{o})}$$

$$\frac{c_{12}-c_{13}}{c_{12}^{o}} = \frac{\Delta_{11}}{c_{11}^{o}}\frac{\Delta_{12}}{c_{12}^{o}} \left(1-\frac{\Delta_{12}}{\Delta_{11}}\right) \frac{1-(\bar{d}_N-\bar{d}_T)^2}{1+(\bar{d}_N-\bar{d}_T)\,(\Delta_{11}/c_{11}^{o})}$$

and

$$\frac{c_{44}-c_{66}}{c_{44}^{o}} = -\left(\frac{\Delta_{44}}{c_{44}^{o}}\right)^2 \frac{1-(\bar{d}_N-\bar{d}_T)^2}{1+(\bar{d}_N-\bar{d}_T)\,(\Delta_{44}/c_{44}^{o})} \tag{68}$$

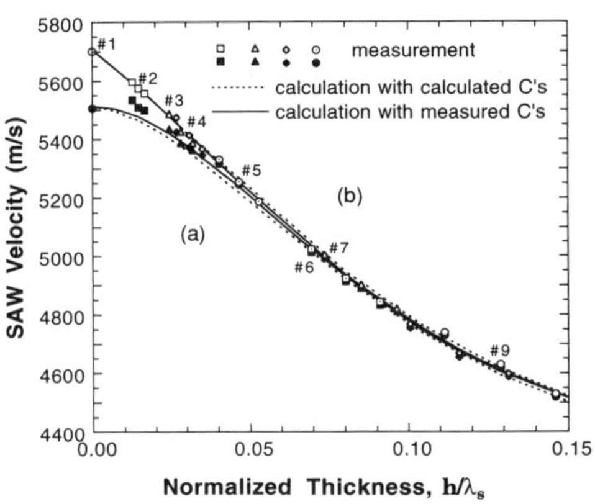

Figure 5.28. Dispersion curves of SAWs on the (001) plane of TiN/NbN superlattice films epitaxially deposited on MgO substrates (a) = [100] direction, (b) = [110] direction.

where $c^o_{ij} = (c^T_{ij} + c^N_{ij})/2$ and $\Delta_{ij} = (c^T_{ij} - c^N_{ij})/2$. We see from Eq. (68) that for small Δ_{ij}/c^o_{ij}, $c_{11} \approx c_{33}$, $c_{12} \approx c_{13}$, and $c_{44} \approx c_{66}$. This means that if the elastic constants of the constituent layers are close to each other, then the symmetry of the superlattice is similar to the symmetry of the constituent layers.

The elastic constants of TiN and NbN are shown in Table 5.1. Since they are very close, we do indeed find that $c_{11} \approx c_{33}$, $c_{12} \approx c_{13}$, and $c_{44} \approx c_{66}$ for the TiN/NbN superlattices. Therefore the TiN/NbN superlattice can be considered a cubic-symmetric structure, which has only three independent elastic constants. With known elastic constants and mass density of the substrate and known mass density of the superlattice, the three independent elastic constants of the superlattice were determined from the experimental SAW dispersion data in Fig. 5.28; they are obtained as $c_{11} = c_{33} = 587$, $c_{12} = c_{13} = 127$, and $c_{44} = c_{66} = 135$ GPa. These measured elastic constants are compared with calculated elastic constants in Table 5.4. Calculated values of c_{11}, c_{33}, c_{44}, and c_{66} agree with measured values within a 2% deviation. Values for c_{12} and c_{13} show however deviations of about 20%. There may be several reasons for these larger deviations. As discussed in Section 5.6.3, the constants c_{12} and c_{13} are generally obtained with less accuracy, in this case, for the individual layers as well as for the superlattice. It is also possible that theoretical expressions for c_{12} and c_{13} are more susceptible to imperfections in the superlattice. To check consistency, dispersion curves calculated from these effective elastic constants are displayed as solid lines in Fig. 5.28. They show good agreement with the measured dispersion data.

The effective elastic constants of TiN/NbN superlattice films were also determined from the directional variation of measured SAW velocities,[60] as discussed in Section 5.6.2 for single-layer films. The open symbols in Fig. 5.26 are the measured SAW velocities as functions of the direction of wave propagation on the (001) plane of TiN/NbN superlattice films (specimen numbers 4, 5, and

Table 5.4. Effective Elastic Constants of TiN/NbN Superlattice Films with TiN Fraction 0.3. Numbers in ( ) Represent the Discrepancy in % Relative to the Effective Elastic Constants Obtained from the Formulas[a]

Determination method	Effective elastic constants (GPa)					
	c_{11}	c_{33}	c_{12}	c_{13}	c_{44}	c_{66}
Formulas	577	575	156	156	134	136
Inversion of dispersion	587 (+2)	587 (+2)	127 (−19)	127 (−19)	135 (+1)	135 (−1)
Inversion of directional variation						
Specimen #4	571 (−1)	571 (−1)	123 (−21)	123 (−21)	133 (−1)	133 (−2)
#5	604 (+5)	604 (+5)	139 (−11)	139 (−11)	129 (−4)	129 (−5)
#6	573 (−1)	573 (−1)	121 (−22)	121 (−22)	132 (−2)	132 (−3)

[a]Specimen numbers correspond to numbers in Table 5.4. (Refs. 60, 68)

Table 5.5. Film Thicknesses and Superlattice Periods of TiN/NbN Superlattice Specimens Used for Experiments

Specimen number	Film thickness, h (μm)	Period, Λ (nm)
1	0	—
2	0.43	9.46
3	0.83	2.37
4	0.91	4.14
5	1.37	8.37
6	2.36	36.3
7	2.5	2.62
8	3.41	5.5
9	3.79	18.9

6 in Table 5.5) epitaxially deposited on MgO substrates. Using the directional variation of the measured SAW velocities, the effective elastic constants of these superlattice films were determined by the inverse method described in Section 5.5. Results obtained for each superlattice film are also listed in Table 5.5 and compared with values determined from the formulas. They show consistency with results obtained from the dispersion data.

5.8. Conclusion

It remains to briefly address the question of accuracy in determining elastic constants by LFAM. Since elastic constants are determined by an inverse method that compares theoretical calculations with measured results, errors may enter because the theoretical measurement model is inadequate, because procedures to measure, process, and interpret data are lacking perfection, or the inverse method is inaccurate.

The theoretical measurement model is based on the assumption that films are homogeneous, interfaces between layers and between films and the substrate are sharp, adhesion is perfect, and residual stresses can be neglected. In the laboratory specimens used in this work, the quality of the films was generally ensured by careful processing. Work is now in progress to process films with interlayer diffusion regions and inadequate adhesion, and attempts are being made to develop theoretical models to account for such deficiencies, see e.g., Refs. 72 and 73. The effect of residual stresses also requires further attention. For small carefully processed specimens, such stresses may be negligible, but they can be of obvious importance in practical thin-film applications. Recent work by Sklar[(10)] discussed an appropriate theory for propagating surface acoustic waves in thin-film/substrate configurations in the presence of uniform static stresses.

Measurement and processing procedures used in the work reported by the present authors are due to Kushibiki and Chubachi.[32] The relative accuracy, which is related to the temperature variation in the environment and mechanical error in the motion of the specimen stage, can be improved by maintaining a constant temperature and careful measurement procedures.[74] As mentioned in Section 5.2.3, a drop of distilled water serves as a couplant between the lens and specimen. The water temperature is controlled by using thermocouples and a 0 °C reference point. Deviations in couplant temperature during the same measurement sequence can give rise to unacceptable errors and must therefore be avoided. The absolute accuracy of the measured velocity and attenuation can be estimated by a calibration method. Other possible problem areas and related improvements in the processing procedure, such as using Teflon reference curves, rectangular windows and $V(z)$ truncation, moving-average filtering, attenuation measurements, and automated analysis of $V(z)$ containing multiple modes, have recently been discussed by Sklar.[10]

The inverse method has potential problems with convergence and uniqueness. As discussed in Section 5.5, these problems can be avoided by a judicious choice of starting points and by repeating the procedure from several initial choices. For example, repeating the $V(z)$ measurement process several times for the same specimen has shown that measurements are repeatable with a small error bar. Comparison with well-established velocity values for a calibration material suggests an absolute accuracy of 0.1%. On this basis, it was concluded that SAW velocities can be measured with an accuracy of a few parts in 10^3. As noted earlier, film thickness and mass density of a film should be available with compatible accuracy for calculating elastic constants.

In summary it is however evident that the experience gained by a number of workers in this developing field shows that quantitative LFAM provides a method of practical feasibility and accuracy for determining thin-film elastic constants.

Acknowledgments

Chapter 5 was completed in the course of research sponsored by the US Department of Energy under Contract No. DE-FG02-86ER13484.

References

1. Weglein, R. D. (1979). A model for predicting acoustic material signatures. *Appl. Phys. Lett.* **34,** 179–81.
2. Weglein, R. D. (1979). SAW dispersion and film thickness measurement by acoustic microscopy. *Appl. Phys. Lett.* **35,** 215–17.
3. Briggs, A. (1992). *Acoustic Microscopy,* Oxford University Press, New York.

4. Kushibiki, J. and Chubachi, N. (1987). Application of LFB acoustic microscope to film thickness measurement. *Electr. Lett.* **23,** 652–54.
5. Kushibiki, J., Ishikawa, T., Chubachi, N. (1990). Cut-off characteristics of leaky Sezawa and pseudo-Sezawa wave modes for thin-film characterization. *Appl. Phys. Lett.* **57,** 1967–69.
6. Behrend, O., Kulik, A., Gremaud, G. (1993). Characterization of thin films using numerical inversion of the generalized Lamb wave dispersion relation. *Appl. Phys. Lett.* **62,** 2787–89.
7. Weglein, R. D. (1980). Acoustic microscopy applied to SAW dispersion and film thickness measurement. *IEEE Trans. Sonics Ultrason.* **SU-27,** 82–86.
8. Weglein, R. D. and Hanafee, J. E. (1985). Nondestructive detection of Rayleigh wave dispersion in beryllium. *Appl. Phys. Lett.* **46,** 347–49.
9. Ishikawa, I., Kanda, H., Katakura, K., Semba, T. (1989). Measurement of a damaged layer thickness with reflection acoustic microscope. *IEEE Trans. Ultrason. Ferroelec. Freq. Contr.* **36,** 587–92.
10. Sklar, Z. (1993). Quantitative Acoustic Microscopy of Coated Materials. Ph.D. diss., University of Oxford.
11. Mirkarimi, P. B., Shinn, M., and Barnett, S. A. (1992). An ultrahigh vacuum, magnetron-sputtering system for the growth and analysis of nitride superlattices. *J. Vac. Sci. Technol. A* **10,** 75–81.
12. Shinn, M., Hultman, L., and Barnett, S. A. (1992). Growth, structure, and microhardness of epitaxial TiN/NbN superlattices. *J. Mater. Res.* **7,** 901–11.
13. Kundu, T. (1992). A complete acoustic microscopical analysis of multilayered specimens. *J. Appl. Mech.* **59,** 54–60.
14. Lee, Y.-C., Kim, J. O., and Achenbach, J. D. (1993). V(z) curves of layered anisotropic materials for the line-focus acoustic microscope. *J. Acoust. Soc. Am.* **94,** 923–30.
15. Somekh, M. G., Bertoni, H. L., Briggs, G. A. D., and Burton, N. J. (1985). A two-dimensional imaging theory of surface discontinuities with the scanning acoustic microscope. *Proc. R. Soc. Lond.* A **49,** 29–51.
16. Li, Z. L., Achenbach, J. D., and Kim, J. O. (1991). Effect of surface discontinuities on V(z) and V(z,x) for the line-focus acoustic microscope. *Wave Motion* **14,** 187–203.
17. Atalar, A. (1978). An angular spectrum approach to contrast in reflection acoustic microscopy. *J. Appl. Phys.* **49,** 5130–39.
18. Achenbach, J. D., Ahn, V. S., and Harris, J. G. (1991). Wave analysis of the acoustic material signature for the line-focus microscope. *IEEE Trans. Sonics Ultrason.* **SU-38,** 380–87.
19. Brekhovskikh, L. M. (1980). *Waves in Layered Media,* 2d ed., Academic, New York, Section 3.
20. Lee, Y.-C. (1994). Line-Focus Acoustic Microscopy for Material Evaluation. Ph.D. diss., Northwestern University.
21. Nayfeh, A. H. (1991). Elastic wave reflection from liquid–anisotropic substrate interface. *Wave Motion.* **14,** 55–67.
22. Chimenti, D. E. and Nayfeh, A. H. (1990). Ultrasonic reflection and guided waves in fluid-coupled composite laminates. *J. Nondestr. Eval.* **9,** 51–69.
23. Nayfeh, A. H. (1991). The general problem of elastic wave propagation in multilayered anisotropic media. *J. Acoust. Soc. Am.* **89,** 1521–31.
24. Bertoni, H. L. (1984). Ray optical evaluation of $V(z)$ in a reflection acoustic microscope. *IEEE Trans. Sonics Ultrason.* **SU-31,** 105–16.
25. Nayfeh, A. H. and Chimenti, D. E. (1984). Reflection of finite acoustic beams from loaded and stiffened half-space. *J. Acoust. Soc. Am.* **75,** 1360–68.
26. Ahn, V. S., Achenbach, J. D., Li, Z. L., Kim, J. O. (1991). Numerical modeling of the V(z) curve for a thin-layer/substrate configuration. *Res. Nondestr. Eval.* **3,** 183–200.
27. Liu, G. R., Achenbach, J. D., Kim, J. O., Li, Z. L. (1992). A combined finite-element method/boundary element method technique for V(z) curves of anisotropic-layer/substrate configurations. *J. Acoust. Soc. Am.* **92,** 2734–40.

28. Ahn, V. S., Harris, J. G., Achenbach, J. D. (1992). Numerical analysis of the acoustic signature of a surface-breaking crack. *IEEE Trans. Ultrason. Ferroelec. Freq. Contr.* **39,** 112–18.
29. Ahn, V. S. and Achenbach, J. D. (1991). Response of line-focus acoustic microscope to specimen containing a subsurface crack. *Ultrason.* **29,** 482–89.
30. Auld, B. (1979). General electromechanical reciprocity relations applied to the calculation of elastic-wave-scattering coefficients. *Wave Motion* **1,** 3–10.
31. Parmon, W. and Bertoni, H. L. (1979). Ray interpretation of the material signature in the acoustic microscope. *Electron. Lett.* **15,** 684–46.
32. Kushibiki, J. and Chubachi, N. (1985). Material characterization by line-focus beam acoustic microscope. *IEEE Trans. Sonics Ultrason.* **SU-32,** 189–212.
33. Tsukahara, Y., Liu, Y., Neron, C., Jen, C. K., Kushibiki, J. (1994). Singularities in acoustic reflection coefficient near the longitudinal critical angle and their effect to V(z) measurement with line-focus beam acoustic microscope. *IEEE Trans. Ultrason. Ferroelec. Freq. Contr.,* submitted.
34. Viktorov, I. A. (1967). *Rayleigh and Lamb Waves,* Plenum, New York, pp. 46–47.
35. Kim, J. O. and Achenbach, J. D. (1992). Line-focus acoustic microscopy to measure anisotropic acoustic properties of thin films. *Thin Sol. Films* **214,** 25–34.
36. Farnell, G. W. and Adler, E. L. (1972). In: *Physical Acoustics IX* (ed. W. P. Mason and R. N. Thurston), pp. 35–127. Academic, New York.
37. Ewing, W. M., Jardetzky, W. S., Press, F. (1957). *Elastic Waves in Layered Media.* McGraw-Hill, New York, Section 4.5.
38. Sezawa, K. (1927). Dispersion of elastic waves propagated on the surface of stratified bodies and on curved surfaces. *Bull. Earthquake Res. Inst. Univ. Tokyo* **3,** 1–18.
39. Weglein, R. D. (1982). Nondestructive film thickness measurement on industrial diamond. *Electron. Lett.* **18,** 1003–1004.
40. Weglein, R. D. (1985). Acoustic micrometrology. *IEEE Trans. Sonics Ultrason.* **SU-32,** 225–34.
41. Kushibiki, J., Maehara, H., Chubachi, N. (1982). Measurements of acoustic properties for thin films. *J. Appl. Phys.* **53,** 5509–13.
42. Kim, J. O., Achenbach, J. D., Mirkarimi, P. B., Shinn, M., Barnett, S. A. (1992). Elastic constants of single-crystal transition-metal nitride films measured by line-focus acoustic microscopy. *J. Appl. Phys.* **72,** 1805–11.
43. Weglein, R. D. and Kim, J. O. (1992). SAW dispersion in diamond films on silicon by acoustic microscopy. *Review of Progress in Quantitative Nondestructive Evaluation,* vol. 11 (D. O. Thompson and D. E. Chimenti, eds.), pp. 1815–22. Plenum Press, New York.
44. Farnell, G. W. (1970). In: *Physical Acoustics VI* (W. P. Mason and R. N. Thurston, ed.), pp. 109–66. Academic Press, New York.
45. Lim, T. C. and Farnell, G. W. (1969). Character of pseudosurface waves on anisotropic crystals. *J. Acoust. Soc. Am.* **45,** 845–51.
46. Achenbach, J. D. and Kim, J. O. (1993). In: *Inverse Problems in Engineering Mechanics* (M. Tanaka and H. D. Bui, eds.), pp. 265–76. Springer, New York.
47. Spendley, W., Hext, G. R., Himsworth, F. R. (1962). Sequential application of simplex designs in optimisation and evolutionary operation, *Technomet.* **4,** 441–61.
48. Nelder, J. A. and Mead, R. (1965). A simplex method for function minimization. *Comput. J.* **7,** 308–13.
49. Caceci, M. S. and Cacheris, W. P. (1984). Fitting curves to data. *Byte* **9** (5), 340–62.
50. Press, W. H., Flannery, B. P., Teukolsky, S. A., Vetterling, W. T. (1986). *Numerical Recipes,* Cambridge University Press, New York, Section 10.4.
51. Anderson, O. L. (1965). In: *Physical Acoustics III B* (W. P. Mason, ed.), pp. 77–83. Academic Press, New York.
52. Karim, M. R., Mal, A. K., Bar-Cohen, Y. (1990). Inversion of leaky Lamb wave data by simplex algorithm. *J. Acoust. Soc. Am.* **88,** 482–91.
53. Kim, J. O. and Achenbach, J. D. (1993). In: *Review of Progress in Quantitative Nondestructive*

Evaluation, vol. 12B (D. O. Thompson and D. E. Chimenti, eds.), pp. 1899–1906. Plenum Press, New York.
54. Holleck, H. (1986). Material selection for hard coatings. *J. Vac. Sci. Technol.* A **4,** 2661–69.
55. Schmid, E. and Boas, W. (1950). *Plasticity of Crystals,* Hughes, London, pp. 14–21.
56. Hertzberg, R. W. (1989). *Deformation and Fracture Mechanics of Engineering Materials,* 3d ed., Wiley, New York, Chap. 1.
57. Mendik, M., Satish, S., Kulik, A., Gremaud, G., Wachter, P. (1992). Surface acoustic wave studies on single-crystal nickel using Brillouin scattering and scanning acoustic microscope. *J. Appl. Phys.* **71,** 2830–34.
58. Kim, J. O. and Weglein, R. D. (1994). Comments on surface acoustic wave studies on single-crystal nickel using Brillouin scattering and scanning acoustic microscope. *J. Appl. Phys.* (in press).
59. Huang, J. and Achenbach, J. D. (1994). Measurement of material anisotropy by dual-probe laser interferometer. *Research in Nondestructive Evaluation* **5,** 225–235.
60. Lee, Y.-C., Kim, J. O., Achenbach, J. D. (1995). Acoustic microscopy measurement of elastic constants and mass density of solids and thin films. *IEEE Trans. Ultrason Ferroelec. Freq. Contr.* (in press).
61. Mal, A. K., Gorman, M. R., Prosser, W. H. (1992). In: *Review of Progress in Quantitative Nondestructive Evaluation,* vol. 11B (D. O. Thompson and D. E. Chimenti, eds.), pp. 1451–8. Plenum, New York.
62. Barnett, S. A. (1993). In: *Physics of Thin Films, vol. 17* (M. H. Francombe and J. L. Vossen eds.), pp. 1–77. Academic Press, New York.
63. Achenbach, J. D. (1975). *A Theory of Elasticity with Microstructure for Directionally Reinforced Composites,* Springer, New York pp. 21–38.
64. Grimsditch, M. (1985). Effective elastic constants of superlattices. *Phys. Rev. B* **31,** 6818–19.
65. Grimsditch, M. and Nizzoli, F. (1986). Effective elastic constants of superlattices of any symmetry. *Phys. Rev. B* **33,** 5891–92.
66. Akcakaya, E. and Farnell, G. W. (1988). Effective elastic and piezoelectric constants of superlattices. *J. Appl. Phys.* **64,** 4469–73.
67. Akcakaya, E., Farnell, G. W., Adler, E. L. (1990). Dynamic approach for finding effective and piezoelectric constants of superlattices. *J. Appl. Phys.* **68,** 1009–12.
68. Kim, J. O., Achenbach, J. D., Shinn, M., Barnett, S. A. (1992). Effective elastic constants and acoustic properties of single-crystal TiN/NbN superlattices. *J. Mater. Res.* **7,** 2248–56.
69. Kim, J. O., Achenbach, J. D., Mirkarimi, P. B., Barnett, S. A. (1993). Acoustic microscopy measurements of the elastic properties of TiN/(V_xNb_{1-x})N superlattice films. *Phys. Rev. B* **48,** 1726–37.
70. Kim, J. O., Achenbach, J. D., Shinn, M., Barnett, S. A. (1994). Effective elastic constants of superlattices determined using acoustic microscopy (to be submitted).
71. Kim, J. O. and Achenbach, J. D. (1993). In: *Dynamic Characterization of Advanced Materials,* NCA, vol. 16 (P. K. Raju and R. F. Gibson, eds.), pp. 163–70. ASME, New York.
72. Nakaso, N., Tsukahara, Y., Kushibiki, J. (1988). Evaluation of adhesion of films by V(z) curve method. *Jpn. J. Appl. Phys.* **28** (Supplement 28–1), 263–65.
73. Mal, A. K. and Weglein, R. D. (1988). In: *Review of Progress in Quantitative Nondestructive Evaluation,* vol. 7B (D. O. Thompson and D. E. Chimenti, eds.), pp. 903–10. Plenum, New York.
74. Kobayashi, T. Kushibiki, J., Chubachi, N. (1992). Improvement of measurement accuracy of line-focus-beam acoustic microscope system. *IEEE Ultrasonics Symposium Proceedings,* New York, pp. 739–42.
75. Kushibiki, J., Wakahara, T., Kobayashi, T., Chubachi, N. (1992). A calibration method of the LFB acoustic microscope system using isotropic standard specimens. *IEEE Ultrasonics Symposium Proceedings,* New York, pp. 719–22.

6

Measuring the Elastic Properties of Stressed Materials by Quantitative Acoustic Microscopy

Z. Sklar, P. Mutti, N. C. Stoodley, and G. A. D. Briggs

6.1. Introduction

Quantitative acoustic microscopy has progressed rapidly since the first observation of oscillations in the acoustic material signature or $V(z)$. The important role of Rayleigh waves was quickly established, and a further crucial step came with the development of line-focus-beam (LFB) lenses, since these allow directional excitation as well as more accurate analysis of the $V(z)$ curve. For some purposes, it is enough to be able to extract the velocity and attenuation of the surface acoustic wave (SAW) from $V(z)$ measurements, and then make qualitative observations about underlying material properties. For example lateral elastic inhomogeneity in piezoelectric wafers used for SAW devices can be detected with high sensitivity. It is often desirable to obtain more quantitative information about the underlying properties. The major challenge is that a change in the measured SAW velocity can result from changes in any number of parameters, such as the elastic constants, density, layer thickness, stress, and strain.

Z. Sklar, P. Mutti, N. C. Stoodley, and G. A. D. Briggs • Department of Materials, University of Oxford, Parks Road, Oxford, OX1 3PH, United Kingdom. *Present address of P.M.*: Dipartimento di Ignegneria Nucleare, Politecnico di Milano, Milano, Italy

Advances in Acoustic Microscopy, Volume 1, edited by Andrew Briggs.
Plenum Press, New York, 1995.

Chapter 6 discusses the principles of an inversion procedure for deducing elastic properties from $V(z)$ measurements and illustrates its use with some specific examples. Similar to the method described in Chapter 5, the strong excitation of surface waves is explicitly acknowledged to identify the most useful information in the $V(z)$. Chapter 6 shows how to incorporate static stresses into the inversion procedure, and the first step is to summarize the effect of stress on acoustic wave propagation. We also demonstrate how to measure the sensitivity and stability of the inversion process to the fitted parameters. A few improvements in the LFB $V(z)$ measurement and analysis technique are also mentioned. Chapter 6 concludes with some examples of the inversion technique applied to apatite single crystals, applied stresses in silicon wafers, amorphous hydrogenated carbon coatings, and ion implanted GaAs wafers.

6.2. Acoustoelasticity and Surface Waves

6.2.1. SAWs in a Stressed Material

It is well-known that the propagation of both bulk and surface acoustic waves is affected by the presence of static stresses, a phenomenon known as the acoustoelastic effect. However the stress cannot be measured directly when using elastic waves but must be inferred from the wave velocity. In certain cases, the presence of stress can be ascertained unambiguously from velocity measurements. For example when there is an anisotropic distribution of stress in the plane of interest, two orthogonally polarized shear horizontal waves propagating in this plane travel at different velocities. Since this difference in velocity vanishes in an unstressed material, the phenomenon provides a direct indication of the presence of stress, and it has been employed in various guises to map stress distributions in materials.[1–5] Considerable progress has been made in describing such effects independent of any particular constitutive relation.[6,7] This is important, since most acoustoelasticity theory assumes a hyperelastic constitutive relation, which must be modified empirically for plastically deformed materials[8] (Hyperelasticity is a special case of elasticity where stress depends only on the rate of change of the internal energy with strain).

Although there are likely to be effects similar to shear wave birefringence for SAWs with predominantly shear horizontal polarization, in most cases the effect of stress is not so clear cut. Since the velocity also depends on second- and third-order elastic constants and the density, the inverse problem, i.e., determining stress from ultrasonic measurements, requires prior knowledge of a number of material parameters. This can lead to a circular problem, since SAW measurements also provide one of the best ways of determining elastic constants, particularly for thin layers. Since the effect of stress on velocity is usually small, we must be sure that a measured velocity change is indeed due to stress, and not simply to inaccurate values for the elastic constants. In practice the material parameters of the natural, unstressed state are often either completely unknown or not known with sufficient accuracy (as is usually the case for residual stresses), or else they cannot be assumed to equal bulk values, as in the case of layered materials. This is a major distinction between situations involving residual as opposed to applied stress, since a reference state of some description is always available in the latter case. This is one reason why experiments on SAWs in stressed materials[9] have nearly always dealt with applied stress. The work described here shows how to improve estimates of the elastic constants of the reference state of a residually stressed layer from SAW measurements. To do this, the stress is taken directly into account when fitting measurements of SAW dispersion.

6.2.2. Theory

In the absence of body forces, Newton's law can be written

$$\frac{\partial \sigma_{ij}}{\partial x_j} = \rho \frac{\partial^2 u_i}{\partial t^2} \tag{1}$$

In linear elastic theory, σ_{ij} corresponds to the traditional stress tensor, and it is related to the infinitessimal strain by Hooke's law (see Ref. 10, Section 6.2). When the deformation is finite, several modifications must be made. Firstly displacements u_i^s and the local displacement gradients $\partial u_i^s/\partial x_j$ associated with material particles are finite, and the change in geometry cannot be ignored. It is usual to define a natural, undeformed reference configuration, a statically deformed or initial configuration, and a current configuration, as shown in Fig. 6.1. In defining physical properties, either the natural or the initial coordinates can be taken to be independent. It is usual to label a particular material particle by its position in the natural state X_i, which leads to the material or Lagrangian description. The complementary description, where initial coordinates are independent, focuses attention on a particular point in space rather than a material particle, and it is therefore referred to as the spatial or Eulerian description.[11] Various measures of the finite static deformation can be defined; the Lagrangian and Eulerian strains η_{ij} and γ_{ij} are particularly useful, since they reduce to the classical strain for an infinitessimal deformation[12]:

$$\eta_{ij} = \frac{1}{2}\left(\frac{\partial u_i^s}{\partial X_j} + \frac{\partial u_j^s}{\partial X_i} + \frac{\partial u_k^s}{\partial X_i}\frac{\partial u_k^s}{\partial X_j}\right) \tag{2}$$

$$\gamma_{ij} = \frac{1}{2}\left(\frac{\partial u_i^s}{\partial x_j} + \frac{\partial u_j^s}{\partial x_i} + \frac{\partial u_k^s}{\partial x_i}\frac{\partial u_k^s}{\partial x_j}\right)$$

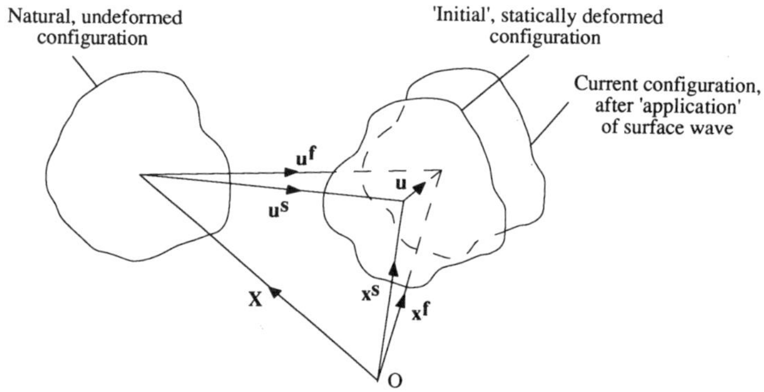

Figure 6.1. The relationship between natural, initial, and current material configurations. The superscript on $\boldsymbol{u}^s$ emphasizes that this variable refers to a static, finite displacement.

Although the Eulerian description has been used by some workers, we restrict ourselves to the Lagrangian formulation here.

When the deformation is finite, Hooke's law is no longer appropriate, and it is usual to postulate a hyperelastic constitutive relation by expanding the internal energy density W as a function of the Lagrangian strain η_{ij}

$$\rho_0 W(\eta) = \frac{1}{2} c_{ijkl}\eta_{ij}\eta_{kl} + \frac{1}{6} c_{ijklmn}\eta_{ij}\eta_{kl}\eta_{mn} + \cdots \tag{3}$$

The nth-order elastic constants are defined as the nth-order derivatives of the energy density with respect to the strain, where the constants are evaluated at zero strain.[13]

Stress is defined in terms of traction components acting on the surface of an elemental volume in the interior of the material (see Ref. 10, Section 6.2). If the traction direction is specified in the same configuration as the surface normal, then the resultant tensor corresponds to the classical or Cauchy stress. When writing equations of motion in terms of the material configurations shown in Fig. 6.1, it is convenient to define alternative measures of stress. The first Piola–Kirchhoff stress P_{ij} relates the components in the x_i directions of the force on a surface that is normal to the X_j axis in the reference configuration. Also in common use is the second Piola–Kirchhoff stress K_{ij}, and while this does not have a simple physical interpretation, it has the advantage of being symmetric unlike P_{ij},[14] therefore simplying certain tensor operations.

In acoustoelasticity displacements u_i associated with the ultrasonic wave are assumed to be infinitessimal. When this is not the case, the equation of motion becomes nonlinear, and wave propagation becomes more complicated.[15] Large-amplitude ultrasonic waves are used to measure third-order and higher elastic constants.[16] We confine our discussion to the case where displacements arising from the passage of the ultrasonic wave are infinitessimal. With this assumption, equations of motion can be linearized with respect to the incremental stress and strain between the initial and current configurations. When written in terms of the current configuration, the linearized equations of motion for a homogeneously stressed anisotropic material can be written as Eq. (1). When written in terms of the initial configuration, which is more useful in practice, Eq. (1) becomes

$$(\delta_{ik}\sigma_{jl} + C_{ijkl})\frac{\partial u_k^2}{\partial x_j \partial x_l} = \rho^i \frac{\partial^2 u_i}{\partial t^2} \tag{4}$$

$$C_{ijkl} = \frac{\rho^i}{\rho^0}\frac{\partial x_i}{\partial X_p}\frac{\partial x_j}{\partial X_q}\frac{\partial x_k}{\partial X_r}\frac{\partial x_l}{\partial X_s}c_{pqrs} \tag{5}$$

$$c_{pqrs} = \rho^0 \frac{\partial^2 W}{\partial \eta_{pq}\partial \eta_{rs}} \tag{6}$$

$$B_{ijkl} = \delta_{ik}\sigma_{jl} + C_{ijkl} \tag{7}$$

where only first-order terms in the displacement gradient $\partial u_i/\partial x_j$ are retained, and C_{ijkl} have the full symmetry of the second-order stiffnesses c_{ijkl}. The form of Eq. (4) arises from the transformation of the Cauchy stress in the current configuration into a Piola–Kirchhoff stress in the initial configuration. It is important to note that since the static stress in Eq. (4) is multiplied by a small displacement gradient term, the stress in this equation has to be specified only by Hooke's law for a consistent level of approximation to be maintained.

Although B_{ijkl} as defined in Eq. (7) can be used to define effective elastic coefficients, these are of a lower symmetry than c_{ijkl} except in the case of a hydrostatic initial stress.[17] We prefer to call C_{ijkl} as defined in Eq. (5) effective elastic constants and keep the dependence on the static stress in Eq. (4) explicit. This enables us to separate contributions from the static stress and strain to the net SAW velocity, as shown later. Thurston also defines wave propagation coefficients with the required symmetry for the application of Voigt's reduced notation by the addition of pairs of the B_{ijkl} in Eq. (7). While useful for obtaining bulk wave velocities, they are of no advantage in SAW problems, since the 6×6 matrix of reduced coefficients is no longer symmetric, and it does not reduce to the standard reduced stiffness matrix in the absence of static stresses.

By retaining terms up to cubic in the strain in Eq. (2), the effective elastic constants C_{ijkl} can be written[18]

$$C_{ijkl} = c_{ijkl}\left(1 - e_{nn}^s\right) + c_{ijklmn}e_{mn}^s \tag{8}$$

$$+ c_{mjkl}\frac{\partial u_i^s}{\partial x_m} + c_{imkl}\frac{\partial u_j^s}{\partial x_m} + c_{ijml}\frac{\partial u_k^s}{\partial x_m} + c_{ijkm}\frac{\partial u_l^s}{\partial x_m}$$

where

$$e^s_{mn} = \frac{1}{2}\left(\frac{\partial u^s_m}{\partial x_n} + \frac{\partial u^s_n}{\partial x_m}\right)$$

In the case of a material that is cubic in the natural configuration and subject only to axial strains e^s_1, e^s_2, and e^s_3, the effective elastic constants C_{IJ} can be expressed neatly in terms of strains and the cubic dilatation Δ as

$$\begin{aligned}
C_{11} &= c_{11}(1 - \Delta + 4e^s_1) + c_{111}e^s_1 + c_{112}(e^s_2 + e^s_3) \\
C_{12} &= c_{12}[1 - \Delta + 2(e^s_1 + e^s_2)] + c_{112}(e^s_1 + e^s_2) + c_{123}e^s_3 \\
C_{13} &= c_{12}[1 - \Delta + 2(e^s_1 + e^s_3)] + c_{112}(e^s_1 + e^s_3) + c_{123}e^s_2 \\
C_{22} &= c_{11}(1 - \Delta + 4e^s_2) + c_{111}e^s_2 + c_{112}(e^s_1 + e^s_3) \\
C_{23} &= c_{12}[1 - \Delta + 2(e^s_2 + e^s_3)] + c_{112}(e^s_2 + e^s_3) + c_{123}e^s_1 \\
C_{33} &= c_{11}(1 - \Delta + 4e^s_3) + c_{111}e^s_3 + c_{112}(e^s_1 + e^s_2) \\
C_{44} &= c_{44}[1 - \Delta + 2(e^s_2 + e^s_3)] + c_{144}e^s_1 + c_{155}(e^s_2 + e^s_3) \\
C_{55} &= c_{44}[1 - \Delta + 2(e^s_1 + e^s_3)] + c_{144}e^s_2 + c_{155}(e^s_1 + e^s_3) \\
C_{66} &= c_{44}[1 - \Delta + 2(e^s_1 + e^s_2)] + c_{144}e^s_3 + c_{155}(e^s_1 + e^s_2) \\
\Delta &= e^s_1 + e^s_2 + e^s_3
\end{aligned} \tag{9}$$

where reduced notation has been used. Note that in the absence of shear strains, the effective elastic constants do not depend on the third-order constant c_{456}.

Finally the density of the material is altered by the deformation. The density in the equation of motion, Eq. (4), refers to the statically deformed state, since this equation is written in terms of the initial configuration. To a consistent level of approximation, the density in the deformed state can be expressed in terms of that in the undeformed state as

$$\rho^i = \rho^0(1 - \Delta) \tag{10}$$

6.2.3. Implementation

The first calculations of the effect of stress on SAW velocity were by Hayes and Rivlin,[19] who derived approximate velocities for an isotropic material. Iwashimizu and Kobori[20] extended this work to propagation directions away from the principal axes, and Hirao and coworkers[9] dealt with stresses that varied as a function of depth. Nalamwar and Epstein[21] calculated velocities for a homogeneously stressed piezoelectric anisotropic layer, but their analysis is incorrect, since they did not maintain a consistent level of approximation for their effective elastic constants. Mase and Johnson[22] presented a calculation for SAWs in a homogeneously stressed half-space, but they do not deal with attenuating modes or layers.

A computer program was written to calculate SAW velocity and attenuation at the boundary between a fluid and an anisotropic, statically stressed, coated half-space.[23] The program allows the calculation of pseudo-SAW-type solutions that decay into the substrate[24,25] and also leaky SAW solutions that decay into the fluid.[26,27] These modes are particularly important in the acoustic microscope (see Ref. 10, Section 10.2). For a description of measurements on stressed materials with the acoustic microscope, see Ref. 10, Section 8.2.4. A number of techniques are available in the program to allow automatic tracking of a single dispersive SAW mode; this is particularly important when fitting calculated curves to experimental results.

In the following, we essentially follow the procedure developed by Farnell and Adler[28] for an unstressed layer. Given an initial guess at the velocity and attenuation of the SAW, we look for displacements of the form

$$u_j = \alpha_j e^{ikbx_3} e^{ik(lx_1 - vt)} \tag{11}$$

where k and v are the wave vector magnitude and velocity, and are both real quantities. The imaginary part of l represents the normalized attenuation of the SAW in the direction of propagation, as shown in Fig. 6.2. Such displacements are substituted into Eq. (4), and the resulting Christoffel equation is solved to obtain values for the decay constants b for both the substrate and the layer. Applying the usual boundary conditions results in a set of simultaneous linear equations. For non-trivial solutions to these equations, we require the accompanying determinant to be zero. This is achieved by numerically iterating the initial guess at the velocity and attenuation until the determinant has been effectively driven to zero

$$\det(v_{saw}, l_{saw}, c_{ij}, c_{ijk}, \sigma_i, \varepsilon_i, t, \rho) = 0 \tag{12}$$

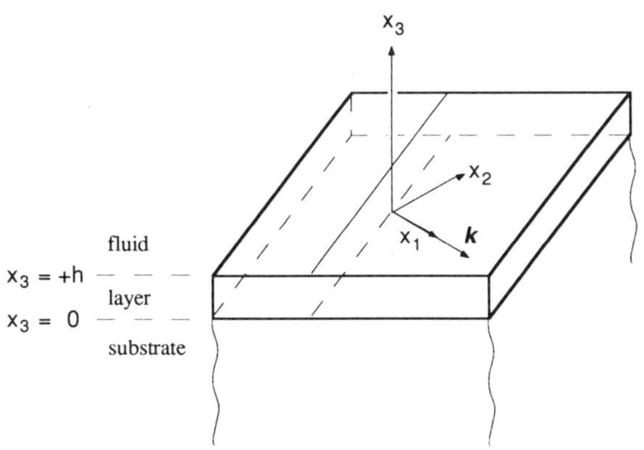

Figure 6.2. A coordinate system for a rotated crystal/layer/fluid system.

where v_{saw} and l_{saw} are the velocity and attenuation of the surface wave, respectively, and t is the layer thickness.

6.2.4. An Example

When modeling a statically stressed material using Eq. (4), three separate effects must be considered: Firstly the strain results in a change in density, as in Eq. (10); secondly the strain modifies the second-order elastic constants according to Eq. (9); lastly the stress is incorporated directly into the equation of motion, Eq. (4). Depending on the particular geometry, these three effects can either enhance or cancel one another's contributions to the overall change in SAW velocity.[21] As a specific example, consider the case of a cubic layer subject to a planar stress and perpendicular strain, bonded to an unstressed infinite substrate.

Figure 6.3 shows SAW dispersion curves calculated along a ⟨110⟩-type direction for a modified GaAs(001) layer on a pure GaAs(001) substrate. The layer is subject to a perpendicular strain of 0.4% and a planar stress of 470 MPa. Its thickness is 1.5 μm, and c_{11}, c_{12}, and c_{44} are decreased by 4%, 0%, and 12%, respectively, from bulk values for GaAs. Figure 6.3 shows that the largest single effect is due to modification of the stiffness by the perpendicular strain. The stress also tends to decrease the velocity, although to a lesser extent. The density decrease, on the other hand, causes an increase in SAW velocity when taken in isolation. The separate contributions to the net SAW velocity have implications for a potential inversion procedure, in that neglecting one or more of them results in a biased fitted set of elastic constants. By the same token, if the stress and strain can be measured and/or calculated independently, then true elastic constants of the reference state of the layer material can be inverted. This approach is demonstrated later for ion-implanted GaAs.

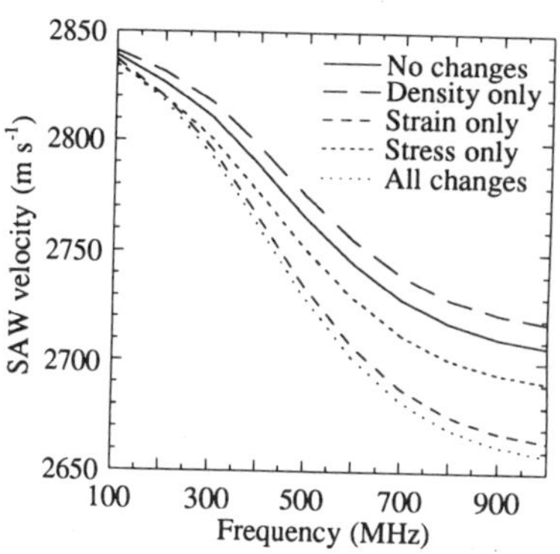

Figure 6.3. Calculated SAW dispersion for a strained GaAs layer on a GaAs(001) substrate along a ⟨110⟩-type direction.

6.3. Line-Focus Beam Acoustic Microscopy

6.3.1. Measurements

The $V(z)$ curves were measured using the line-focus beam acoustic microscope described by Kushibiki and Chubachi,[29] with a cylindrical lens operating at 225 MHz. To improve the accuracy and repeatability of measurements, the quasi-monochromatic radio frequency circuit described in Ref. 10, Section 5.2 was employed. The precision in the inferred SAW velocity was estimated to be of the order of 0.05% by Kushibiki and others.[30] This figure is realistic only when a single, well-defined mode with many oscillations is present in the $V(z)$. When fast substrates, layers, or thin plates are measured, oscillations due to various modes are often present over the entire range of defocus. Truncation effects in the analysis can give rise to a substantial loss of accuracy. The maximum usable range of defocus is limited by the lens opening angle and more fundamentally by attenuation in the coupling fluid, which determines the maximum focal length of the lens. The limitation of 560 μm of negative defocus for the standard 225-MHz lens actually arises because the lens geometry causes the second specimen reflection to begin to enter the usual fixed receive gate at about this value. For the present work, the second reflection was avoided by automatically moving the position of the gate during $V(z)$ acquisition, resulting in a maximum usable negative defocus of 660 μm. The extra 100 μm of defocus thus obtained can make a substantial difference to the analysis when more than one mode is present.

6.3.2. Analysis

The standard analysis procedure for LFB acoustic microscope data was given by Kushibiki and Chubachi,[29] and a description is found in Ref. 10, Section 8.2.1. When dealing with large numbers of complicated $V(z)$s, it is impractical to analyze each one manually. An automated analysis procedure was therefore developed, which locates all the maxima and minima within a specified range of the spectrum produced by an initial trial Fourier transform. Three adjustable criteria are then employed to eliminate the vast majority of these peaks. Firstly all peaks below a threshold level relative to the highest peak are discarded. Then all peaks that are either too shallow or too closely spaced are ignored; such peaks often arise from rippling effects in the initial transform. Approximate values of Δz and attenuation are then calculated for each of the selected peaks and passed as reference values to a subsequent Fourier analysis. This may either be the standard analysis or an enhanced version, whereby each of the local peaks in the spectrum can be Fourier-filtered and then inverse-transformed, with a linear fit applied to the inverse transformed waveform to obtain the attenuation.

The method relies rather heavily on a good initial transform. The presence of multiple modes in the $V(z)$ can generate interference between the side lobes of peaks in the power spectrum and produce errors in determining SAW propagation parameters. This effect, called leakage, is reduced if the $V(z)$ is first multiplied by a smooth window function. A Hamming window gives a good trade-off between the side lobe falloff and the width of the main lobe.[31] The result is an improvement in the power spectrum, and determining the peak position is less subject to interference effects. The disadvantage lies in a broader main lobe as compared to the spectrum of the unwindowed $V(z)$.[32] With due care, it is usually possible to extract most of the information in a multiple-mode $V(z)$ with little loss of accuracy over a fully blown manual analysis. This procedure is demonstrated in the following example.

Figure 6.4 shows a typical $V(z)$ on a Si(111) wafer, and a number of weak contributions can be identified in the Fourier spectrum. Figure 6.5 shows the

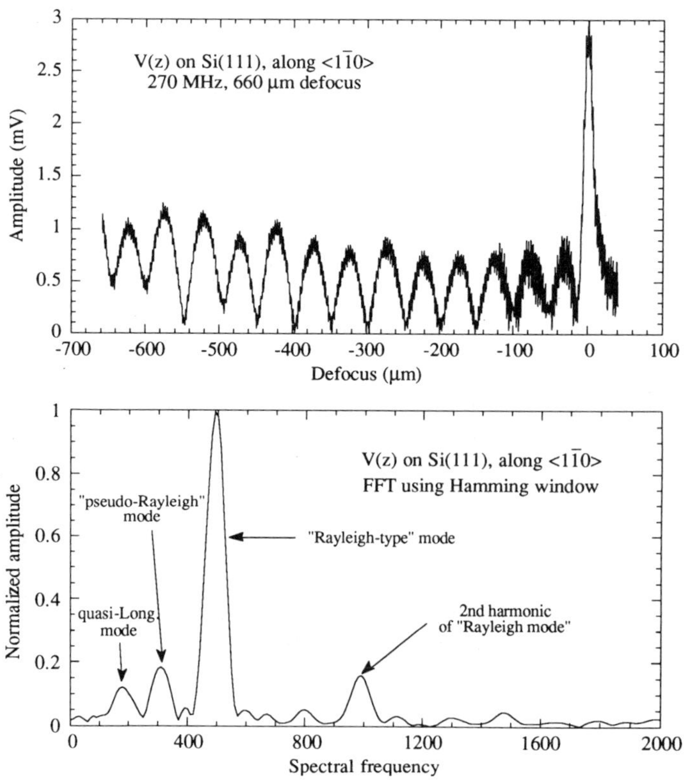

Figure 6.4. An analysis of a $V(z)$ taken along a $\langle 1\bar{1}0\rangle$-type direction on Si(111); (a) 270-MHz, 660-μm Defocus, (b) fast fourier transform (FFT) using Hamming Window.

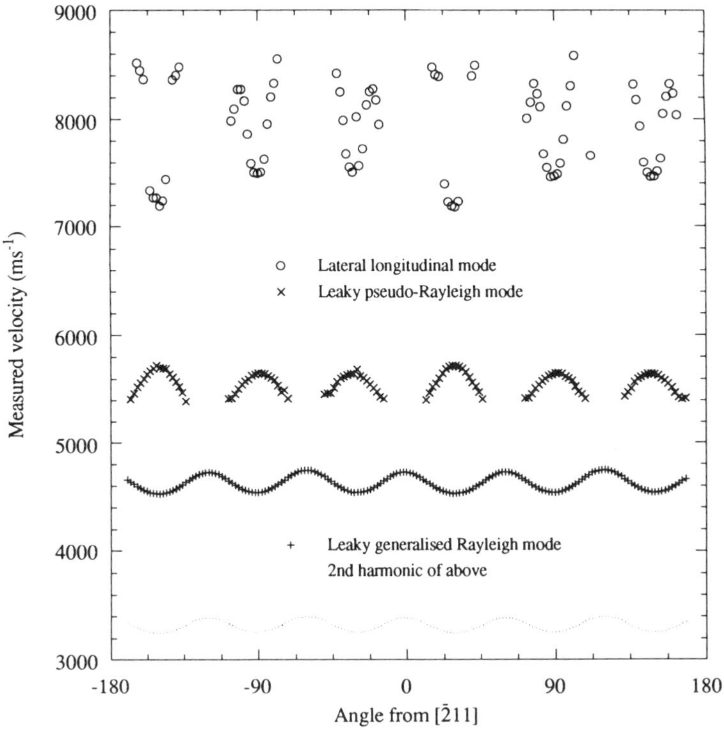

Figure 6.5. Results of automatic multiple-mode analysis for 600-μm defocus measurements at 270 MHz on silicon (111).

results of an automatic multiple-mode analysis using a Hamming window for a 360° rotational scan on a Si(111) wafer. Measurements were taken at 270 MHz using 660 μm of defocus. The utility of the automatic multiple-mode analysis method is immediately apparent in the appearance of the leaky pseudo-Rayleigh mode. This mode is not excited significantly around ⟨$\bar{2}$11⟩-type directions, since its character degenerates toward shear horizontal. With the threshold for peak rejection set to 10% of maximum, two other modes are selected besides the main peak due to the leaky generalized Rayleigh wave. Firstly a high-velocity mode corresponding to the lateral quasi-longitudinal wave is clearly identified. Absolute velocity values for this mode are only accurate to about 5%, since the defocus available with the standard lens limits the number of oscillations to three or four at most. This is an example of when it is not a good idea to remove the first oscillation, since the analysis suffers significantly as a result. The lowest velocity mode in Fig. 6.5 is spurious, arising from the second harmonic of the strong leaky Rayleigh wave on silicon. Other spurious modes can also arise when there is a rippling effect in the Fourier spectrum. The azimuthal asymmetry, which is most noticeable for the longitudinal mode but also present for the other modes, is due to the cut of the wafer being slightly off (111). The degree of asymmetry depends on the amount of misalignment, and in this case, it was specified by the manufacturers as 3°. When comparing results for vicinal wafers, the direction along which velocities are measured must be the same.

The preceding example indicates some of the advantages and new problems inherent in analyzing results with a good automatic procedure. The method becomes particularly useful for layered materials, such as the thin amorphous hydrogenated carbon coatings considered later.

6.4. Determining Material Properties

6.4.1. Inverting the V(z)

Several attempts have been made to invert reflectance functions from a measured $V(z)$.[10,33,34] This is more ambitious than using the ray visualization to extract surface wave velocities, since the lens response is dominated by only one or two aspects of the reflectance function, and the pupil function of a real lens is difficult to characterize accurately. The most accurately reproduced feature in the inverted reflectance function is the phase change at the Rayleigh angle, from which the Rayleigh velocity can be obtained. Kinks in the magnitude at the longitudinal and transverse wave numbers are less reliable, but these enable an approximate determination of bulk velocities.

Yu and Boseck[35] obtain material parameters by attempting to fit the $V(z)$ itself, modeled using the standard Fourier method.[36] There are many potential advantages to this technique, the principal one is avoiding directly calculating the position of complex poles in the reflectance function. For layered materials, such poles, or equivalently the highly attenuating SAWs they represent, can be very difficult to locate. Much effort has been expended trying to reproduce accurately all the features of a real $V(z)$. The absence of a good characterization of the lens pupil function is a major problem. There is then a danger in attempting to fit a whole $V(z)$ that the least-squares routine will be unduly influenced by parts of the curve that are not well accounted for by the model in use. While this is probably the best that can be done for a single $V(z)$, a greater degree of confidence is possible if the combined information in a number of $V(z)$s is taken as a function of an angle or frequency.

6.4.2. Fitting SAW Dispersion

Fitting theory to experimental SAW data presents a number of practical problems, and most of these stem from the absence of an analytic expression for the velocity of a surface wave. In this respect, Lamb waves are much easier to fit because of the availability of such an expression for their dispersion. To locate a surface wave, the traditional method is first to bracket the anticipated position of the velocity, and then search until the boundary conditions imposed at the surface and interfaces are satisfied. This happens when the determinant of the matrix formed by applying the boundary conditions is effectively driven to zero [see Eq. (12)]. Since all calculations take place numerically, there is no way of telling to which branch in the dispersion curve a particular solution belongs. Furthermore boundary conditions can be satisfied or nearly satisfied at or near points that do not lie on the dispersion curves, such as the limiting bulk velocities of the substrate or layer. Problems multiply further when attenuating solutions are sought, since a two-dimensional search is now required to locate the zero in the boundary condition determinant. All these problems can make life difficult for a prospective fitting routine, since the correct theoretical point must be automatically chosen over a wide range of material parameters.

We now examine two approaches to inverting material parameters from measurements of SAW velocity dispersion. In the first approach, which was developed here, an initial guess at the parameters is used to calculate the SAW velocity corresponding to each experimental point by driving the determinant in Eq. (12) to zero. The overall discrepancy between theory and experiment is then minimized by improving the initial guess using a nonlinear least squares routine. Minimize

$$\sum_{i=1}^{n}\left(\Delta v_i\right)^2 = \left(v_i^{\text{measured}} - v_i^{\text{theoretical}}\right)^2$$

with respect to the material parameters.

In the second approach, referred to as determinant fitting, the measured velocities are substituted into Eq. (12) and the magnitude of the determinant calculated at each of the n experimental points. The sum of these determinants is then minimized by improving the guess at the material parameters. Minimize

$$\sum_{i=1}^{n}\left|\det_i\right|^2$$

with respect to the material parameters.

Determinant fitting has been used by a number of workers,[37–40] to invert elastic constants of substrates and coatings from measurements of SAW dispersion. The most attractive aspect of the procedure is that it eliminates most of the numerical requirement for calculating SAW velocities, and as a result it is much faster to use and considerably easier to implement than the first approach. Determinant fitting also has a serious disadvantage, stemming from the fact that it does not involve a true fit of theory to experiment, and hence statistical procedures used to provide a quantitative assessment of the quality of the fit cannot be used.

This is particularly important for SAW problems, since the relative complexity of calculating the SAW velocity in an anisotropic material can tend to mask an underdefined or poorly defined problem. A curve, fitted by eye or determinant fitting, may look good, but the parameters used to create the fit are not of much use without a proper estimate of their uncertainty. A further advantage of the approach developed here over that of determinant fitting is that the former is not affected by the local shape of the boundary determinant.

Another technique in recent use analyzes a calculated $V(z)$ to obtain a theoretical value for the measured SAW velocity, and then proceeds with a nonlinear least squares fit much as for the first technique discussed (see Chapter 5). This is particularly useful when the microscope resolution is insufficient to distinguish between two adjacent modes, so that the measured velocity lies between the true velocities of the pair.

All of the three preceding methods share the advantage of ray analysis, namely, the ray model concentrates on the strongest interaction with the sample and therefore on the features of the $V(z)$ data that most reliably depend on the elastic properties of the sample.

6.4.3. Solution Stability and Uniqueness

The most difficult aspect of many inversion problems is the stability of the system when only a limited amount of experimental data is available. Dispersion measurements on the LFB acoustic microscope are limited to a very narrow range of frequency. It is asking a great deal of any fitting routine to extrapolate the entire dispersion characteristic reliably from such a small segment, no matter to what accuracy that segment is available. A least squares fitting routine applied to such a system may not realize that it is in trouble. The problem is that the routine has no reliable way of telling whether the minimum it has been pointed to or else has stumbled on is a true global minimum or merely a local feature. A good way of picturing this is to suppose the experimental data are perfect, but only a very narrow range of dispersion is available. The sum of squares of residuals then contains a large number of local minima, many of similar form and magnitude, and one of these happens to drop to zero. If the routine is steered in the right direction by giving it an initial guess close to the true parameters, then it usually works its way to the correct minimum and returns a good result. Otherwise it picks the nearest minimum, does the best it can with it, and then looks in a few other places to see if it can do better. If it happens to stumble on the true zero, it is by accident rather than design. Introducing experimental error ensures that the true minimum no longer goes to zero. This can make the problem truly unstable, in that there is no one local minimum significantly better than the others. Such a scenario is best exposed by trying to fit the data from many different starting points, since significant variation in the final values indicates an unstable problem. As the range of available dispersion is increased, the global minimum becomes deeper while local minima becomes shallower. The global minimum also becomes broader at the mouth, which helps it to swallow any wandering minimization routines.

There are two other considerations that affect the question of stability. A small change in the experimental values may cause a relatively large change in the fitted parameters, which is clearly unsatisfactory. Also when fitting data for thin layers, it is crucial for the substrate to be well-characterized, since the layer has only a small perturbing effect on it. Fitting to narrow ranges of dispersion for very thin films is likely to be unstable even to small changes in substrate constants.

6.4.4. Quality of Fit

Having obtained a best fit using the method developed here, some indication of its quality and errors in the fitted parameters needs to be given. When the experimental errors are normally distributed with a known variance, the χ^2 goodness-of-fit measurement can be applied. When SAW velocities are measured in the acoustic microscope, the variance is not usually known with any accuracy due to possible errors in the analysis. It is assumed in the following discussion that errors are distributed approximately normally, with the same unknown variance for each point. In this case, the root mean square (rms) deviation of the residuals provides an initial indication of the goodness of fit. If this value is greater than the expected typical error in velocity, then the fit must be called into question immediately. A good value is not necessarily sufficient however, in that the fitting routine may distort the curve in some manner to minimize the discrepancy. The best way of checking for this is by examining residuals as a function of the independent variable. Residuals should change sign in a largely random manner, since they reflect the experimental errors. Since residuals are not unbiased estimates of the random errors, some trends may well be expected, particularly for data with very low scatter. However long runs where the residuals do not change sign are grounds for suspicion.[41] If an acceptable fit cannot be found, then either the errors are much larger than expected and are probably due to systematic errors in the analysis of the $V(z)$, or the theoretical model is incorrect.

Once a fit has been deemed acceptable, the covariance matrix can be estimated from the Hessian matrix evaluated by using the best fit parameters. Large values of the diagonal elements indicate that the minimum in the sum of residuals is shallow, and hence there is little sensitivity to one or more of the parameters. If the error distribution is taken to be normal, then these diagonal values represent the actual variances of the fitted parameters.[42]

6.5. Examples

6.5.1. Fluoroapatite and Hydroxyapatite

As a first example of the application of the inversion procedure just described, we consider LFB $V(z)$ data taken on the $(10\bar{1}0)$ plane of hexagonal single crystals of fluoroapatite and hydroxyapatite, which are constituents of tooth enamel.[43] Angular scans on these materials are shown in Figs. 6.6 and 6.7 with curves obtained using literature values of elastic constants for the crystals and best fits to the data using the code we developed. The velocities calculated using literature constants for fluoroapatite are in good agreement with the measured values, whereas agreement for hydroxyapatite is very poor.

Twenty-eight experimental points were fitted, using a fixed offset of 8.0 m s^{-1} to account for fluid loading. This is the theoretical value obtained using the literature constants, and it is almost independent of direction in the plane. The literature values of the five elastic constants and density were used for the initial guess, and an attempt was made to fit all five constants with the density held fixed. Results are summarized in Table 6.1, with all elastic constants in GPa.

The rms deviation is measured between the experimental points and the best fitted curve. As previously discussed, the rms deviation gives a good indication of the quality of fit when experimental error is the same for all points.

The fitted curves in Figs. 6.6 and 6.7 are therefore of good quality, since the rms deviation is similar in magnitude to the experimental error, and there are no significant trends in the residuals. The variances of the fitted parameters

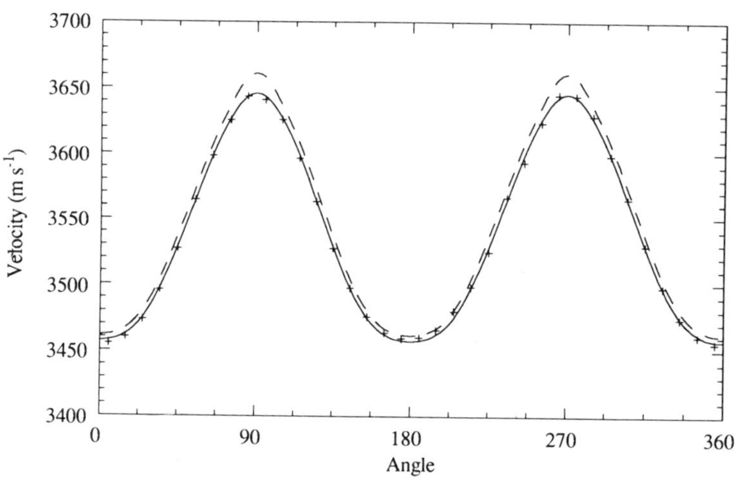

Figure 6.6. Measured and calculated leaky SAW velocities on single-crystal fluoroapatite. – – – –, calculated using literature values; ———, fitted using SAW theory; ++++, LFAM experiment.

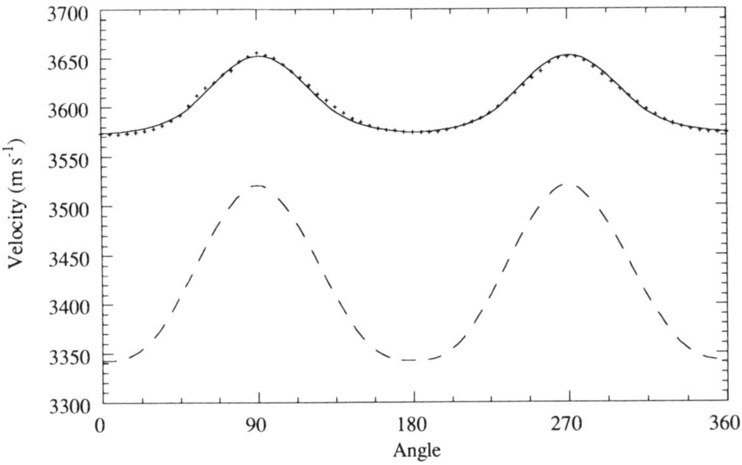

Figure 6.7. Measured and calculated leaky SAW velocities on single-crystal hydroxyapatite. – – – –, calculated using literature values; ————, fitted with SAW theory; ++++, LFAM experiment.

indicate the sensitivity with which each parameter has been determined. Large values of the variance indicate that a parameter can be changed significantly without greatly altering the sum of squares of residuals. Since the distribution of errors in SAW velocity is not assumed to be normal, variance values in Table 6.1 are not to be taken as formal error bounds on elastic constants. Despite the good fit, variances in Table 6.1 demonstrate that most of the constants are ill-determined, which is a consequence of measuring in only one plane of symmetry. Only c_{44} has been determined with any sensitivity. Variances in the other constants, particularly for hydroxyapatite, make their values almost worthless. This demonstrates how a good fit cannot be used as the sole justification for accepting fitted constants.

Table 6.1. Estimated Elastic Constants (in GPa) for Fluoroapatite and Hydroxyapatite Crystals

	Fluoroapatite			Hydroxyapatite		
Constants	Literature	Fitted	Variance	Literature	Fitted	Variance
c_{11}	150.5	133.9	13.4	137.0	144.3	40.5
c_{12}	48.8	30.9	18.9	42.5	43.0	51.3
c_{13}	62.2	31.4	13.2	54.9	65.1	64.2
c_{33}	185.0	140.0	18.7	172.0	171.9	88.9
c_{44}	42.5	43.7	0.07	39.6	47.6	0.1
RMS deviation		0.33 m s^{-1}			0.52 m s^{-1}	

6.5.2. Amorphous Hydrogenated Carbon Coatings

When a thin coating is present, determining material parameters becomes more difficult, since we are usually dealing with small perturbations to the SAW velocity of the substrate. As an example, we consider LFB acoustic microscope measurements on amorphous hydrogenated carbon (a-C:H) coatings on silicon and fused quartz wafers.

There has been considerable interest in hard-carbon-based films over the last two decades. Their hardness, low friction, and chemical inertness makes them attractive as protective coatings, while their relatively low infrared absorption and variable refractive index fosters their use in optics. Over the last three or four years, attention has increasingly focused on the production of true polycrystalline diamond films by various forms of chemical vapor deposition. However this is still usually a high-temperature process, so hard a-C:H coatings produced at room temperature by plasma or ion beam deposition remain important for many substrate materials.[44]

Amorphous hydrogenated carbon films were produced on silicon (001) and (111) wafers and fused quartz substrates using a standard capactively coupled, water cooled, radio frequency (r.f.) glow discharge system,[45] using cyclohexane as the source gas. The gas pressure and r.f. power were varied over a wide range, while the substrate was kept at room temperature. The six different sets of conditions used (denoted A–F) are summarized in Table 6.2. Coatings from 0.1–6.0 μm were deposited under each of these sets of conditions. Coating densities were measured using a sink float technique[46]; these are shown in Fig. 6.8. Compressive stresses present in the coatings were inferred from the curvature of thin circular glass cover slips,[47] and these are shown in Fig. 6.9. The hardness of the coatings, determined by nanoindentation, was found to range from 2–15 GPa.

Figures 6.10 and 6.11 summarize LFB acoustic microscope results for a-C:H coatings on fused quartz and silicon (001), respectively, along a ⟨100⟩-type direction. Frequency scans for five or six coatings produced under identical conditions are combined in each case. It is necessary to use a range of samples

Table 6.2. Process Conditions for a-C:H Coatings

Set	Pressure (mTorr)	r.f. power (W)	Residual stress (GPa)	Density (kg m^{-3})
A	17	300	1.2	1650 ± 50
B	100	125	0.17	1250 ± 25
C	480	300	0.24	1230 ± 25
D	100	75	0.15	1230 ± 25
E	100	200	0.52	1325 ± 25
F	100	300	0.71	1420 ± 20

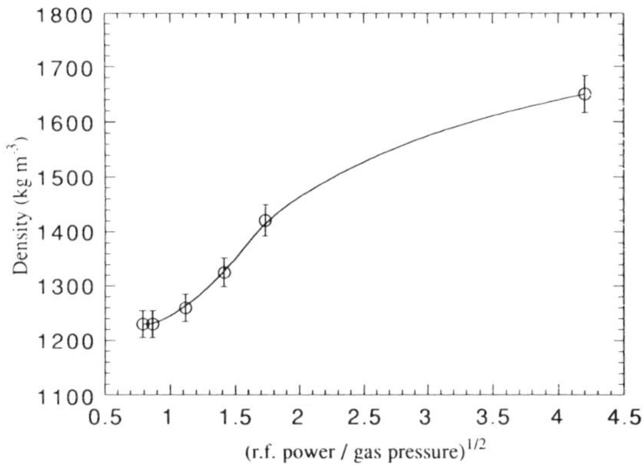

Figure 6.8. Measured densities of a-C:H coatings as a function of approximate bias voltage.

with coatings of various thicknesses, since the inversion of elastic constants is unstable when limited to thin coatings. The fact that there is a high degree of overlap between frequency scans on coatings of different thicknesses produced under identical conditions shows that the elastic constants of these coatings do not change by a large factor as a function of thickness.

In fitting such dispersion curves, the variance in c_{11} was always a factor of 2–5 higher than for c_{44}. Measured values of residual compressive stress were entered directly into the fitting process, but the lack of knowledge about third-order elastic constants means that strain effects are embedded in the estimated second-order constants.

Only the lowest velocity modes are shown in Figs. 6.10 and 6.11. When more than one mode was significantly excited, these were fitted simultaneously, since this gives a much more stable inversion. An example is shown in Fig. 6.12 for coatings in Set C on silicon. For the fused quartz substrates, and also along

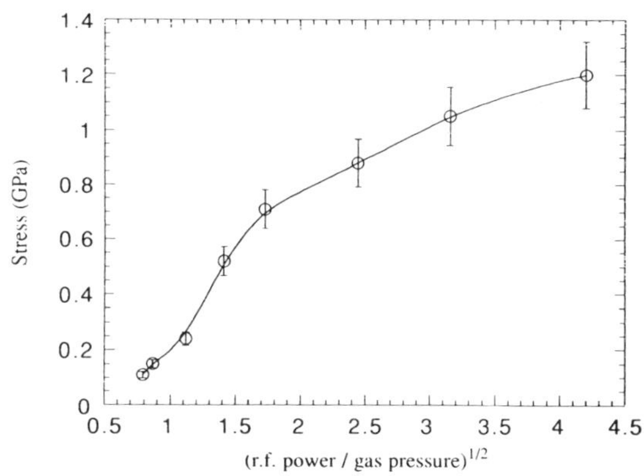

Figure 6.9. Compressive stress in a-C:H coatings as a function of approximate bias voltage.

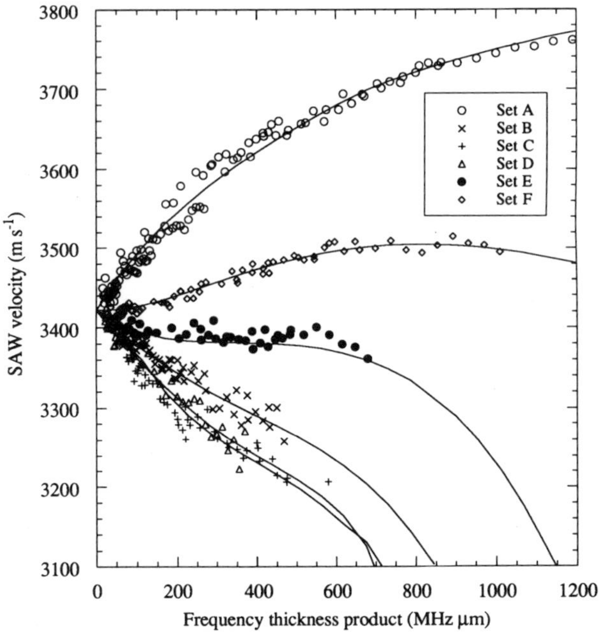

Figure 6.10. Fitted SAW dispersion for the lowest velocity modes measured for a-C:H coatings on fused quartz.

most symmetry directions in silicon, the shear horizontal (SH) components of particle displacement are uncoupled from the shear vertical (SV) and longitudinal (L) components. Two independent sets of dispersive SAW modes result, and are known as Love type and Rayleigh type, respectively.[28] Since the acoustic microscope cannot couple to SH waves, it can detect only Rayleigh type modes. Away from symmetry directions, the SH, SV, and L components all couple together, and two new sets of modes are present, both containing vertical displacements. Along a $\langle 1\bar{1}0\rangle$-type direction on the (111) plane, the degree of coupling is sufficient to allow both sets of modes to be detected using the acoustic microscope, and this is a bonus for the fitting process. This effect can be seen by comparing measured curves along the $\langle 1\bar{1}0\rangle$ and $\langle \bar{2}11\rangle$ directions in silicon (111) for coatings in Set C, as shown in Figs. 6.12(a) and (b). This is an interesting result, since it shows that more information about an isotropic layer can potentially be obtained when the substrate is anisotropic rather than isotropic. The automatic selection of the correct mode to fit to becomes very awkward near regions of mode pinching, such as that in Fig. 6.12(a), where two branches exchange polarizations.[23] Fitting results are summarized in Figs. 6.13 and 6.14, which show the inverted values of c_{11}, c_{44}, and Poisson's ratio for the coatings. Results

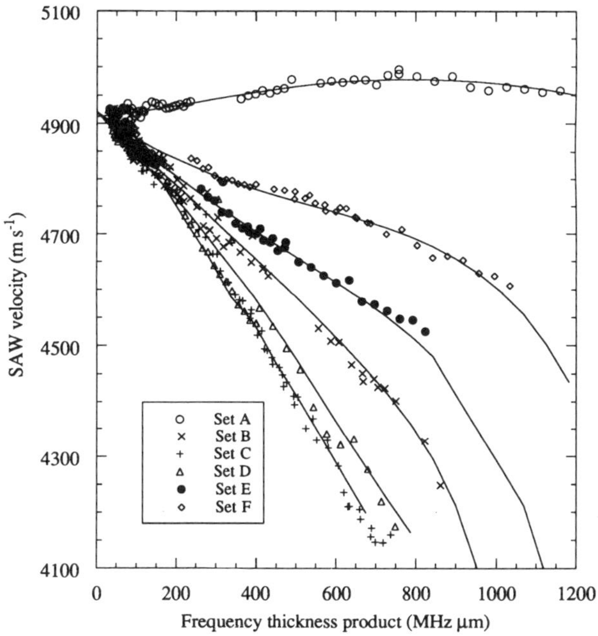

Figure 6.11. Fitted SAW dispersion for the lowest velocity modes measured for a-C:H coatings on silicon (001) along a ⟨100⟩-type direction.

along the two symmetry directions have been averaged for the (001)- and (111)-oriented silicon wafers.

The single most important process parameter is the negative self-bias voltage V_b, which accelerates the bombarding ions. Although this could not be measured directly, it can, under certain approximations,[48] be related to the r.f. power P and the source gas pressure p by $V_b \propto \sqrt{P/p}$. All figures are therefore plotted as a function of $\sqrt{P/p}$, a quantity that should be proportional to the bias voltage. Both c_{11} and c_{44} increase with decreasing gas pressure and increasing r.f. power, since both of these effects increase the ionic bombardment. Stiffnesses vary by almost an order of magnitude, ranging from values typical of vitreous carbon at the low end and approaching a diamondlike hydrocarbon material at the other extreme. Similarly Poisson's ratio increases from a value typical of soft polymeric materials to something more akin to a hard, brittle substance. From Figs. 6.13 and 6.14, the rate of increase in c_{44} is slightly faster than c_{11}, consistent with a slowly decreasing Poisson's ratio. The rate of increase in the density is much lower, and this ties in with bulk and SAW velocities that rise quickly with the bias. Qualitative comparisons of these results with theoretical studies of the elastic properties of random covalent networks have been made.[49]

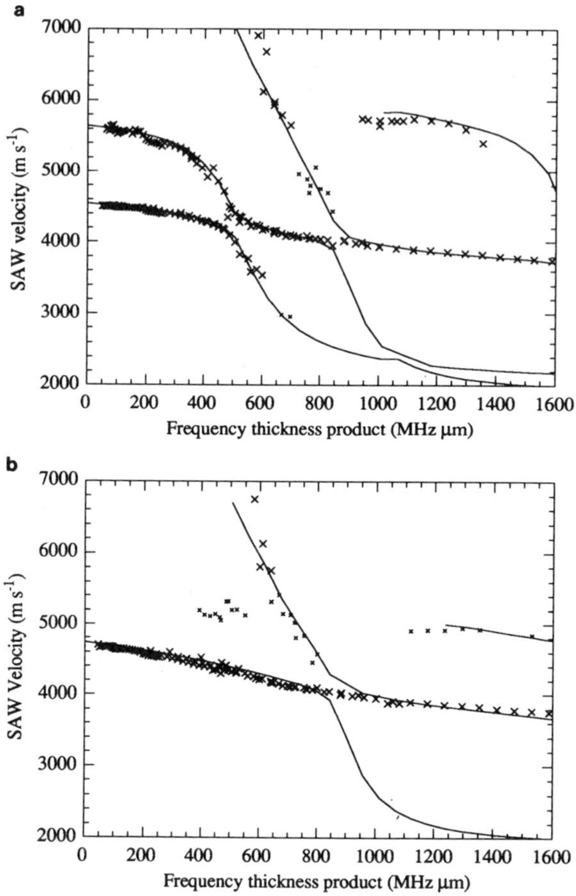

Figure 6.12. Fitted SAW dispersion for a-C:H coatings of Set C on silicon {111}; (a) a-C:H Set C on Si (111) along $\langle 1\bar{1}0 \rangle$, (b) a-C:H Set C on Si (111) along $\langle \bar{2}11 \rangle$.

6.5.3. Applied Stresses and Third-Order Elastic Constants in Silicon

The dependence of SAW velocity on applied stress was investigated here for silicon wafers using the LFB acoustic microscope, and results were compared with theoretical calculations. To generate a strain field in the sample, the vacuum pump normally used to hold specimens to the stage provided a bending force to one side of the wafer. The wafer was placed on a brass ring on the stage, and the volume under the wafer was evacuated using the vacuum pump. A needle valve was fitted between the pump and the stage to vary the extent of the evacuation, thereby changing the net force on the top surface of the wafer from

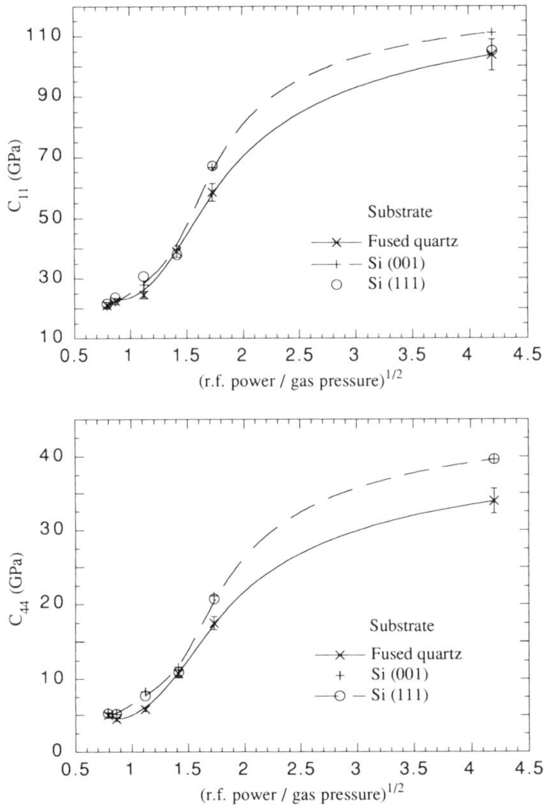

Figure 6.13. Stiffness of a-C:H coatings as a function of approximate bias voltage.

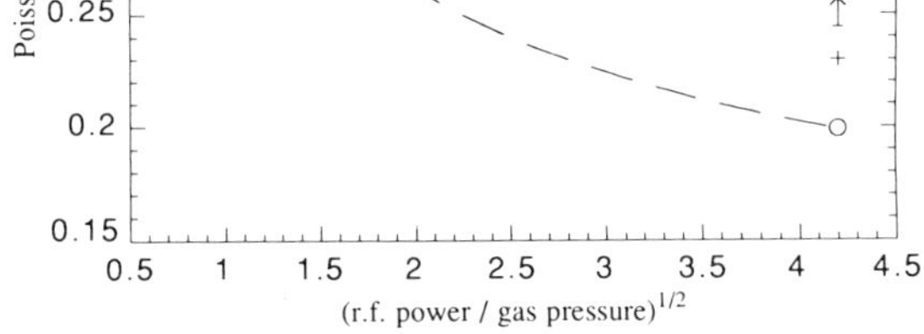

Figure 6.14. Poisson's ratio of a-C:H coatings as a function of approximate bias voltage.

the atmospheric pressure. A stress analysis could then be performed from knowing the overpressure, and from this the strain present in the surface of the wafer could be calculated. The system is shown in Fig. 6.15.

Using this arrangement, the pressure difference between the top and bottom of the wafer was incrementally increased, with a 45-minute pause at each new pressure to allow the system to reach equilibrium. The lowest point on the wafer was found by plotting the point of focus along two mutually perpendicular lines on the surface of the wafer and adjusting the position of the lens until it was over the lowest point throughout the rotation.

Wafers used were p-type silicon (dopant 10^{14} cm^{-3}), 125 mm in diameter and 500 μm thick, with an optical polish on the top surface. Attempts were also made to use wafers that had been etched on both surfaces to remove surface defects and hence give the wafers greater strength, but the surface finish of these wafers was not good enough for the precise measurements needed in these experiments. Results were obtained along ⟨100⟩ and ⟨110⟩ directions on (001) wafers and along $\langle 1\bar{1}0\rangle$ directions on (111) wafers.

Calculating the stress and strain in the top surface of the wafer due to the overpressure is an anisotropic elasticity problem. To simplify the problem, an approximate isotropic model was used that involved calculating effective Poisson's ratios for the three directions ⟨100⟩, ⟨110⟩, and $\langle 1\bar{1}0\rangle$ by using the thin shell method outlined in Ambartsumyan.[50] The effective Poisson's ratios calculated for the three directions are 0.279 along ⟨100⟩, 0.362 along ⟨110⟩, and 0.181 along $\langle 1\bar{1}0\rangle$. These were then used in the theory of flexural bending of isotropic plates

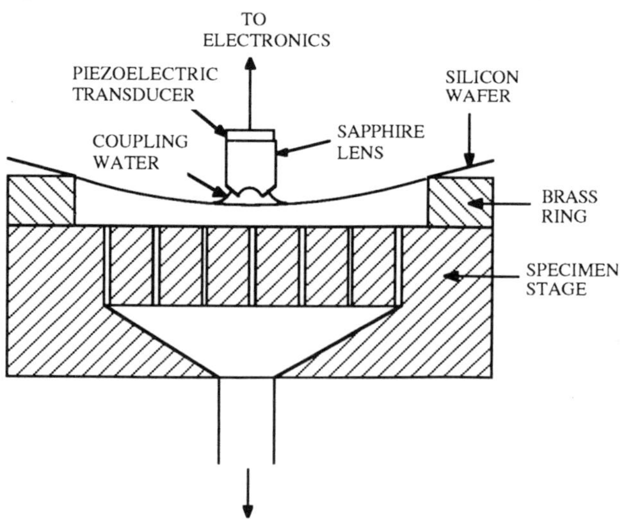

Figure 6.15. Schematic diagram of the lens and wafer.

given in Timoshenko and Goodier,[51] which gives the stress in the plate as a function of the Poisson's ratio and the pressure on the top surface as

$$(\sigma_r)_{r=0,\, z=-c} = -q\left(\frac{2+\sigma}{8} - \frac{3}{8}\frac{2+\sigma}{5} + \frac{3(3+\sigma)}{32}\frac{a^2}{c^2}\right) \tag{13}$$

where c = half-thickness of the wafer,
a = radius of the wafer,
σ = Poisson's ratio, and
q = overpressure on the top surface of the wafer.

At the center of the wafer, the tangential stress σ_ϕ and the radial stress σ_r are equal, and it can be shown by a transformation of axes that they are equivalent to the stress components σ_{11} and σ_{22} referred to a Cartesian coordinate system. The value of $(\sigma_r)_{r=0}$ at a depth of 20 μm, approximately equal to 1 SAW wavelength, is 96% of its surface value, and so it is assumed that the radial stress is constant with depth. For (001) samples, the top surface can be assumed under a biaxial compressive plane stress, since the stress perpendicular to the plane of the wafer is negligible in comparison. In the (111) case, orthogonal stresses are not identical, but since the difference is small, the stress field has been approximated to the equibiaxial case. The strain, calculated by Hooke's law, is negative in the plane of the wafer and positive perpendicular to this. To check the validity of our model, an experimental determination of the strain along ⟨110⟩ was made by gluing a strain gauge onto the surface of the wafer and recording the change in resistance as the pressure difference was changed. The strain gauge behaved linearly up to an overpressure of 33 kPa, corresponding to a stress in the wafer of around 400 MPa. Above this value, the small deformation theory used here breaks down, since the deflection of the center of the wafer becomes comparable with its thickness. As a result, all values above a pressure difference of 33 kPa were discarded. A straight line fit to the strain gauge electrical resistance with the overpressure in the linear region gave a mean gradient that agreed with the theoretical dependence of the strain on pressure difference to within 2%.

Experiments were also done to measure the leaky surface-skimming compressional wave (SSCW) velocity on perspex in an attempt to reproduce results obtained by Obata and others[52] using tensile test specimens under the LFB acoustic microscope. To account for the different stress situation used in the bending arrangement, the superposition principle proposed by Obata and others[52] was used, so that results from the two experiments were directly comparable. Results of one set of readings for perspex are shown in Fig. 6.16. To obtain the average gradient of SSCW velocity versus applied stress for many sets of data, a weighted mean of the gradients obtained by fitting data with a straight line was calculated by using the correlation factor of the fit as a weighting factor. This weighted mean and the standard weighted error are shown in Table 6.3.

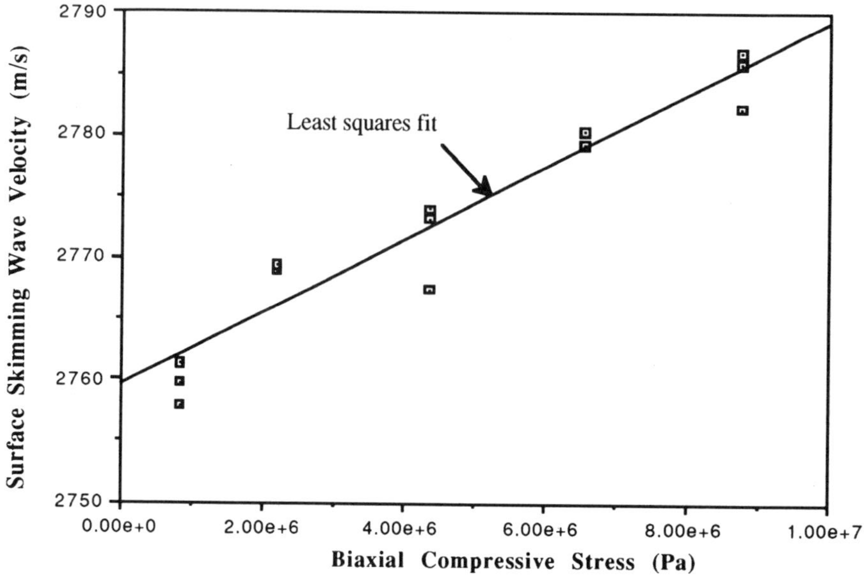

Figure 6.16. A typical set of results for perspex.

Obata and others' theoretical gradient[52] for the dependence of the leaky SSCW velocity on applied stress, calculated for the biaxial stress field used in these experiments, is also shown. The agreement is within the error for this experiment, and although no error is quoted by Obata for agreement between his theoretical results and his experimental findings, the agreement between theory and experiment appears to be good in his paper.[52]

For silicon wafers, sets of readings were taken along the symmetry directions, then straight line fits were performed to give the best fit gradient. Again the correlation factor for each set of results was used as a weighting factor, and a weighted mean for each direction was calculated. These are shown in Table 6.4 with the standard weighted errors. A set of results for a typical loading sequence is shown in Fig. 6.17, with three $V(z)$s taken at each pressure increment to show the amount of scatter at each pressure level.

Table 6.3. Dependence of SSC Wave Velocity on Stress for Perspex

Research	Gradient (m s^{-1}/MPa)
Obata and others	2.56
This experiment	2.42 ± 0.24

Table 6.4. Experimental and Fitted Dependences of Surface Wave Velocity on Stress for Silicon

Manually fitted constants	Weighted average of Gradients (m s^{-1}/MPa)	Standard weighted Error (m s^{-1}/MPa)	Fitted gradient (m s^{-1}/MPa)
(001)⟨100⟩	0.027	0.005	0.028
(001)⟨110⟩	0.030	0.003	0.029
(111)⟨1$\bar{1}$0⟩	0.037	0.008	0.030

A theoretical calculation of third-order elastic constants of Si was performed using the Keating model,[53,54] whereby the elastic constants of diamondlike crystals are calculated from a knowledge of the lattice parameter and two constant terms α and β, which are constants in the expansion of the strain energy in terms of microscopic strain. The microscopic strain is defined such that the energy of the crystal is invariant under rigid rotations. The parameters α and β are found by fitting to experimental data of second-order elastic constants in the linear approximation of Hooke's law. Good agreement for diamond, silicon, and germanium is found. In an extension of this theory, Keating[54] shows how three additional constants (γ, δ, and ε) can be used to describe the six third-order elastic constants. Fitting these three constants to experimental values of third-order elastic constants for silicon and germanium also gives reasonable results, despite the fact that the theory is strictly valid only at 0 K, since it deals only with first-

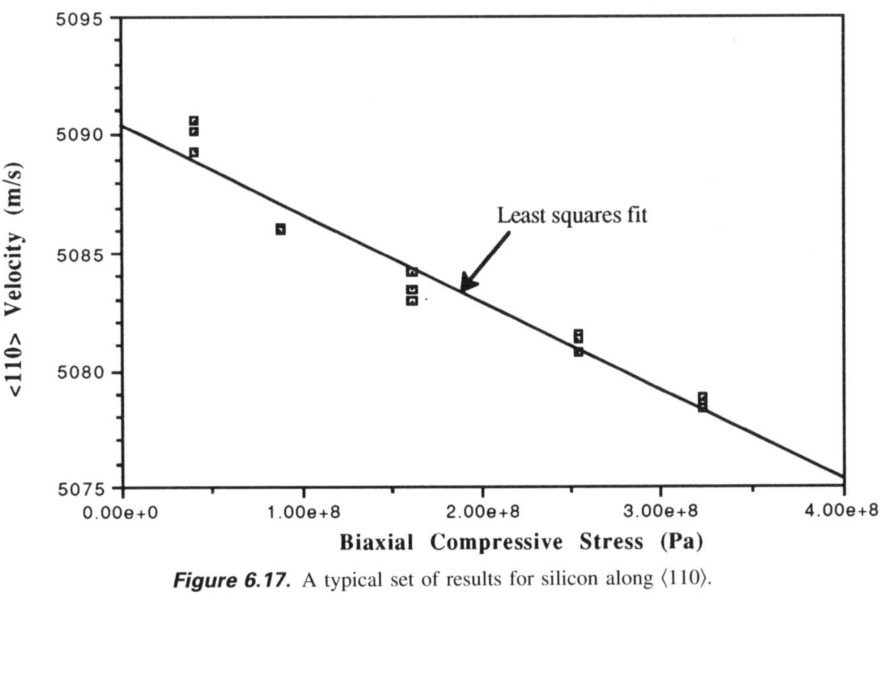

Figure 6.17. A typical set of results for silicon along ⟨110⟩.

and second-neighbor interactions. This method has also been used to separate linear combinations of third-order elastic constants obtained using the second harmonic generation and hence to obtain values for all six constants.[55]

In light of the agreement between fitted and experimental values of the third-order constants just described, it was decided to use Keating's method to fit only three parameters γ, δ, and ε to experimental values for the dependence of surface wave velocity on stress, and then to extract values for the six third-order constants from these three parameters via expressions given in Keating.[54] Values for second-order constants were obtained with a fixed density of 2329.08 kg m^{-3} by the fitting method described earlier, using unstressed velocities along the three symmetry directions. With this procedure, we obtained the following values: c_{11} = 164.0 GPa, c_{12} = 63.5 GPa, and c_{44} = 79.7 GPa. These data were then used in expressions for modified second-order constants in Eq. (9), and values of γ, δ, and ε were tried to give the best fit to experiment.

Results of SAW velocity calculations using manually fitted constants are shown in Table 6.4, the third-order elastic constants that were used are shown in Table 6.5. The means and standard deviations of the literature values of third-order elastic constants, calculated from data taken from Landolt-Bornstein,[56] are also shown for comparison.

Experimental errors in velocity gradients versus stress are significantly reduced by averaging velocity measurements to obtain a weighted mean and error for the gradients. Each gradient used in the weighted average was typically made up of 15 velocity measurements, and for each crystallographic direction, from 5–10 gradients were found and then averaged. Those data sets with a lot of scatter were given a low weighting factor to increase the accuracy of the process.

Table 6.5 shows that all third-order elastic constants lie within the range of the literature values except for c_{123}. The differences between the fitted constants and the literature values may be simply due to the fitting method: There may be a different combination of constants that gives as good a fit, and all six values lie within the literature range. However it may also be due to the effects of doping on third-order constants,[57] but the size of this effect has not yet been quantified.

Table 6.5. Fitted and Literature Values of Third-Order Elastic Constants of Silicon

Elastic constants	Fitted value (GPa)	Mean literature value (GPa)	Standard deviation of Literature value (GPa)
c_{111}	−638	−663	272
c_{112}	−382	−428	115
c_{123}	−154	−13	50
c_{144}	−17	+14	58
c_{166}	−246	−366	130
c_{456}	−28	−99	94

In conclusion the effect of stress and strain on the surface wave velocity of silicon has been measured subject to errors of 0.008 m s^{-1}/MPa or less along three crystallographic symmetry directions. Agreement between experimental findings and theoretical predictions can be found in all three directions studied. A possible set of third-order elastic constants has been found by using a lattice dynamic model to relate third-order constants to each other and fit these constants to experimental parameters. It is acknowledged that these third-order elastic constants represent only one set of constants that could fit the experimental data. It is possible that these values give unphysical values for the dependency of velocity on stress away from the three symmetry directions investigated in this experiment. To obtain a valid set of constants, at least six directions must be studied, and then a minimization routine must be run to achieve the best fit between fitted values and experimentally determined parameters.

6.5.4. Ion-Implanted GaAs

In this last example, we use LFB acoustic microscope measurements to extract longitudinal and shear elastic constants during the amorphization of ion-implanted GaAs samples. In this problem, we must determine the elastic constants of the residually stressed implanted region, given the thickness of this region and second- and third-order elastic constants of the unperturbed material.

Experimentally determining the softening of elastic constants during amorphization is of particular interest in view of the possibility that the crystal-to-amorphous (c-a) transition may be initiated by a mechanical instability.(58,59) In this regard, molecular dynamic simulations of the elastic properties of mono-atomic(60) and biatomic(61) lattices at the (c-a) transition indicated a different behavior of the shear and longitudinal elastic constants, i.e., c_{44} and c_{11} in cubic materials, and showed a greater softening of c_{44} at amorphization. Experimentally a sharp variation in elastic properties at the onset of the c-a transition was observed in several materials, such as semiconductors(62–64) and intermetallic compounds.(65) In all of these experimental studies, only variation in the shear elastic constant was measured, and no indication of change in the longitudinal modulus was given.

Two sets of GaAs wafers implanted with Si^{+} ions were studied. Samples in the first set were implanted at room temperature (labeled RT-samples) at a fixed dose of 10^{14} ions cm^{-2}, with ion energy from 0.5–4 MeV. The second set of samples was implanted at liquid nitrogen temperature (labeled LN-samples) to avoid annealing while varying the incident ion dose (5×10^{12}, 5×10^{13}, 5×10^{14}, and 5×10^{15} ions cm^{-2}) at a fixed energy of 1.5 MeV. Pieces were cleaved from the same wafer and chemically polished to eliminate residual processing damage. A reference sample was left as polished (see sample REF). A summary of implantation conditions is presented in Table 6.6.

Table 6.6. Summary Table of Si^+ Implanted GaAs Samples

Sample	Implantation temperature	Ion energy (MeV)	Ion dose (ions cm^{-2})
REF	as polished	//	//
RT1	RT	0.5	1×10^{14}
RT2	RT	1.5	1×10^{14}
RT3	RT	4.0	1×10^{14}
LN1	LN	1.5	5×10^{12}
LN2	LN	1.5	5×10^{13}
LN3	LN	1.5	5×10^{14}
LN4	LN	1.5	5×10^{15}

The presence of amorphous material was determined by Rutherford backscattering (RBS) and channeling with both a 1.5-MeV He^+ and 1-MeV H^+ beam. Almost no damage was observed for the RT samples and the low-dose LN implant (5×10^{12} ions cm^{-2}). The 5×10^{13} ions cm^{-2} LN implant causes a roughly Gaussian damage profile with depth, with a FWHM of about 1 μm. The 5×10^{14} ions cm^{-2} LN implant gave a fully amorphous layer about 1.3 μm thick buried about 0.3 μm below the surface. The crystalline near-surface layer for this implant is of good quality at the surface, but it has a progressively higher number of defects nearer the buried layer. The high-dose LN implant (5×10^{15} ions cm^{-2}) gave a fully amorphized surface layer about 1.85 μm thick. Double-crystal X-ray rocking curves of the (004) Bragg reflection for the set of RT implants were recorded to determine the depth distribution of the residual strain left after implantation (see Fig. 6.18).

To understand the influence of ion implantation damage on the elastic properties of the crystal, we can consider a simple representation of the solid where lattice atoms are interconnected by springs, and ion implantation damage can be viewed as the rupture or variation of a certain number of springs. Elastic constants represent the average of the strength of these springs over all lattice atoms, and therefore these vary with the concentration of defects. In addition the

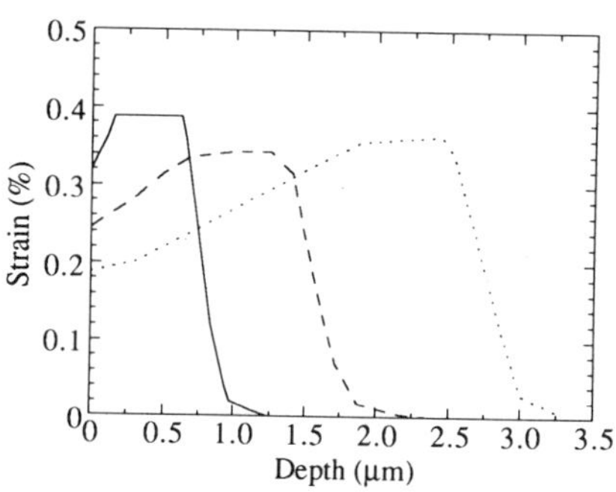

Figure 6.18. Perpendicular strain depth profile measured by double crystal diffractometry in RT ion-implanted GaAs(001) samples. Solid line, 0.5 MeV; dashed line, 1.5 MeV; dotted line, 4 MeV.

volume expansion created in the ion-implanted material by the accumulation of lattice damage sets up lateral stresses in the implanted layer. The nature and the amount of stress can be calculated by modeling the ion-implanted sample as a strained layer on top of an unperturbed substrate. With good approximation, we can assume that the depth of the implanted layer is much smaller than the lateral dimensions and thickness of the sample. We therefore neglect any functional dependence of the strain and stresses on lateral coordinates. Disregarding the negligible bending of our specimens, we assume that the volume expansion $\Delta V/V$ induced by the accumulation of lattice damage initially generates an isotropic dilatation in the layer, and then we treat the problem in an analogous manner to a thermoelastic phenomenon.[66] Equilibrium conditions in the layer written using the Voigt notation are $\partial\sigma_3/\partial x_3 = 0$, where the x_3-axis is taken normal to the surface. Combining with this the boundary condition for a stress-free surface (i.e., $\sigma_3 = 0$ at $x_3 = 0$) implies that $\sigma_3 = 0$ everywhere in the layer. The constraint imposed by the unperturbed substrate on the lateral expansion of the implanted layer is such that $\varepsilon_1 = \varepsilon_2 = 0$ everywhere in the layer. From these relationships, we find the perpendicular strain ε_3

$$\varepsilon_3 = \frac{1}{3}\frac{\Delta V}{V}\frac{2c_{12} + c_{11}}{c_{11}} \tag{14}$$

and lateral stresses σ_1 and σ_2

$$\sigma_1 = \sigma_2 = -\varepsilon_3 \frac{c_{11}^2 + c_{11}c_{12} - 2c_{12}^2}{c_{11} + c_{12}} \tag{15}$$

Measurements were made as a function of azimuthal angle at a fixed frequency of 225 MHz; these measurements are shown as the discrete points in Figs. 6.19 and 6.20. Fitting was restricted to points within 10–15° of the symmetry

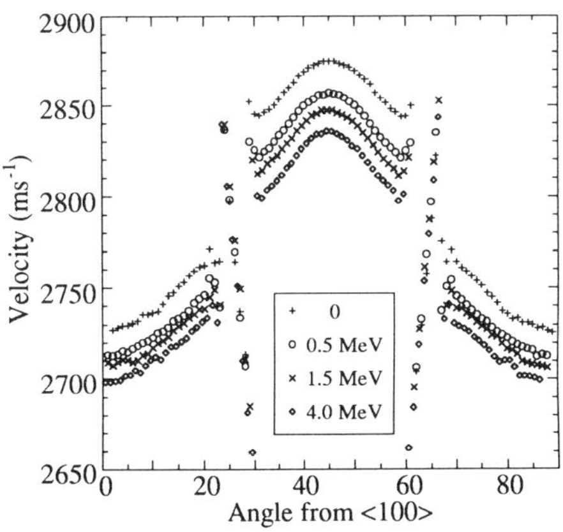

Figure 6.19. Dependence of SAW velocity with propagation direction as measured by LFB acoustic microscopy for the RT-implanted GaAs(001) wafers.

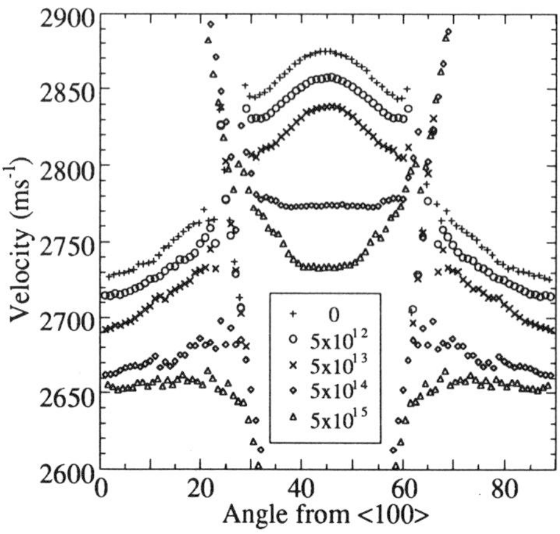

Figure 6.20. Dependence of SAW velocity with propagation direction as measured by LFB acoustic microscopy for the LN-implanted GaAs(001) wafers.

directions, since significant distortion is introduced into the analysis when the SAW and pseudo-SAW branches are both significantly excited. A fixed offset of 4.9 m s^{-1} was used to take account of fluid loading, since this value is almost constant over the range of angles considered. With fitting restricted to 10–15° degrees of ⟨110⟩ type directions, the velocity of the unattenuated local minimum in the boundary determinant in Eq. (12) lies very close to that of the true pseudo-SAW. These two simplifications help reduce computation time by avoiding the use of two-dimensional minimization as required for the exact calculation of attenuating SAWs.

The fitting procedure begins with an initial guess at the three elastic constants for the cubic implanted material. The measured perpendicular strain is then used to calculate modified constants using Eq. (9). The actual tetragonal nature expected of the strained material arises naturally in these modified constants. The requirement for the third-order elastic constants in Eq. (9) presents difficulties, since these may well be completely unknown in general. The wide variability in third-order elastic constants of silicon was discussed in Section 6.5.3. For GaAs implanted at low doses, implantation produces a damaged but still crystalline layer, and third-order constants were kept fixed at the literature values for unimplanted GaAs.[67] Compressive planar stress is then calculated by Eq. (15). The justification for using Hooke's law to determine the static stress was presented in Section 6.2.2. The planar stress is entered directly into the equation of motion, Eq. (4), with the new density given by Eq. (10). The SAW velocities are then calculated as a function of azimuthal angle in the (001) plane and the sum of squares of residuals is minimized by improving the guess at the unmodified elastic constants.

Since the implanted region has elastic constants similar to the substrate's, and the implanted region is thin compared to the SAW wavelength (1.6 μm compared to 12 μm), the layer causes a relatively small perturbation in the SAW velocity of the substrate. Inverting the layer's elastic constants is therefore very sensitive to values used for the substrate. The analysis procedure can also introduce distortion into the shape of angular scans.[23] Hence rather than use literature values for substrate second-order stiffnesses, elastic constants were fitted to the rotational scan for unimplanted GaAs, and these values were used in all subsequent fitting for implanted specimens.

Large values of diagonal elements of the covariance matrix (see Section 6.3.2.) indicate a shallow minimum in the sum of squares of residuals and hence insensitivity of the fit to one or more parameters. The variance in c_{12} was found to be an order of magnitude or more higher than for c_{11} or c_{44}, which indicates that the SAW velocity in the (001) plane is relatively insensitive to c_{12}. The fitting routine can therefore easily distort this constant without altering the residual sum by a great deal. We therefore present the anisotropy ratio $A = 2c_{44}/(c_{11}c_{12})$ instead, since this is subject to a smaller error.

The results of applying the fitting procedure to experimental SAW phase velocity dispersion curves are taken for the set of samples implanted at different doses as in Fig. 6.21. A summary of results for the specimens implanted at constant dose is shown in Table 6.7. Values obtained for c_{11} and c_{44} are constant

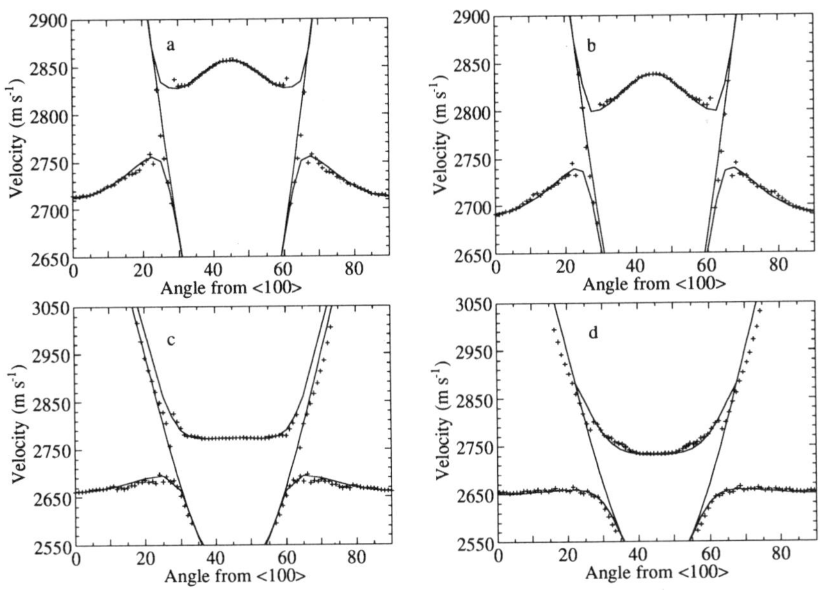

Figure 6.21. Theoretical curves fitted to LFB acoustic microscope rotational scans for GaAs(001) substrates implanted with increasing doses of Si^{+} ions; (a) dose = 5×10^{12}, (b) dose = 5×10^{13}, (c) dose = 5×10^{14}, (d) dose = 5×10^{15}.

Table 6.7. Values of c_{11} and c_{44} and Anisotropy Ratio A Obtained by Fitting the Two Sets of Ion-Implanted Samples

Sample	c_{11} (GPa)	c_{44} (SPa)	A
RT1	116.6 (95.9%)	52.9 (88.3%)	1.78
RT2	118.0 (97.0%)	52.5 (87.7%)	1.73
RT3	119.8 (98.5%)	53.9 (90.0%)	1.72
LN1	119.1 (97.9%)	55.6 (92.8%)	1.80
LN2	113.4 (93.2%)	51.8 (87.2%)	1.84
LN3	107.8 (88.6%)	35.9 (60.0%)	1.42
LN4	101.0 (83.0%)	35.4 (59.0%)	1.01

to within the errors associated with fitting, representing an average decrease of 3.5% and 12.5%, respectively. The difference in measured velocities in Fig. 6.19 is thus almost entirely a thickness effect, i.e., implantation at a fixed dose creates a layer with elastic constants approximately independent of energy. The sample implanted at 4 MeV does not quite fit this trend: If elastic constants fitted for other materials are used to calculate velocities for this sample, these are about 15 m s^{-1} lower than those actually measured. The measured velocity occurs for a thickness of 2.1 μm rather than the value of 2.7 μm actually used. This is consistent with the buried nature of the layer at 4 MeV, as seen in the strain profile in Fig. 6.18. Using the value of 2.7 μm for fitting is an overestimation, which therefore resulted in an underestimated decrease in elastic constants.

The softening of c_{11} and c_{44} found with an increasing ion dose is represented in Fig. 6.22, where estimated elastic constants of the implanted material $(c_{ij})'$ normalized to the elastic constants of the as-polished sample $(c_{ij})^{o}$ are shown with the anisotropy factor $A=2c_{44}/(c_{11}-c_{12})$. In the two lower dose implants (5×10^{12} ions cm^{-2} and 5×10^{13} ions cm^{-2}), the noticeable feature is a progressively smooth softening of c_{11} and c_{44}. For these two implants, the anisotropy factor has approximately the value of the unperturbed GaAs substrate. In the implant at 5×10^{14} ions cm^{-2}, a drop in c_{44} appears with a corresponding decrease in the anisotropy ratio; the change in c_{11} is smaller. The highest dose implant (5×10^{15} ions cm^{-2}) exhibits a further decrease in c_{11}, and it is characterized by an anisotropy ratio approaching unity, indicating that the implanted region has become isotropic. The difference in the anisotropy ratio between the 5×10^{14} ions cm^{-2} and 5×10^{15} ions cm^{-2} implants is due to the approximation of the implanted region as a single layer, whereas there is a damaged but still crystalline surface layer for the 5×10^{14} ions cm^{-2} implant, as shown by RBS.

As observed by other authors,[(65)] the decrease in elastic constants for amorphization of GaAs is larger than for Si. This is in agreement with molecular dynamics (MD) simulations[(60,61)] of a defect-induced amorphization, and it may be attributed

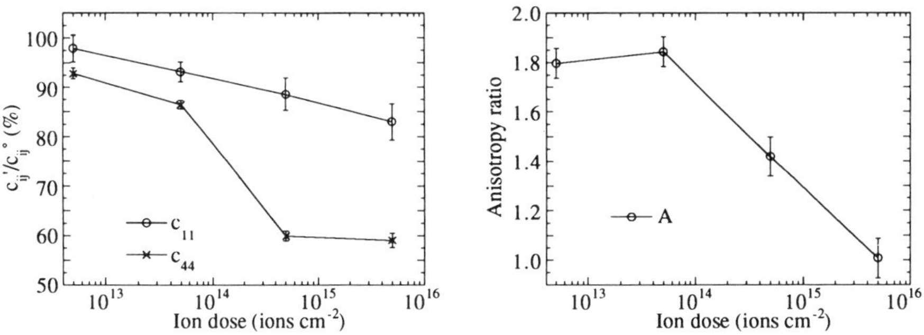

Figure 6.22. Elastic properties of the liquid nitrogen-implanted specimens as a function of the ion dose; (left) extracted elastic constants of the implanted layer $(c_{ij})'$ normalized to the fitted elastic constants of the reference sample $(c_{ij})^{o}$ (right) anisotropy ratio A of the implanted layer.

to chemical disorder generated by wrong bonds between like atoms in the amorphous structure. The decrease at amorphization in c_{44} is larger than that in c_{11}, which is also in agreement with MD studies.

6.6. Conclusions

Using the methods described in Chapters 5 and 6, it is possible to deduce the elastic properties of materials from measurements by quantitative acoustic microscopy. In Chapter 6, acoustoelasticity theory is used to include the effect of static stresses in the forward calculation of SAW velocities. It is then possible to invert the elastic constants of the natural, unstressed state of a residually stressed material. If the unstressed material is available in its own right, information about third-order elastic constants or alternatively the static stress can be obtained.

In any multiparameter problem, if you start with zero knowledge you are liable to end with infinite ignorance. A major motivation for this work was therefore to obtain a quantitative assessment of the sensitivity of the inversion process for each of the fitted parameters. In the first example given, the density and crystal symmetries of fluoroapatite and hydroxyapatite are assumed, so it is possible to invert the value of only a single elastic constant, the shear modulus, with any confidence. In the second example, the elastic properties of stress-free silicon are known, so it is possible to present a consistent (though not unique) set of nonlinear elastic parameters. In the third case, the change in elastic constants of amorphous hydrogenated carbon coatings is measured as a function of the processing parameters. These examples illustrated the sensitivity of the method to three different properties—anisotropic elastic constants, applied stress, and a surface layer. The final example employs elements of all the examples to evaluate elastic constants of the natural state of a residually stressed layer produced by ion implantation in GaAs.

Acknowledgments

We wish to thank Dr. C. B. Scruby for invaluable discussions; Professor D. K. Bowen and Dr. J. Bradler for X-ray measurements; Dr. C. Jeynes for preparing the ion-implanted GaAs samples and RBS measurements; Dr. R. Falster of MEMC for providing silicon wafers; Dr. R. E. W. Jansson, Dr. K. Price, Dr. F. S. Delk, Dr. B. K. Daniels, and Dr. R. H. Petrmichl of Monsanto Chemical Company, St. Louis, Missouri, and Monsanto Europe, Belgium, for all their help and support; Professor J. Kushibiki for establishing the line-focus system at Oxford and for providing the lenses; the Science and Engineering Research Council, Monsanto Chemical Company, Logitech Ltd. and British Nuclear Fuels plc for funding.

References

1. Thompson, R. B., Smith, J. F., Lee, S. S. (1983). In: *Review of Progress in Quantitative Nondestructive Testing,* vol. 2B (D. O. Thompson and D. E. Chimenti, eds.), p. 1339. Plenum Press, New York.
2. King, R. B. and Fortunko, C. M. (1983). Determination of in-plane residual stress states in plates using horizontally polarised shear waves. *J. Appl. Phys.* **54**, 3027–35.
3. Kino, G. S., Hunter, J. B., Johnson, G. C., Selfridge, A. R., Barnett, D. M., Hermann, G., Steele, C. R. (1979). Acoustoelastic imaging of stress fields. *J. Appl. Phys.* **50**(4), 2607–13.
4. Dike, J. J. and Johnson, G. C. (1990). Residual stress determination using acoustoelasticity. *J. Appl. Mech.* **57**, 12–7.
5. Drescher–Krasicka, E. (1993). Scanning acoustic imaging of stress in the interior of solid materials. *J. Acoust. Soc. Am.* **94**, 453–64.
6. Man, C.-S. and Lu, W.-Y. (1987). Towards an acoustoelastic theory for measurement of residual stress. *J. Elast.* **17**, 159–82.
7. Man, C.-S., Lu, W.-Y, Gu, Q., Tang, W. L. (1992). Ultrasonic measurements of stress in weakly anisotropic thin sheets. *J. Acoust. Soc. Am.* **91**, 2643–53.
8. Johnson, G. C. (1981). Acoustoelastic theory for elastic-plastic materials. *J. Acoust. Soc. Am.* **70**, 591–95.
9. Hirao, M., Fukuoka, H., Hori, K. (1981). Acoustoelastic effect of Rayleigh surface wave in isotropic material. *J. Appl. Mech.* **48**, 119–23.
10. Briggs, G. A. D. (1992). *Acoustic Microscopy,* Oxford University Press, Oxford, UK.
11. Spencer, A. J. M. (1980). *Continuum Mechanics,* Longman, London.
12. Thurston, R. N. (1964). In: *Physical Acoustics,* vol. IA (W. P. Mason, ed.). Academic Press, London, pp. 1–110.
13. Brugger, K. (1964). Thermodynamic definition of higher order elastic coefficients. *Phys. Rev.* **133A**, 1611–12.

14. Truesdell, C. (1961). General and exact theory of waves in finite elastic strain. *Arch. Ration. Mech. Anal.* **8**, 263–304.
15. Breazeale, M. A. (1972). Ultrasonic studies of the nonlinear properties of solids. *Int. J. Nondestr. Test* **4**, 149–66.
16. Breazeale, M. A. (1991). In: *Reviews of Progress in Quantitative Nondestructive Evaluation,* vol. 10B (D. O. Thompson and D. E. Chimenti, eds.), pp. 1797–1803. Plenum Press, New York.
17. Thurston, R. N. (1965). Effective elastic constants for wave propagation in crystals under stress. *J. Acoust. Soc. Am.* **37**, 348–56.
18. Pao, Y. H., Sasche, W., Fukukoa, H. (1984). In: *Physical Acoustics,* vol. XVII (W. P. Mason and R. N. Thurston, eds.), pp. 61–143. Academic Press, New York.
19. Hayes, M. and Rivlin, R. S. (1961). Surface waves in deformed elastic materials. *Arch. Ration. Mech. Anal.* **8**, 358–80.
20. Iwashimizu, Y. and Kobori, O. (1978). The Rayleigh wave in a finitely deformed isotropic elastic material. *J. Acoust. Soc. Am.* **64**, 910–16.
21. Nalamwar, A. L. and Epstein, M. (1976). Surface acoustic waves in strained media. *J. Appl. Phys.* **47**, 43–48.
22. Mase, G. T. and Johnson, G. C. (1987). An acoustoelastic theory for surface waves in anisotropic media. *J. Appl. Mech.* **54**, 127–35.
23. Sklar, Z. (1993). Quantitative acoustic microscopy of coated materials. D. Phil. thesis, Oxford University.
24. Lim, T. C. and Farnell, G. W. (1969). Character of pseudosurface waves on anisotropic crystals. *J. Acoust. Soc. Am.* **45**, 845–51.
25. Adler, E. L. and Sun, I. H. (1971). Observation of leaky Rayleigh waves on a layered half-space. *IEEE Trans.* **SU-16**, 181–84.
26. Viktorov, I. A., Grishchenko, E. K., Kaekina, T. M. (1963). An investigation of the propagation of ultrasonic surface waves at the boundary between a solid and a liquid. *Sov. Phys. Acoust.* **9**, 131–37.
27. Campell, J. J. and Jones, W. R. (1970). Propagation of surface waves at the boundary between a piezoelectric crystal and a fluid medium. *IEEE Trans.* **SU-17**, 71–76.
28. Farnell, G. W. and Adler, E. L. (1972). In: *Physical Acoustics,* vol. IX (W. P. Mason and R. N. Thurston, eds.), Academic Press, New York, pp. 35–127.
29. Kushibiki, J. and Chubachi, N. (1985). Material characterisation by line-focus beam acoustic microscope. *IEEE Trans.* **SU-32**, 189–212.
30. Kushibiki, J., Takahashi, H., Kobayashi, T., Chubachi, N. (1991). Characterization of lithium niobate crystals by line focus beam acoustic microscopy. *Appl. Phys. Lett.* **58**, 2622–25.
31. Nuttall, A. H. (1981). Some windows with very good sidelobe behaviour. *IEEE Trans.* **ASSP-29**, 84–94.
32. Mutti, P. (1993). Surface acoustic waves for semiconductor characterization. D. Phil. thesis, Oxford University.
33. Smith, I. R., Sinclair, D. A., Wickramasinghe, H. K. (1980). Acoustic microscopy of elastic constants. *IEEE 1980 Ultrasonics Symposium,* IEEE, New York, pp. 677–82.
34. Liang, K. K., Kino, G. S., Khuri–Yakub, B. T. (1985). Material characterisation by the inversion of $V(z)$. *IEEE Trans.* **SU-32**, 213–24.
35. Yu, Z. and Boseck, S. (1991). Inversion of $V(z)$ data in the scanning acoustic microscope to determine material properties of a layered solid. *Optik* **88**, 73–79.
36. Atalar, A. (1979). A physical model for acoustic signatures. *J. Appl. Phys.* **50**, 8237–39.
37. Biebl, E. M., Russer, P. H., Anemogiannis, K. (1989). SAW propagation on protonexchanged lithium niobate. *IEEE 1989 Ultrasonics Symposium,* IEEE, New York, pp. 281–84.
38. Behrend, O. (1991). Mesure des constantes elastiques de couches de surface par ondes de Lamb. Diplôme d'ingenieur-physicien, Ecole Polytechnique federale de Lausanne.

39. Kim, J. O. and Achenbach, J. D. (1992). Line-focus acoustic microscopy to measure anisotropic acoustic properties of thin films. *Thin Sol. Fi.* **214**, 25–34.
40. Kim, J. O., Achenbach, J. D., Mirkarimi, P. B., Shinn, M., Barnett, S. A. (1992). Elastic constants of single-crystal transition metal nitride films measured by line-focus acoustic microscopy. *J. Appl. Phys.* **72**, 1805–11.
41. Bard, Y. (1974). *Nonlinear parameter estimation,* Academic Press, London.
42. Press, W. H., Flannery, B. P., Teukolsky, S. A., Vetterling, W. T. (1986). *Numerical Recipes,* Cambridge University Press, Cambridge.
43. Gardner, T. N., Elliott, J. C., Sklar, Z., Briggs, G. A. D. (1992). Acoustic microscope study of the elastic properties of fluoroapatite and hydroxyapatite, tooth enamel and bone. *J. Biomech.* **25**, 1265–77.
44. Lettington, A. H. (1993). Applications of diamondlike carbon thin films. *Philos. Trans. Roy. Soc. London* **A342**, 287–96.
45. Bubenzer, A., Dischler, B., Brandt, G., Koidl, P. (1983). Rf-plasma-deposited amorphous hydrogenated hard-carbon thin films: preparation, properties, and applications. *J. Appl. Phys.* **54**, 4590–95.
46. Banks, B. A. and Rutledge, S. K. (1982). Ion beam sputter-deposited diamondlike films. *J. Vac. Sci. Technol.* **21**, 807–14.
47. Grill, A. Patel, V., Meyerson, B. S. (1990). Optical and tribological properties of heat-treated diamondlike carbon. *J. Mater. Res.* **5**, 2531–37.
48. Catherine, Y. (1991). In: *Diamond and Diamondlike Films and Coatings* (R. E. Clausing, L. L. Horton, J. C. Angus, and P. Koidl, eds.), 193–227. Plenum Press, New York.
49. Sklar, Z. and Briggs, G. A. D. (1994). Elastic constants of amorphous hydrogenated carbon coatings measured by line focus acoustic microscopy. Submitted to *Thin Sol. Fi.*
50. Ambartsumyan, S. A. (1964). *Theory of anisotropic shells.* NASA Technical Translation F-118, Washington, D.C.
51. Timoshenko, S. P. and Goodier, J. N. (1970). *Theory of Elasticity,* McGraw-Hill, New York.
52. Obata, M., Shimada, H., Mihara, T. (1990). Stress dependence of leaky surface wave on PMMA by line-focus beam acoustic microscope. *Exp. Mech.* **30**, 34–39.
53. Keating, P. N. (1966a). Effect of invariance requirements on the elastic strain energy of crystals with application to the diamond structure. *Phys. Rev. B* **145**, 637–45.
54. Keating, P. N. (1966b). Theory of the third-order elastic constants of diamondlike crystals. *Phys. Rev. B* **149**, 674–78.
55. Philip, J. and Breazeale M. A. (1983). Third-order elastic constants and Gruneisen parameters of silicon and germanium between 3 and 300K. *J. Appl. Phys.* **54**, 752–57.
56. Landolt, H. and Bornstein, R. (1983). *Numerical Data and Functional Relationships in Science and Technology,* 17a, 22a (Series III), Springer, Berlin.
57. Hall, J. J. (1967). Electronic effects in the elastic constants of n-type Silicon. *Phys. Rev.* **161**, 756–61.
58. Koike, J. (1993). Elastic instability of crystals caused by static atom displacements—a mechanism for solid-state amorphization. *Phys. Rev. B* **47**, 7700–04.
59. Massorbio, C. and Pontikis, V. (1992). Percolation model for elastic softening in intermetallic compounds during solid-state amorphization. *Phys. Rev. B* **45**, 2484–87.
60. Hsieh, H. and Yip, S. (1987). Defect-induced crystal to amorphous transition in an atomistic simulation model. *Phys. Rev. Lett.* **59**, 2760–63.
61. Hsieh, H. and Yip, S. (1989). Atomistic simulation of defect-induced amorphization of binary lattices. *Phys. Rev. B* **39**, 7476–91.
62. Burnett, P. J. and Briggs, G. A. D. (1986). The elastic properties of ion-implanted silicon. *J. Mat. Sci.* **21**, 1828–36.
63. Bhadra, R., Pearson, J., Okamoto, P., Rehn, L. E., Grimsditch, M. (1988). Elastic properties of silicon during amorphization. *Phys. Rev. B* **38**, 12656–59.

64. Sharma, R. P., Bhadra, R., Rehn, L. E., Baldo, P. M., Grimsditch, M. (1989). Crystalline to amorphous transformation in GaAs during Kr ion bombardment—a study of elastic behaviour. *J. Appl. Phys.* **66**, 152–55.
65. Rehn, L. E., Okamoto, P., Pearson, J., Bhadra, R., Grimsditch, M. (1987). Solid-state amorphization of Zr_3Al—evidence of an elastic instability and first-order phase transformation. *Phys. Rev. Lett.* **26**, 2987–90.
66. Nowacki, W. (1962). *Thermoelasticity.* Pergamon Press, Oxford, UK.
67. Lardner, R. W. and Tupholme, G. E. (1986). Nonlinear surface waves on cubic materials. *J. Elast.* **16**, 251–65.

7

Surface Brillouin Scattering—Extending Surface Wave Measurements to 20 GHz

P. Mutti, C. E. Bottani, G. Ghislotti, M. Beghi, G. A. D. Briggs, and J. R. Sandercock

7.1. Introduction

Brillouin light scattering is generally referred to as the inelastic scattering of an incident optical wave field by thermally excited elastic waves (elastic waves of thermal origin are usually called *acoustic phonons*) in a sample. This subject was first investigated early in the century by Brillouin[1] and Mandelshtam[2] in the case of scattering from transparent materials. Since the advent of the laser as a powerful source of monochromatic light, Brillouin scattering has received considerable interest for characterizing elastic and optoelastic bulk properties of materials.[3,4] More recently with the introduction of high-contrast spectrometers,[5] scattering from opaque materials can be studied, thereby permitting considerable advances in the study of surface acoustic waves in solids. In the last decade,

P. Mutti, G. A. D. Briggs • Department of Materials, University of Oxford, Oxford, England; *present address of P.M.:* Dipartimento di Ignegneria Nucleare, Politecnico di Milano, Milano, Italy. • G. Ghislotti • Dipartimento di Ignegneria Nucleare, Politecnico di Milano, Milano, Italy. C. E. Bottani, M. Beghi • Dipartimento di Ingegneria Nucleare, Politecnico di Milano, Milano, Italy and Consorzio Interuniversitario Nazionale per la Fisica della Materia, Unitá di Ricerca Milano Politecnico, Milano, Italy. and J. R. Sandercock • JRS, Zurich, Switzerland.

Advances in Acoustic Microscopy, Volume 1, edited by Andrew Briggs.
Plenum Press, New York, 1995.

Brillouin scattering from surfaces, more often called *surface Brillouin scattering* (SBS), has been widely used to investigate elastic properties of thin films, interfaces, and layered materials.

Chapter 7 focuses on SBS as a tool for studying the propagation of *surface acoustic waves* (SAW) in solid materials.

Section 7.2 discusses the kinematics of inelastic scattering of light and the basic equations used in the theory of Brillouin scattering, which are taken from bulk acoustic phonons (see Section 7.3.1). The presence of surfaces and the basic interactions in scattering from opaque materials are briefly described to clarify the nature of SAWs that can be investigated by using SBS (see Section 7.3.2). The instrumentation currently used for SBS is presented in Section 7.4. A major emphasis is given to assessing the accuracy and reproducibility of SAW velocity and attenuation measurements. The performance of quantitative acoustic microscopy and SBS are also compared. Section 7.5 reviews various experimental observations of SAWs obtained by SBS in semi-infinite media and layered structures. In Section 7.5.1 more fundamental results are considered to illustrate the basic characteristics of surface acoustic waves detected using Brillouin light scattering. Section 7.5.2 presents applications to materials science where SAWs are used to obtain relevant information about solid materials.

It is worth mentioning that the theory presented in the following section is probably unfamiliar to the acoustic microscopist. Some general aspects must therefore be kept in mind when approaching the subject of Brillouin scattering. The main difference of the latter technique from more traditional acoustic techniques is the nature of the acoustic modes investigated. Whereas in traditional techniques used in physical acoustics, elastic waves are excited by a transducer at a fixed frequency, in a Brillouin scattering experiment, the whole population of thermal acoustic modes which are allowed by the scattering configuration and sample geometry is probed. This fact enables us to study acoustic waves at much higher frequencies, since there is no need to excite a particular monochromatic acoustic mode. The difficulty of most acoustic microscopes used to study SAW having frequencies from 1–2 GHz, due to the large attenuation of high-frequency sound waves in water, is no longer a problem for Brillouin scattering. It is clear that surface acoustic phonons from 30–40 GHz can be routinely studied by this technique. Another useful feature of Brillouin scattering is that the phonon density of states can be studied. This information, which cannot be obtained with transducer-generated acoustic modes, sheds a new light, as it will appear later, on the physics of *pseudosurface waves*. The spectral density of phonon states can be also used to extract material parameters. At thermal equilibrium, the energy of acoustic phonons is very weak, and scattered light over a finite solid angle must be collected to produce a measurable Brillouin signal. This causes an indeterminacy in the wave vector resolution with a consequent lack of resolution for studying the propagation parameters of SAW when using this technique. This is probably the major limitation of SBS.

7.2. Kinematics of Brillouin Scattering

The kinematics of Brillouin scattering can give quite a direct and intuitive picture of the scattering process between photons and phonons. For the sake of simplicity, we assume that the incident wave is already traveling inside an isotropic medium, and we discard all problems connected with refraction and reflection of electromagnetic waves at the boundary.

Let us assume that a single thermal elastic wave (an acoustic phonon) is propagating within the medium with wavevector $\mathbf{q}$ and frequency $\omega(\mathbf{q})$. This sound wave sets up a modulation in the dielectric constant ε, which is viewed as a moving diffraction grating by an incident light wave. Then Brillouin scattering can be explained by the familiar concepts of *Bragg reflection* and *Doppler shift.* This is equivalent to saying that, in a particle description, the laws of conservation of total momentum and energy must be obeyed in the scattering process.

The first law is a vector equation, and it corresponds to conservation of total wave vector in the scattering process

$$\mathbf{k}_s - \mathbf{k}_i = \pm\mathbf{q} \tag{1}$$

where $\mathbf{k}_i$ and $\mathbf{k}_s$ are wave vectors of the incident photon and scattered photon. The second law is a scalar equation that reads

$$\omega_s - \omega_i = \pm\omega(\mathbf{q}) \tag{2}$$

In Eqs. (1) and (2), the plus sign refers to *anti-Stokes* events when a phonon is annihilated in the process and the scattered photon is more energetic, while the minus sign corresponds to *Stokes* events when a new phonon is created in the process and the scattered photon has lost energy.

The photon wave vector, on the order of 10^5 cm^{-1}, is much smaller than the dimension of the Brillouin zone ($\approx 10^8$ cm^{-1}), and information is provided only about phonons near the center of the Brillouin zone ($\mathbf{q} \approx 0$). In this region, the dispersion relation is linear for acoustic phonons, and $\omega(\mathbf{q}) = v|\mathbf{q}|$, where v is the velocity of sound inside the material. In a classical picture, Eq. (2) represents the Doppler shift of an electromagnetic wave diffracted by an elastic wave moving either in the direction of vector $\mathbf{q}$ or in the direction of vector $-\mathbf{q}$.

Brillouin scattering can be viewed as the Bragg reflection of the incident wave by the moving diffraction grating generated by the elastic wave that modulates the index of refraction. From Eq. (2), with the aid of the dispersion relations for photons $\omega_i = (c/n)\mathbf{k}_i$ and $\omega_s = (c/n)\mathbf{k}_s$ (where c is the velocity of light in vacuum and n is the index of refraction), we have

$$\frac{|\mathbf{k}_s| - |\mathbf{k}_i|}{n|\mathbf{q}|} = \frac{v}{c} \tag{3}$$

The order of magnitude of the ratio v/c is typically 10^{-5} so that we can take $|\mathbf{k}_s| \approx |\mathbf{k}_i| = k$; we write therefore

$$2k\sin(\phi/2) = |\mathbf{q}| \tag{4}$$

where ϕ is the scattering angle [see Fig. 7.1(a)].

This equation can be given the form of Bragg law

$$2d\ \sin(\psi) = \lambda \tag{5}$$

where $\psi = \phi/2$ is the Bragg angle. The grating spacing d is equal to the phonon wave length

$$\frac{2\pi}{|\mathbf{q}|} = 2\pi v/|\omega_s - \omega_i|$$

and $\lambda = \lambda_o/n$ is the wavelength of the incident photon in the medium (λ_o is the corresponding quantity in a vacuum). The measurement of the Brillouin shift $\Delta\omega = |\omega_s - \omega_i|$ in a fixed scattering geometry leads to the possibility of measuring the sound velocity by the following formula:

$$v = \frac{\lambda_o|\Delta\omega|}{4\pi n\sin(\phi/2)} \tag{6}$$

In backscattering $\phi = \pi$, and Eq. (6) yields

$$v = \frac{\lambda_o|\Delta\omega|}{4\pi n} \tag{7}$$

Surface acoustic phonons are completely characterized by their wave vector $\mathbf{q}_x$ parallel to the surface [see Fig. 7.1(b)]. The wave vector conservation law for surface Brillouin scattering therefore becomes

$$\mathbf{k}_{s_x} - \mathbf{k}_{i_x} = \pm\mathbf{q}_x \tag{8}$$

where $\mathbf{k}_{i_x}$ and $\mathbf{k}_{s_x}$ are the projection of the incident and scattered photon wave vectors parallel to the surface, respectively. For backscattering from SAWs, the SAW phase velocity v_{SAW} can be obtained by

$$v_{SAW} = \frac{\lambda_o \sin\theta_i|\Delta\omega|}{4\pi} \tag{9}$$

where θ_i is the angle between the incident photon wave vector and the normal to the surface [the angle of incidence; see Fig. 7.1(b)]. With scattering from surface phonons, it is not necessary to know the index of refraction of the medium to determine the SAW phase velocity, but the angle of incidence must be known.

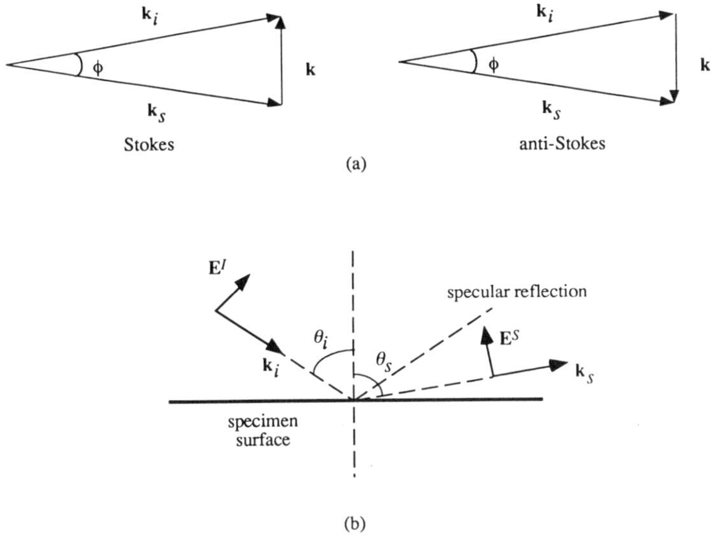

Figure 7.1. Kinematics of a light-scattering experiment; (a) Stokes and anti-Stokes events occurring in Brillouin scattering, (b) Brillouin scattering from a surface.

7.3. Theory of Brillouin-Scattering Cross Section

Equations (1) and (2) are sufficient to determine acoustic wave velocity from the position of a peak in the Brillouin spectrum. To understand the intensity of the scattered wave, e.g., the scattering cross section and hence the form of a spectrum, we must be able to describe the strength of the interaction between photons and phonons. Selection rules arise in connection with the scattering angle, the polarizations of both photons and phonons, and their relative orientations with respect to the scattering plane (defined by $\mathbf{k}_i$ and $\mathbf{k}_s$). In the case of crystalline bodies, it is also necessary to take elastic anisotropy into account. Before describing the basic equations for Brillouin scattering from bulk and surface acoustic phonons, we make some general remarks on the subject of inelastic light scattering.

In the following paragraphs, we consider only electromagnetic radiation with wavelength λ of the order of hundreds of nanometres (e.g., visible light) and always understand that all fields represent statistical averages with respect to thermal fluctuations of the sources at thermodynamic equilibrium (e.g., *thermal averages*). Averages are taken over an infinitesimal but already macroscopic volume much smaller than λ^3. The medium is therefore considered a *continuum*, with fluctuating optical properties represented by the polarization vector $\mathbf{P}$.

When a monochromatic plane electromagnetic wave of frequency ω_i and wave vector $\mathbf{k}_i$ impinges on the free surface of an isotropic dielectric medium, the electromagnetic field in the medium (the *refracted wave*) exhibits a very simple behavior if the frequency of the incident wave is far from internal resonances. The medium appears transparent, and the transmitted wave is a monochromatic plane wave propagating with a phase velocity $c/n(\omega_i)$ [$n(\omega_i)$ is the real part of the index of refraction of the medium at frequency ω_i] and wave vector $\mathbf{k}_i' = n(\omega_i)\mathbf{k}_i$. The frequency of the refracted wave is equal to that of the incident one. Despite the simplicity of this well-known and elementary phenomenon, its microscopic explanation is quite intricate. It was only around 1915 that within the framework of Maxwell equations, a satisfactory but very complex mathematical formulation, the so-called *extinction theorem*(6) was given. The transmitted field is the sum of the incident field and the dipole field radiated by excited atomic dipoles present within the medium. The dipole field is in turn the sum of two terms: The first completely cancels the incident wave, whereas the second corresponds to the refracted wave.

The last result is true only if we consider the time average of the dipole field with respect to the motion of all source charges induced by thermal agitation. Indeed additional scattered waves, with frequencies and directions differing from those of the incident wave, are radiated by thermal fluctuations of the polarization driven by the incident electric field. Inelastic light scattering is a manifestation of the existence of such fluctuations around thermodynamic equilibrium states. In the usual classical description of electromagnetic wave propagation in matter, scattering effects are neglected, since the total scattered power is usually a very small fraction of the incident power. An important exception is critical opalescence in fluids about the critical point.

Even though scattering is due to microscopic effects and should therefore be interpreted in the framework of quantum mechanics, in the case of visible light scattering, a simple classical explanation can be given: Brillouin light scattering is described in terms of the modulation of the dielectric constant of the medium due to the presence of thermally excited elastic waves (or acoustic phonons). Frequencies of acoustic phonons are always much smaller than the incident light frequency and characteristic frequencies of electronic motions, so that we can assume the fluctuating dielectric constant of the medium to depend adiabatically on the phonon coordinates.

7.3.1. Bulk Brillouin Scattering from Transparent Materials

The key ingredient of bulk Brillouin-scattering theory is the instantaneous dielectric susceptibility of the medium at frequency ω_i, e.g., the tensor

$$\chi_{ij} = \chi\delta_{ij} + \delta\chi_{ij} \tag{10}$$

where χ is the time-independent isotropic susceptibility, while $\delta\chi_{ij}$ is the fluctuating part of the susceptibility due to the presence of thermal phonons; δ_{ij} is the Kroenecker delta. If the static dielectric constant is written as $\varepsilon = \varepsilon_o (1 + \chi)$, then the corresponding fluctuating part of the dielectric constant is simply

$$\delta\varepsilon_{ij} = \varepsilon_o \delta\chi_{ij} \tag{11}$$

The fluctuating part of the polarization vector radiating the scattered waves can be written as $\delta P_i = \delta\varepsilon_{ij} E_j$, where summation over repeated indexes is understood.

$$E_j = E_j^I + E_j^S \tag{12}$$

is the sum of the incident field and the scattered field. The scattered electric field $\mathbf{E}^S$ can be computed by means of first-order perturbation theory (*Born approximation*), by writing

$$\delta P_i \approx \delta\varepsilon_{ij} E_j^I \tag{13}$$

Equation (13) is justified by the smallness of both the scattered field and $\delta\varepsilon_{ij}$. In this approximation, the excess polarization $\delta\mathbf{P}$ is driven only by the incident electric field $\mathbf{E}^I$. The expression for the scattered electric field in the far-field approximation, appropriate for most Brillouin-scattering experiments,[7] is

$$\mathbf{E}^S(\mathbf{r}, t) \approx \mathbf{R}\left[-\frac{E_o e^{i(\mathbf{k}_s \cdot \mathbf{r} - \omega_s t)}}{4\pi\varepsilon r} \mathbf{k}_s \times (\mathbf{k}_s \times \mathbf{G})\right] \tag{14}$$

In Eq. (14), the observation point $\mathbf{r}$ is a great distance from the scattering volume V where the origin of the coordinate system was chosen; $\mathbf{k}_s$ and ω_s have the same definitions as in Section 7.2.

The vector-scattering integral $\mathbf{G}$ in Eq. (14) is given by the following expression:

$$G_i = \int_V (\delta\varepsilon_{ij} e_j) e^{-i(\mathbf{Q}\cdot\mathbf{r}')} d\mathbf{r}' \tag{15}$$

where the time dependence of $\delta\varepsilon_{ij}$ was removed and e_i are the components of the polarization unit vector of the incident electromagnetic wave. The transferred wave vector $\mathbf{Q}$ is defined as

$$\mathbf{Q} = \mathbf{k}_s - \mathbf{k}_i \tag{16}$$

For bulk phonons in transparent materials, $\mathbf{Q}$ must be equal to the phonon wave vector $\mathbf{q}$ [see Eq. (1)]; while this is demonstrated by direct inspection of the integral in $\mathbf{G}$, it does not necessarily apply to SBS from opaque materials. Equations (14) and (15) express the scattered electric field as a function of the random variables $\delta\varepsilon_{ij}$. Near thermodynamic equilibrium, the thermal average $\langle \mathbf{E}^S \rangle_{th} = 0$, but this is not true for the mean square value $\langle |\mathbf{E}^S|^2 \rangle_{th}$, which depends on the mean square value of the phonon normal coordinate implicit in $\delta\varepsilon_{ij}$ (see the following paragraphs).

To illustrate the main conclusions that can be drawn from Eqs. (14) and (15), we take the simplest case of an elastically and optically isotropic material. In this case, the coupling between $\delta\varepsilon_{ij}$ and fluctuations of the strain tensor u_{ij}, induced by the presence of long-wave-length thermal acoustic phonons, can be written in terms of two elastooptic constants a_1 and a_2

$$\delta\varepsilon_{ij} = a_1 u_{ij} + a_2 u_{ll}\delta_{ij} \tag{17}$$

Sometimes a_1 and a_2 are written in terms of the Pockels coefficients p_{ij} as

$$a_1 = -\varepsilon^2(p_{11} - p_{12}) \qquad a_2 = -\ \varepsilon^2 p_{12}$$

For a cubic crystal, three independent coefficients are necessary to express the elastooptic coupling, and different selection rules are obtained.

The smallness of u_{ij} near thermodynamic equilibrium at normal temperatures allows one coefficient, using linear elasticity; strains can therefore be expressed as

$$u_{ij} = \frac{1}{2}\left(\frac{\partial u_i}{\partial x_j} + \frac{\partial u_j}{\partial x_i}\right) \tag{18}$$

For a bulk sound wave of wave vector $\mathbf{q}$ and frequency ω, the spatial part of the fluctuating displacement vector field $\mathbf{u}$ is then given by

$$\mathbf{u} = \mathcal{R}[\mathbf{u}^\circ \exp(i\mathbf{q} \cdot \mathbf{r})] \tag{19}$$

where $\mathbf{u}^\circ = Q\,\mathbf{e}$ is the product of the phonon normal coordinate Q and the phonon unit polarization vector $\mathbf{e}$. For longitudinal elastic waves (longitudinal acoustic, LA, phonon), $\mathbf{e}$ is parallel to $\mathbf{q}$, while for transverse elastic waves (transverse acoustic, TA, phonon), $\mathbf{e}$ is perpendicular to $\mathbf{q}$.

The scattering integral $\mathbf{G}$ can now be written as

$$\mathbf{G} = \mathbf{g}\int_V e^{-i(\mathrm{Q}-\mathrm{q})\cdot \mathrm{r}'}\, d\mathbf{r}' \tag{20}$$

where

$$g_i = i\left[\left(\frac{a_1}{2}\right)\left(u^\circ{}_i q_k + u^\circ{}_k q_i\right) + a_2 u^\circ{}_l q_l \delta_{ik}\right] e_k \tag{21}$$

If $V \gg |\mathbf{q}|^{-3}$, the integral in Eq. (27) turns out to be

$$\int_V e^{-i(\mathrm{Q}-\mathrm{q})\cdot \mathrm{r}'}\, d\mathbf{r}' \propto \delta(\mathbf{Q} - \mathbf{q}) \tag{22}$$

where δ is the Dirac delta function. This means that scattering with the phonon $\mathbf{q}$ can occur only if the conservation of the total wave vector [see Eq. (1)] is satisfied. Taking this into account, the integral in Eq. (20) can be replaced by the scattering volume V.

To complete our result, we note that the mean square value of Q at temperature T is

$$\langle Q^2 \rangle_{th} = \frac{4k_B T}{V\rho v^2 |\mathbf{q}|^2} \tag{23}$$

which is obtained by equating the mean kinetic energy per acoustic mode to k_B T. In Eq. (23), ρ is the mass density, and k_B is the Boltzmann constant.

Several general conclusions can be drawn from Eqs. (14), (20), and (21).

Light scattered by transverse phonons is completely depolarized; that is it does not keep the polarization of the incident wave.

There is no scattering by transverse phonons polarized in the scattering plane.

The intensity of scattering from transverse phonons goes to zero in backscattering (in materials elastically isotropic).

Scattering by longitudinal phonons is fully polarized.

The power spectrum $S_E^S(\omega)$ of the scattered light, which is essentially the measured outcome of a Brillouin-scattering experiment, is defined as

$$S_E^S(\omega) = \int_{-\infty}^{\infty} \langle E^S(t+\tau)^* E^S(t) \rangle_{th}\, e^{-i\omega\tau} d\tau \tag{24}$$

where E^S is the complex amplitude of the scattered field. From the preceding discussion, it turns out that the $S_E^S(\omega)$ for an isotropic material generally contains two doublets around the central elastic peak of the incident light: at lower frequency shifts, peaks due to transverse horizontal phonons; at higher shifts, peaks due to longitudinal phonons, which propagate at a higher velocity v_l v_t [see Eq. (6)]. The theory just outlined predicts delta function shapes for these peaks. If phonons have finite lifetimes, theoretical peaks are Lorentzian line shapes whose widths are directly related with the time decay (damping) of acoustic phonons.

7.3.2. Surface Brillouin Scattering in Opaque Materials

Since the formal theory of SBS involves cumbersome computations for both the phonon density of states and the scattering cross section, we shall outline the main aspects without going into too much detail; for a more detailed discussion, see Refs. 8–16.

Although some general conclusions about bulk scattering also apply in the case of SBS by opaque or semiopaque materials, such as metals or semiconductors, new features do appear. The finite penetration depth of light in the material (e.g., about 1000 nanometres in Si and only a few nanometres in Al at $\lambda_o = 514$ nm, the wave length of an argon ion laser typically used in Brillouin-scattering experiments) is such that the wave vector component normal to the surface is no longer conserved, and thus the wave vector conservation law [see Eq. (1)] is no longer valid.

Let us consider the effect of opacity on scattering by bulk phonons. Calling δ the optical skin depth of the material, the effective scattering volume is limited in the direction z perpendicular to the surface within a length on the order of δ. Optical absorption can be taken into account, introducing a complex transferred wave vector

$$\mathbf{Q} = Q_x\hat{\mathbf{e}}_x + (Q_z - i\delta^{-1})\hat{\mathbf{e}}_z \tag{25}$$

in the integral in Eq. (20), which now becomes approximately

$$\int_{-\infty}^{+\infty} \exp[-i(Q_x - q_x)x]dx \int_{0}^{+\infty} \exp[-i(Q_z - q_z)z]e^{-z/\delta}dz = -2\pi\delta(Q_x - q_x)\frac{1}{\delta^{-1} - i(Q_z - q_z)} \tag{26}$$

Two main conclusions can be drawn from Eq. (26):

The total parallel wave vector is conserved. Let θ_s and θ_i be the angles (positive in the clockwise direction) between the outgoing surface normal and the incident and scattering direction. The first integral in Eq. (26) can be replaced by the dimension of the illuminated area and the condition

$$|\mathbf{k}_s|(\sin\theta_s - \sin\theta_i) = \pm q_x \tag{27}$$

In Brillouin scattering from opaque materials, the parallel wave vector component is conserved; this is the fundamental kinematic relation for SBS.

The scattered intensity I_s is proportional to the squared modulus of the integral in Eq. (26), so

$$I_s = \frac{1}{\delta^{-2} + (Q_z - q_z)^2} \tag{28}$$

The shape of the Brillouin peak in an opaque material is a Lorentzian function whose width depends strongly on opacity. If $\delta|Q_z| \gg 1$, the line shape is a rather sharp peak centered at $q_z = Q_z$. In this limit, a situation very similar to bulk scattering in transparent materials is found. Yet the bulk line is broadened by the presence of absorption, which partially destroys conservation of a wave vector perpendicular to the surface. In the opposite limit $\delta|Q_z| \ll 1$, we have a rather flat broad band receiving contributions from all modes, with q_z ranging from 0 (bulk waves propagating parallel to the surface) to δ^{-1} (modes propagating almost perpendicular to the surface). In strongly opaque materials, bulk peaks are not symmetric, and the peak position depend on δ, that is on the imaginary part of the refraction index. The peak asymmetry can be derived theoretically if the reflection at the surface of phonons having wave vectors $\mathbf{q}$ and $-\mathbf{q}$ is taken into account.[17]

In addition to peaks corresponding to bulk acoustic phonons, the form of the spectrum in an opaque material also shows features at much lower frequencies corresponding to the population of surface acoustic phonons.[18,19] We assume a backscattering geometry where the sagittal plane, defined by the phonon propagation direction and the surface normal, coincides with the scattering plane, and we consider a *p*-polarized (polarized in the scattering plane) incident electromagnetic wave with incidence angle θ_i. From dispersion relations for transverse and longitudinal waves, we have

$$q_{t_z} = \sqrt{\frac{\omega^2}{v_t^2} - q_x^2} \tag{29a}$$

$$q_{l_z} = \sqrt{\frac{\omega^2}{v_l^2} - q_x^2} \tag{29b}$$

where q_z and q_x are the normal and parallel wave vector components, respectively. Once q_x is fixed, as in the case of SBS experiments, we see from Eq. (29) a lower *transverse* and a *longitudinal frequency threshold,* respectively, at $\omega_t = v_t q_x$ and $\omega_l = v_l q_x$ for acoustic phonons having real normal component of the wave vector $\mathbf{q}$. The following frequency regimes can therefore be distinguished.

$\omega > \omega_l$: *continuous spectrum of bulk modes;* q_{t_t} and q_{l_z} are real; the spectrum is a rather flat continuous band starting at the longitudinal threshold. All modes are composed of propagating bulk waves reflected by the surface. As shown by Eq. (28), peaks corresponding to shear horizontal and longitudinal bulk phonons appear in the cross section of semiopaque materials when the normal component Q_z of the transferred wave vector is such that $Q_z = q_{t_z}$ and $Q_z = q_{l_z}$.

$\omega_t < \omega < \omega_l$: *continuous spectrum of mixed modes;* q_{t_z} is real, but q_{l_z} is imaginary. The spectrum is a continuous shoulder that is zero at the transverse threshold. The modes, called *mixed modes,* are composed of propagating bulk transverse waves and evanescent longitudinal waves. These mixed modes can give rise around $\omega = \omega_l$ to a pronounced peak in the scattering cross section corresponding to the so-called *longitudinal resonance* or *high-frequency pseudo-surface wave* (HFPSW).[20–22] The latter can be seen as a leaky surface wave radiating energy into the bulk with a pronounced x polarization or longitudinal character. The HFPSW is not a true surface wave but only a packet of bulk waves with a displacement field strongly localized immediately below the surface.

$\omega < \omega_t$: *discrete spectrum;* q_{t_z} and q_{l_z} are imaginary. In this frequency range, the spectrum shows only discrete lines corresponding to true surface waves characterized by a wave vector $\mathbf{q}_x = q_x\hat{\mathbf{e}}_x$ parallel to the surface. In a semi-infinite medium, the discrete spectrum is composed solely by the *Rayleigh wave* (RW) peak at frequency $\omega_R = v_R q_x$; v_R is the RW phase velocity. The scattering integral for SAW in the discrete spectrum can be derived by writing the displacement field associated with a surface wave as

$$\exp|iq_x x) \sum_k a_k \exp(-\alpha_k z) \tag{30}$$

where the sum runs over the number of discrete modes and the decay constants α_k have a magnitude comparable to q_x. The scattering integral is proportional to

$$\delta(Q_x - q_x) \sum_k \frac{ak}{\alpha_k + \delta^{-1} + iQ_z} \tag{31}$$

Due to the absence of q_zs in the denominator, only parallel wave vector conservation can occur even in the case of low opacity. Hence scattering by SAW gives rise to discrete lines in the spectrum in the low-frequency region $\omega < \omega_t$, and Brillouin shifts of these lines scale purely as q_x.

In the presence of a surface, besides the bulk *elasto-optic effect,* a new scattering mechanism, called the *ripple effect,* is operative.[9,10,14] The phonons modulate the instantaneous shape of the free surface so that the latter differs from that of a perfectly flat plane. The instantaneous corrugated surface scatters the incident light with a Doppler effect, giving scattered photons with frequencies shifted from ω_i. The strength of this scattering process depends on the reflectivity of the surface, and it is active only for the normal component u_z of the phonon displacement field. The ripple scattering cross section is proportional to the power spectrum of the u_z displacement field computed at the surface ($z = 0$)[9,10]

$$S_{u_z|0)}(\omega, q_x) = \frac{k_B T}{\pi\rho v_t^3 q_x^2} \mathbf{R}\left\{\frac{\omega v_t^3 q_x^2 q_{ez}}{4v_t^4 q_x^2 q_{e_z} q_{t_z} + (\omega^2 - 2v_t^2 q_x^2)^2}\right\} \tag{32}$$

where A is the illuminated area of the sample surface and q_{l_z}, q_{t_z} are given by Eq. (29). The proportionality factor in the ripple cross section is represented by the electric field reflection coefficient per unit of solid angle at the corrugated surface. Explicit expression of the reflection coefficients for different polarizations of the incident and scattered fields can be found in Ref. 23. Plots of the backscattering reflection coefficient for *p-p* (incident and scattered light polarized in the scattering plane) and *s-s* polarization (incident and scattered light polarized normal to the scattering plane) are given in Fig. 7.2. The ripple cross section is null for *p-s* or *s-p* scattering (incident light polarized in the scattering plane and scattered light polarized normal to the scattering plane, and viceversa). For RW and many other sagittal SAWs, the ripple effect is the strongest contribution to the Brillouin-scattering cross section in opaque materials.

When the sample geometry is more complicated, as in the case of supported films or other layered structures, the main features of SBS just presented remain.[24–28] The spectrum of scattered light is in general the union of a discrete and a continuous part. In the discrete spectrum, lines associated with true *SAWs* are allowed by the sample geometry are observable. These are the *generalized RW,* the *generalized Lamb wave* or *Sezawa waves,* and the *Stoneley wave.*[18,29] Shear horizontal surface waves (*generalized Love waves*),[18,29] not observable by acoustic microscopy, can also be detected via the elasto-optic scattering mechanism. In the continuous part of the spectrum, beyond the transverse threshold of substrate material, the spectrum can exhibit rather defined peaks corresponding to scattering by packets of bulk waves (or *pseudowaves*) with a strong surface character. Pseudo-Sezawa waves, pseudo-Love wave in layered structures, and pseudo-SAWs in anisotropic materials fall into this frequency region. In the high-frequency part of the spectrum, above the longitudinal resonance, bulk transverse and longitudinal peaks can be observed.

We summarize the results found for Brillouin scattering from acoustic phonons in solid materials as follows:

Both surface and bulk acoustic thermal phonons produce inelastic scattering of light. In the former case, peaks in the spectrum appear at frequencies Ω given by the relation $\Omega = v_{SAW}Q_x$, where Q_x is the parallel component of the transferred wave vector and v_{SAW} is the phase velocity of the SAW.

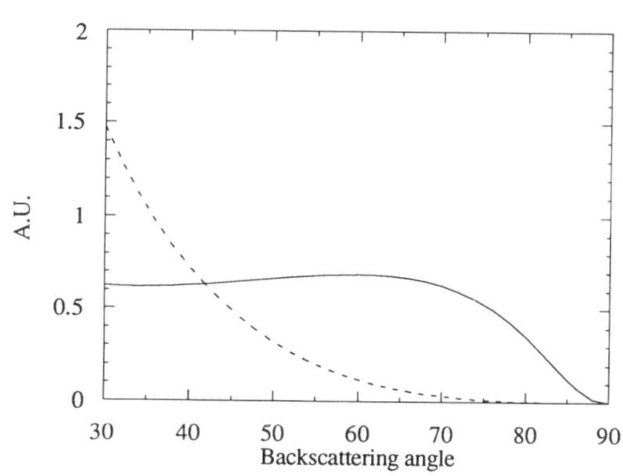

Figure 7.2. The dependency of the ripple cross section with the angle of incidence for *p-p* (solid line) and *s-s* (dashed line) backscattering. (from Ref. 45)

In scattering from bulk phonons, the total transferred wave vector is conserved only in transparent or semitransparent materials. In backscattering bulk phonon peaks appear at frequencies $\Omega = 2nk_o v$, where k_o is the wave vector of the incident light.

In transparent materials, Brillouin scattering is caused by the strain-induced modulation of the dielectric constant (elasto-optic effect). The strength of the scattering depends on the elasto-optic constants, and it is proportional to the scattering volume. Both SAWs and bulk acoustic waves can scatter the incident light by this mechanism. The line shape for bulk phonons is Lorentzian with width proportional to the phonon lifetime.

In opaque materials, the major contribution to inelastic light scattering is due to surface corrugation produced by the propagation of SAWs with normal displacement components (ripple effect). The scattering cross section does not depend on the scattering volume on elasto-optic constants. In strongly opaque materials, such as metals, this is the only scattering contribution.

In materials that are partially opaque, such as semiconductors, both ripple and elasto-optic effects are present, and the total Brillouin cross section is given by the combination of both contributions. The opacity of material distorts the shape of the bulk phonon peaks, which appears shifted and shows increased broadening.

7.4. Surface Brillouin Scattering Setup

The experimental setup currently used for SBS is illustrated in Fig. 7.3. The key elements are the scattering geometry, the laser source, and the high-contrast spectrometer. Figure 7.3 shows the most widely used scattering geometry for backscattering on reflection from a surface. This configuration is most suitable when scattering from opaque and semiopaque materials such as metals and semiconductors, is studied. Other scattering geometries, which involve electromagnetic waves guided parallel to the surface and detected by wedge optical prisms, have also been used to enhance Brillouin-scattering efficiency in thin films.[12,13,30,31]

The major requirement for the laser source is a stable continuous wave (cw) operation in a single longitudinal cavity mode. In a typical SBS experiment, surface phonons with frequencies in a range from 30–40 GHz are measured. Since the scattered light lies so close in frequency to the incident light, the spectral width of the laser line must be kept as narrow as possible. This is achieved only when the laser oscillates in single longitudinal mode. Another important characteristic of the laser source is the availability of linearly polarized

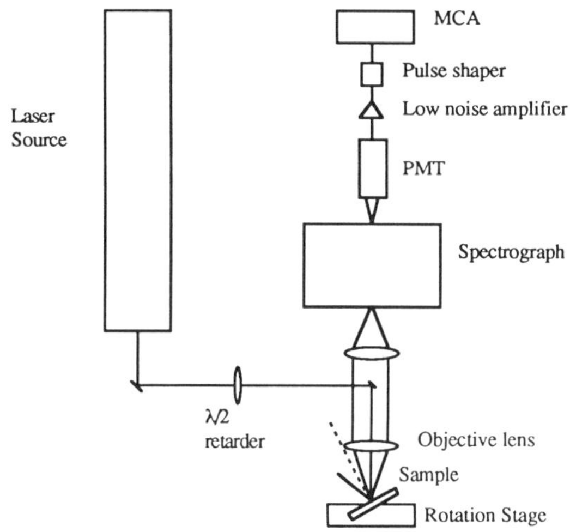

Figure 7.3. An instrumental setup for Brillouin-scattering spectroscopy. (from Ref. 45)

output light, since the Brillouin-scattering cross section depends strongly on the polarization of the incident optical wave. The typical incident laser power depends on the material under examination, and it can range from 50–250 mW. The most widely used laser source is the cw argon ion laser oscillating at a 488- or 514.5-nm wavelength, but experiments can also be performed using the new generation of cw solid-state lasers, such as the diode-pumped frequency-doubled Nd:YAG lasers or the Ti:sapphire tuneable lasers. The model used in the Poltecnico of Milano laboratory is a cw Ar^+ laser (Coherent Innova 300™), where single longitudinal-mode oscillation is achieved by means of an etalon inserted into the laser cavity. The temperature of the intracavity etalon is regulated in a feedback control loop to track the oscillating mode. Temperature drifts in the cavity cause variation in the cavity spacing, with a consequent broadening of the laser line or even mode hopping in the laser output. The output laser power is kept fixed by means of a control loop inserted on the input laser current. A $\lambda/2$ retarder is used to select the incident polarization.

The incident laser beam is directed by a small prism onto the sample surface, and it is focused by a 50-mm objective lens. The same lens is also used for collecting the scattered light from the sample. The angle between the incident light beam and the normal to the specimen surface is a measurement parameter that must be varied in a range from 0°–90° with great accuracy and repeatability. We use a 360° range DC motor-driven rotation stage (Physik Instrument M-038™) having a repeatability of ±0.3 arc sec. The possibility of rotating around the surface normal must be considered for the sample stage if anisotropic materials or single crystals have to be examined. If polarization of scattered light is analyzed, a dichroic polarizer is inserted before the spectrograph.

The spectrometer consists of a couple of plane parallel Fabry–Perot (PFP™) interferometers mounted in tandem multipass configuration. Details of the spectrometer are given in the following section.

Given the low intensity of the Brillouin signals, single-photon counting is necessary for the detection system. Low-noise photomultiplier tubes (PMT) having dark counts not higher than 1–2 counts per second are generally used for this purpose. A drawback of this type of detector is the low quantum efficiency of the photocathode. Quite recently a new generation of low-background noise, single-photon avalanche photodiodes with a higher quantum efficiency than a typical PMT, have also been available off-shelf. In our system, we used a room-temperature-operated PMT (Hamamatsu R-464 S™) having less than one dark count per second. The electronics that follows is standard when PMT single-photon signals are passed to an amplifier, a pulse shaper, and then counted and stored by an multichannel analyzer (MCA).

7.4.1. Tandem Multipass Fabry–Perot Interferometer

In Brillouin-scattering spectroscopy from surfaces, acoustic phonons with frequencies in the range of 30–40 GHz are measured. The frequency shift of the scattered light is therefore 10^4–10^5 times smaller than the central frequency of the laser light, and therefore a high-resolution spectrometer, such as the plane Fabry–Perot interferometer, is used; for a detailed description of the Fabry–Perot interferometer, see Ref. 32. The PFP interferometer consists of two partially reflecting plane mirrors mounted parallel to each other at a distance L. For normal incidence, the instrument transmits light of wavelength λ only if the distance L is equal to a multiple of half the value of the incident wavelength, $L = m\lambda/2$, where m integer. This can be understood by considering the interference between the ray directly transmitted by the two mirrors and the ray that experiences a double reflection inside the interferometer cavity. Neighboring orders of interference are therefore separated in frequency by an amount $\Delta\nu$ equal to $L/2$ wave numbers ($\Delta\nu = c/(2L) = 150/L$ GHz mm^{-1}). The interorder spacing is called the *free spectral range* (FSR) of the interferometer. If we take into account multiple reflections inside the interferometer cavity, then the transmitted intensity is given by the *Airy function* $T(L/\lambda)$[32] (see Fig. 7.4).

$$T(L/\lambda) = \frac{T_o}{1 + \left(\frac{4F^2}{\pi^2}\right)\sin^2(2\pi L/\lambda)} \tag{33}$$

where T_o is the maximum transmission and the parameter F, the *finesse* of the interferometer, determines the width of the fringes and depends on the flatness and the reflectivity of the mirrors. The resolving power $\delta\nu$ of the PFP interferometer is

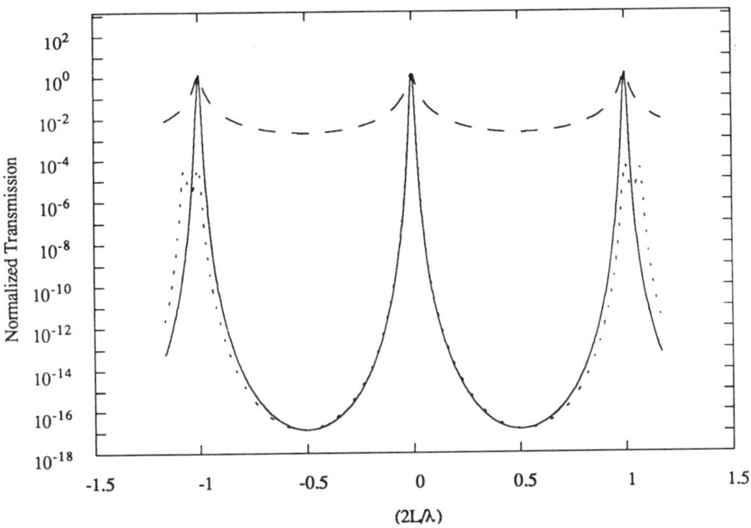

Figure 7.4. Normalized transmission curves for single-pass and multipass interferometry ($F = 40$); dashed line, 1 pass; dotted line, 3 + 3 passes; solid line, 6 passes. (from Ref. 45)

defined as the full width at half-maximum of the Airy function, and it equals the ratio between the free spectral range and the finesse

$$\delta\nu_{FWHM} = FSR/F \qquad (34)$$

In our experiments, we used mirrors with reflectivity $R = 93\text{–}94\%$ and $\lambda/500$ flatness over the whole area (a 37-mm diameter), corresponding to a nominal finesse $F = 40$.

When mirror spacing is scanned, interferometer transmission can be tuned to analyze spectral components of the incident signal. In PFP interferometry, a tilt-free scanning stage must be used.[(33)] The scanning stage used in the experiments consists of a deformable parallelogram platform mounted on top of a roller translation stage for fine and coarse scanning. A screw made of low-thermal-expansion material positions the mirror spacing while the scan is performed with a piezoelectric transducer. Mirror spacing is measured by a plane capacitor, and an active loop is used for regulating the scan voltage to the transducer. The PFP interferometer is extremely sensitive to mirror misalignment, and an active feedback must be used to regulate the parallelism between the mirrors.[(34)] The scanning stage is mounted on top of an active vibration isolation table. This is necessary, since mirror displacements as small as 20 Å are necessary to scan the PFP interferometer through a transmission peak. Spectrum distortions can occur for any mechanical disturbance that changes the mirror spacing by as little as a few angstroms.

In a Brillouin spectrum, elastically scattered light from surface roughness and irregularities is usually several orders of magnitude more intense than Brillouin signals, and situations where the elastic peak is 10^6 times higher than the Brillouin signals are often encountered. This makes a surface Brillouin-scattering experiment feasible only if very high-contrast interferometers are used. High contrast is even more important when the small scattering volume associated with surface phonons, usually limited to a few acoustic wavelengths, is considered. Signals of surface acoustic phonons are therefore very weak compared with Brillouin scattering from bulk phonons in transparent materials. In these cases, instruments whose contrast is better than 10^6-10^7 must be used. The contrast C of the interferometer can be defined as the ratio between maximum and minimum transmission of the Airy function; and hence it is given by $C = 1 + (2F/\pi)^2$. Since the finesse of an interferometer cannot be much higher than 100, a single-pass interferometer cannot have a contrast higher than 10^4. The contrast is greatly enhanced if multiple passes of the light through the interferometer are performed[35–37] (see Fig. 7.4). In this case, the form of the transmission coefficient is ideally obtained by raising the single-pass Airy function to the nth power, where n is the number of passes. The contrast for a multiple-pass interferometer becomes $C_n = [1 + (2F/\pi)^2]^n$, which can be raised to 10^8-10^9. This happens only if separate passes are truly decoupled from one another. In practice coupling between passes in the PFP interferometer is present, leading to a loss in the contrast of the multipass with respect to its theoretical value. The full width of the transmission curve of a multipass PFP interferometer can be readably obtained from

$$\frac{2F}{\pi \sin(\pi\delta\nu/2FSR)} = \sqrt{2^{1/n} - 1} = f(n) \tag{35}$$

The spectral resolution of the multiple-pass interferometer is equal to the value of a single-pass interferometer with a finesse $F/f(n)$. The function $f(n)$ varies from 0.5 for $n = 3$ to 0.35 for $n = 6$. The spectral resolution of a multipass interferometer is therefore slightly improved with respect to that of a single pass. Various schemes of multipass interferometers were proposed in the literature. In our measurements, we used the 3 + 3 multipass configuration proposed by Mock and others.[38]

Another problem encountered in PFP interferometry is the overlapping of interference orders of different spectral features of the scattered signal. When scattered light is composed of frequencies extending to a maximum frequency ν_{max}, an unambiguous interpretation of the spectrum can be given only if the FSR is chosen such that $FSR > 2\nu_{max}$. This sets a limit on the spectral resolution, since by using a large value of FSR, low-energy peaks in the spectrum cannot be resolved from the central elastic peak. To increase the FSR without degrading the spectral resolution involves using a cascade of two or more PFP interferometers having different spacing (*tandem interferometer*).[5,39–41] If two interferometers with mirror distances L_1 and L_2 are placed in cascade to transmit two different interference orders $m_1 = 2L_1/\lambda$ and $m_2 = 2L_2/\lambda$ of the incident wavelength λ, the transmitted intensity is composed of the product of the Airy function of both interferometers. The total transmission curve is composed of a central elastic peak and small ghosts representing weakly transmitted interference orders of the two interferometers (see Fig. 7.4). The overall free spectral range can be increased 10–20 times over the *FSR* of a single interferometer. Usually distances L_1 and L_2 are chosen such that $L_2/L_1 = k/n$, where k and n are incommensurate prime numbers. The effective *FSR* becomes k times the free spectral range of the first interferometer.

The major difficulty of tandem interferometry is represented by the synchronization of the two interferometer scans. The relation $\delta L_1/L_1 = \delta L_2/L_2$ between increments in the mirror distances L_1 and L_2 must be satisfied during the scan with Å accuracy. A simple way to make a synchronously scanned tandem interferometer[33] is realized involves mounting the scanning mirrors of the two interferometers on the same scanning stage at an angle α (see Fig. 7.5). In this way, $L_2 = L_1\cos\alpha$, and $\delta L_1/L_1 = \delta L_2/L_2$ is satisfied. In the tandem interferometer used for the measurements, $L_2/L_1 = 19/20$, corresponding to $\alpha \approx 18°$. The overall spectral resolution of the system was $\delta\nu_{FWHM} \approx FSR/100$, obtained by fitting the

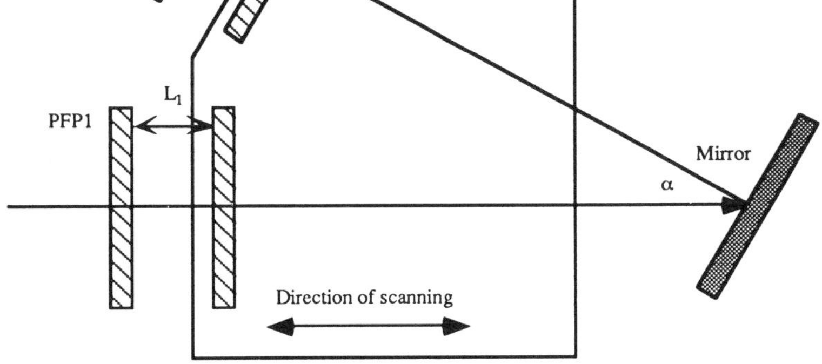

Figure 7.5. The tandem PFP interferometer used in the experiments. (from Ref. 5)

central elastic peak of the interferometer response using a Gaussian function to determine the FWHM. This result in agreement with the overall finesse of the 3 + 3 multipass interferometer when the single-pass finesse is $F = 35 - 45$, $f(n) = 0.35$, and we consider the degrade in resolution associated with the cross talk between the two interferometers.

7.4.2. Calibration and Errors

Determining acoustic wave velocity and attenuation from Brillouin spectroscopy depends on measuring frequency shifts in the phonon peaks in the spectrum. To accomplish this, it is necessary to calibrate the scale of the MCA on a sample of known acoustic characteristics. The calibration can be performed by using the first ghost of the interferometer (see Fig. 7.4) and knowing the distance of one of the two mirrors. From the position of the corresponding ghost peak, it is possible to determine the *FSR* and calibrate the MCA scale. Some difficulties occur when the intensity of the ghost is either too low or too high. In the first case, the shape of the ghost is rather distorted, so that fluctuations in the counts can misdirect positioning the peak. If the intensity of the ghost is higher than the dynamic range of the detection system, this can produce a flat and featureless ghost spectral shape. In this case, it is impossible to determine the ghost peak position. An alternative way uses the Stokes and anti-Stokes longitudinal bulk peaks of a sample of known optical and acoustic properties as markers in the spectrum. We used a sample of high-purity fused silica with longitudinal acoustic wave velocity $v_l = 5970$ m s^{-1} (literature value, see Ref. 42, Chap. 6) and optical refractive index $n = 1.458$. The index of refraction was measured with a precision of one part in a thousand. Taking several spectra in backscattering, changing the distance of the mirrors and the scan amplitude, and measuring each time the position of the Stokes and anti-Stokes longitudinal phonon peaks, it is possible to calibrate the interferometer. The relationship for calculating the frequency shift of a phonon peak from positions in the MCA of the anti-Stokes and Stokes lines N_{AS} and N_S is

$$\Delta\nu = \frac{(N_{AS} - N_S)\gamma x N_{cal}}{2DN_{ch}} [GHz] \tag{36}$$

where γ is the calibration constants expressed in [GHz mm], x is the reading of the scan amplitude (usually a potentiometer), N_{cal} is the number of MCA channels used in the calibration, N_{ch} is the number of channels used in the acquisition, and D is the mirror distance of the PFP interferometer. Using the ghosts of the interferometer to calibrate the instrument, we had an error in determining the fused quartz longitudinal velocity of the same order as the indetermination in the refractive index. It can be argued that readings of x and D are themselves affected by errors. While the position of the scan amplitude potentiometer can be kept fixed for all of the measurements, the spacing necessarily has to be changed. In our instrument, the reading of D is made by watching a clock gauge. An error of approximately ± 10 μm may therefore occur in positioning the mirror distance. Since a typical mirror distance for observing surface acoustic phonons is on the order of 5–10 mm, an error on the order of 0.1% can take place in setting the *FSR*. If the peak position is taken without fitting, a source of error in determining frequency shifts is given by the discreteness of the frequency axis in the MCA. When $N_{ch} = 1024$ and $D = 10$ mm, an error of one MCA channel in the position of the peak corresponds to approximately 30 MHz. For a SAW, velocity $v_{SAW} = 3000$ m s^{-1} and an angle of incidence $\theta_i = 60°$, the relative error in the velocity is $\Delta v_{SAW}/v_{SAW} \approx 0.3\%$. When measuring frequency shifts, it is to position correctly the center ($\Delta\nu = 0$) in the MCA scale. To eliminate the error in $\Delta\nu = 0$ positioning, caused by jitter in the elastic peak position, it is always necessary to consider the arithmetical average of the Stokes and anti-Stokes frequency shifts.

The second aspect that must be considered when determining the SAW velocity is the angle of incidence measurement, since this parameter enters directly into the estimation of the transferred parallel wave vector [see Eq. (9)]. Although a rotation stage with very good precision can be used, an error can still be introduced when fixing the parallelism between the incident laser beam and the normal to the surface ($\theta_i = 0°$). This is usually done by visually checking that the laser beam reflected from the sample is returned parallel to the incident beam. It is difficult to assess the precision in setting $\theta_i = 0°$, but we calculate that it may be on the order of $\pm 0.1°$. For a typical measurements at $\theta_i = 60°$, this indetermination in the angle of incidence yields a relative error in the velocity on the order of $[\sin(60.1°)-\sin(60°)]/\sin(60°) \approx 1 \times 10^{-3}$. A worst case hypothesis for the relative error in estimating SAW velocity may therefore be on the order of 0.5%.

A final point to consider concerns precision in measuring surface phonon velocity for anisotropic materials such as single crystals. In single crystals, the acoustic wave velocity is usually anisotropic, and it is necessary to align the crystallographic axes of the sample with respect to the propagation direction. Given the long measurement time that is necessary in Brillouin-scattering spectroscopy, it is not feasible to check crystal alignment using the acoustic wave velocity itself, and sample alignment is made visually, using in most cases the cleavage plane of the specimen as an indicator. We calculate that misalignment as large as $\pm 5°$ may occur with this procedure.

7.4.3. Geometrical Aperture Effects (Measurement Accuracy)

Accuracy in determining SAW velocity and attenuation using Brillouin-scattering spectroscopy is influenced by the finite aperture of the collection lens.

Given the low intensity of signals, it is often necessary to use a large-aperture collection lens to allow enough scattered radiation to be analyzed. This broadens the phonon line shape and introduces a systematic shift in the position of the peak.

The reason for the broadening is quite clear: It arises because scattered light is not collected at a single scattering angle θ but over a range of angles $\theta \pm \Delta\theta$. Thus a range of surface phonons are probed with a consequent distribution of frequencies. In backscattering if $\Delta\theta \ll \theta_i$, the spread in parallel transferred momentum Q induced by the finite collection aperture is $\Delta Q = 2k_o \cos\theta_i \sin\Delta\theta$. Broadening is minimized by reducing the collection lens aperture and increasing the angle of incidence. For a collection lens diameter $D = 24$ mm and focal length $F = 50$ mm, the subtended semiangle is $\Delta\theta \approx 13.5°$. Taking a typical Rayleigh velocity $v_R = 3000$ m s^{-1} and an angle of incidence $\theta_i = 60°$, the width of the spectral line shape induced by the geometrical aperture is $\Delta\omega = \Delta Q v_R \approx 1.2$ GHz. The geometrical aperture broadening can be orders of magnitude higher than the FWHM associated with the surface phonon lifetime.

There are two primary reasons for the systematic shift in the position of the surface phonon peak introduced by the finite collection aperture. Firstly when a finite collection aperture is used in backscattering, instead of the transferred parallel momentum $Q = 2k_o \sin\theta_i$, its mean value on the distribution of θ, $\langle Q \rangle = 2k_o \langle \sin\theta_i \rangle$, should be considered. The factor $\langle \sin\theta_i \rangle$ is given by

$$\langle \sin \theta_i \rangle = \frac{\int_{\theta_i-\Delta\theta}^{\theta_i+\Delta\theta} \sin \theta \, d\theta}{\int_{\theta_i-\Delta\omega}^{\theta_i+\Delta\theta} d\theta} = \sin \theta_i \frac{\sin \Delta\theta}{\Delta\theta} \tag{37}$$

The mean value of the transferred parallel momentum is therefore smaller than the ideal backscattering value, and for a semiangle $\Delta\theta \approx 13.5°$ and $\theta_i = 60°$, we have that $\langle Q \rangle / Q \approx 99\%$.

The other aspect to be considered is the angular dependence of the Brillouin-scattering cross section. This can be done more easily if only the *p-p* ripple contribution is considered. For surface sagittal modes in opaque materials, this is by far the dominant term in the cross section. As shown in Fig. 7.2, variation with the angle of incidence of the *p-p* reflection coefficient is quite uneven and drops sharply to zero beyond 70°. For backscattering at $\theta_i > 70°$, the reflection coefficient weighs more heavily contributions centered at smaller angles, with a consequent shift in the position of the peak in the spectrum.

Previous arguments considered only a finite aperture in the in-plane angle of scattering θ. In a real experiment, the finite collection aperture also allows out-of-plane-scattering events to contribute to the total cross section. In backscattering transferred parallel wave vectors become $Q = k_o(\sin\theta_i + \sin\theta\cos\phi)$, where angle ϕ is illustrated in the geometry of the collection (see Fig. 7.6).

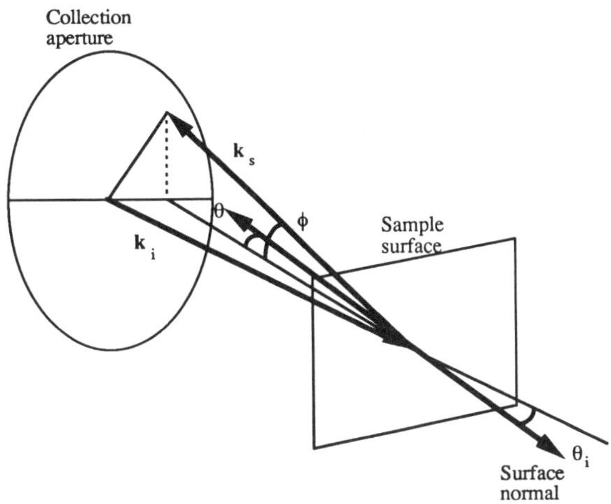

Figure 7.6. The geometry of the collection in a surface Brillouin-scattering experiment. (from Ref. 45)

The effect of the finite collection aperture on the Brillouin spectrum can be calculated as an integral over the collection solid angle. Assuming that the line shape of the SAW is a Lorentzian curve $L[\omega - \Omega(\theta, \phi); \Gamma]$ centered at a frequency $\Omega(\theta, \phi) = Qv = k_o v(\sin\theta_i + \sin\theta\cos\phi)$, with half-width Γ proportional to the acoustic attenuation, then the total Brillouin-scattering spectrum $S_B(\omega)$ is given by the following expression:

$$S_B(\omega) \propto \int_{\Omega} L[\omega - \Omega(\theta, \phi); \Gamma] R(\theta, \phi) d\Omega \tag{38}$$

where $d\Omega = 2\pi\sin\theta d\theta d\phi$ is the elemental solid angle and $R(\theta, \phi)$ is the reflection coefficient for the scattered field.[23] The integration domain is illustrated in Fig. 7.7. The central part of the integration domain is occupied by the shadow of the small prism that necessarily has to be used in backscattering to direct incident light onto the sample surface. The result of this integral for a RW with velocity $v_R = 2720$ m s^{-1}, angle of incidence 60°, and $\Gamma = 200$ MHz is given for the

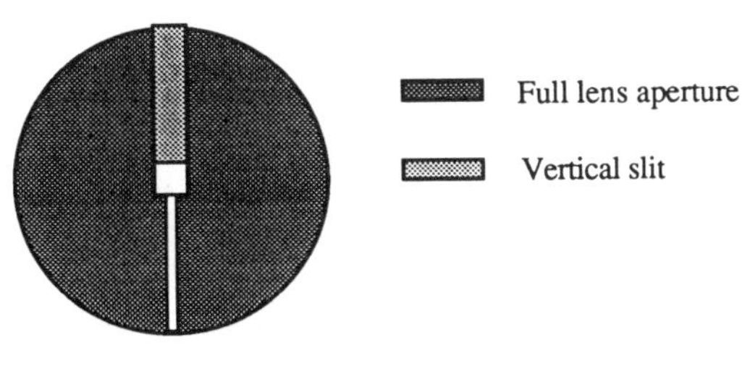

Figure 7.7. The integration domain for Eq. (37). (from Ref. 45)

full-lens aperture in Fig. 7.8 and compared with experiments. The shadow of the prism causes the appearance of a spurious double peak in the line shape. The dip can be used as a marker for identifying the position of the Rayleigh peak. In Fig. 7.8, the intrinsic Lorentzian line shape is also given.

If a slit is used to limit the aperture in the collection lens, the spurious double peak can be removed, and contrast for determining the peak can be increased. For those experiments where the highest resolution in determining SAW velocity is the main task, we use a central slit with the same width as the prism ($\approx$ 5 mm). Care must be exerted in aligning the incident laser beam in the center of the prism to minimize errors in positioning the scattering angle. The presence of the prism holder shadows the lower part of the slit aperture with a consequent shift in the position of the phonon peak in the spectrum as compared to the true backscattering configuration.

To obtain the measured Brillouin spectrum, the function $L[\omega - \Omega(\theta, \phi); \Gamma]$ must be convoluted with the instrumental response function $I(\omega)$ of the laser spectrograph system. The laser instrumental function can be approximated as a delta function, while the spectrograph response is given by the appropriate transmission function of the interferometer. For our purposes, it can be approximated as a Lorentzian or a Gaussian curve having a FWHM equal to the *FSR* divided by the overall finesse. The effect of the instrumental response on the measured spectral phonon line shape introduces a further broadening, which may be, according to the selected *FSR*, the dominant term in the line shape FWHM.

For the most accuracy, we used Eq. (37), evaluated in the light-gray-shadowed domain in Fig. 7.7, to fit the experimental line shape of the surface phonon taken in backscattering using a central slit in the collection. To test our fitting procedure, we performed a series of measurements along $\langle 110 \rangle$ on a Si(001) sample at different angles of incidence, and we fitted the experimental line shapes using both a Lorentzian curve and the integral in Eq. (38). The results are summarized in Table 7.1. The FWHM of the Lorentzian curve decreases with an increasing incidence angle, and it agrees with a broadening caused by the finite collection aperture. From $\theta_i = 60°$, the line shape width is limited by the instrumental response of the interferometer. For these measurements, we used $FSR = 30$ GHz, which corresponds to a spectral width of the interferometer transmission curve of approximately 300 MHz. Velocity values found by the fitting procedure are slightly higher than those found by taking positions of the peaks and assuming a true backscattering configuration. Corrected values are in better agreement with the literature value for the $\langle 110 \rangle$ Rayleigh wave velocity in Si(001) ($v_R = 5080$ m s^{-1}). Regarding extracted FWHM values, data are quite disperse, and these vary from 140–200 MHz. These measured widths are not realistic, since they are smaller than the broadening induced by the interferometer response line shape ($\approx$300 MHz). We see therefore that an accurate measurement of SAW velocity must take into account the effect of using a finite aperture in

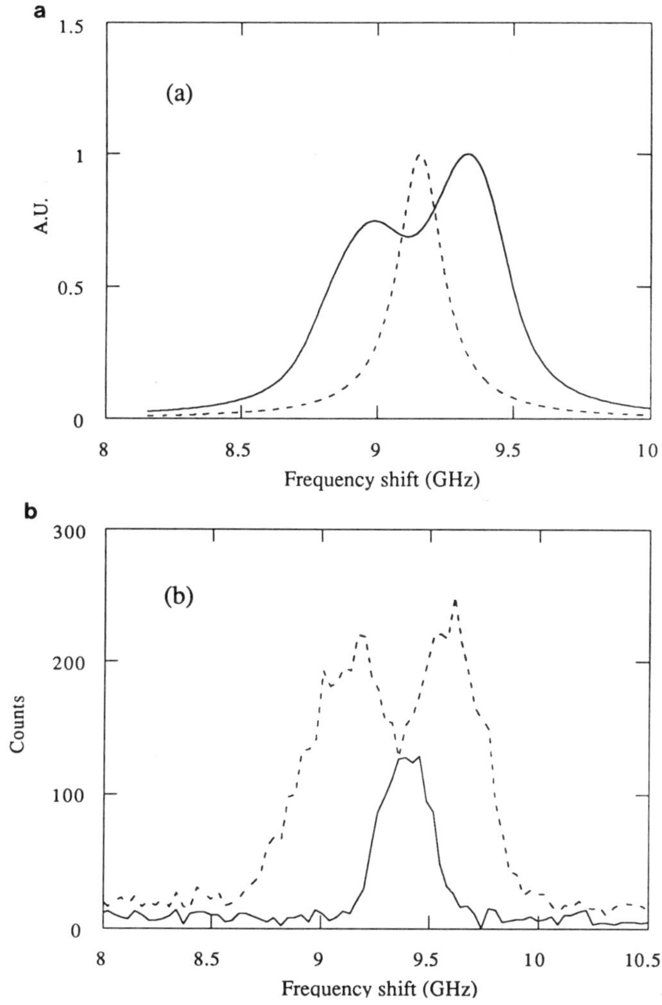

Figure 7.8. A surface phonon spectrum measured in a Brillouin-scattering experiment; (a) line shape determined by Eq. (38) using the full-lens aperture and compared with the intrinsic Lorentzian curve (v_R = 2700 m s^{-1}, angle of incidence 60°, and Γ = 200 MHz); (b) measured spectrum of Rayleigh wave using full-lens aperture (*solid line*) and a vertical slit (*dashed line*) [GaAs(001) along ⟨110⟩, angle of incidence 60°, *FSR* = 15 GHz]. (from Ref. 45)

Table 7.1. Velocity and Width (FWHM) of an RW in a Series of Measurements at Different Angles of Incidence along ⟨110⟩ on a Si(001) Sample[a]

Angle of incidence	Lorentzian		Eq. (37)	
	Velocity (m s^{-1})	FWHM (MHz)	Velocity (m s^{-1})	FWHM (MHz)
30°	5045	630	5058	200
40°	5053	580	5066	190
50°	5040	470	5055	140
60°	5036	350	5056	135
70°	5036	300	5054	145

Source: From Ref. 45.
[a]Experimental line shapes were fitted using both a Lorentzian curve and Eq. (37).

the collection. This can be done if the experimental surface phonon line shape is fitted using Eq. (37) as trial function.

Broadening caused by the spread in scattered wave vectors and the spectral resolution of the interferometer may prevent determining the phonon lifetime. The surface phonon lifetime has a negligible effect on the overall line shape unless it is larger than the geometric broadening. Since in practice slits with an angular width $\Delta\theta$ less than $\pm 2°$ cannot be used without deteriorating the *S/N* ratio and typical *FSR* are 15–30 GHz, phonon lifetimes smaller than 200–300 MHz cannot be measured by the present system. The combined effect of the intrinsic phonon lifetime and parasitic aperture broadening on the surface phonon peak is illustrated in Fig. 7.9, where the line shape was calculated by assuming 60° incidence, Rayleigh velocity $v_R = 2720$ m s^{-1}, and angular width $\Delta\theta = 2.5°$. The intrinsic phonon line width Γ was taken to be 100 MHz [see Fig. 7.9(a)], 200 MHz [see Fig. 7.9(b)], 400 MHz [see Fig. 7.9(c)], and 800 MHz [see Fig. 7.9(d)]. In Fig. 7.9 the Lorentzian curve corresponding to the intrinsic phonon lifetime is also shown. The difference between the curves becomes negligible when the intrinsic line width has the same order as the geometrical aperture broadening [see Figs. 7.9(c)(d)].

7.4.4. Comparison with Quantitative Acoustic Microscopy

To clarify advantages and drawbacks of SBS in measuring the propagation parameters of SAWs, we compare the performance of the former technique with quantitative acoustic microscopy.[(42)]

The main difference between the two techniques is the nature of the acoustic modes investigated. When using quantitative acoustic microscopy, the propagation of a monochromatic SAW excited by a transducer is analyzed; in Brillouin scattering, the population of acoustic phonons present in the sample at thermal equilibrium is probed. Thus acoustic phonons can be studied at much higher

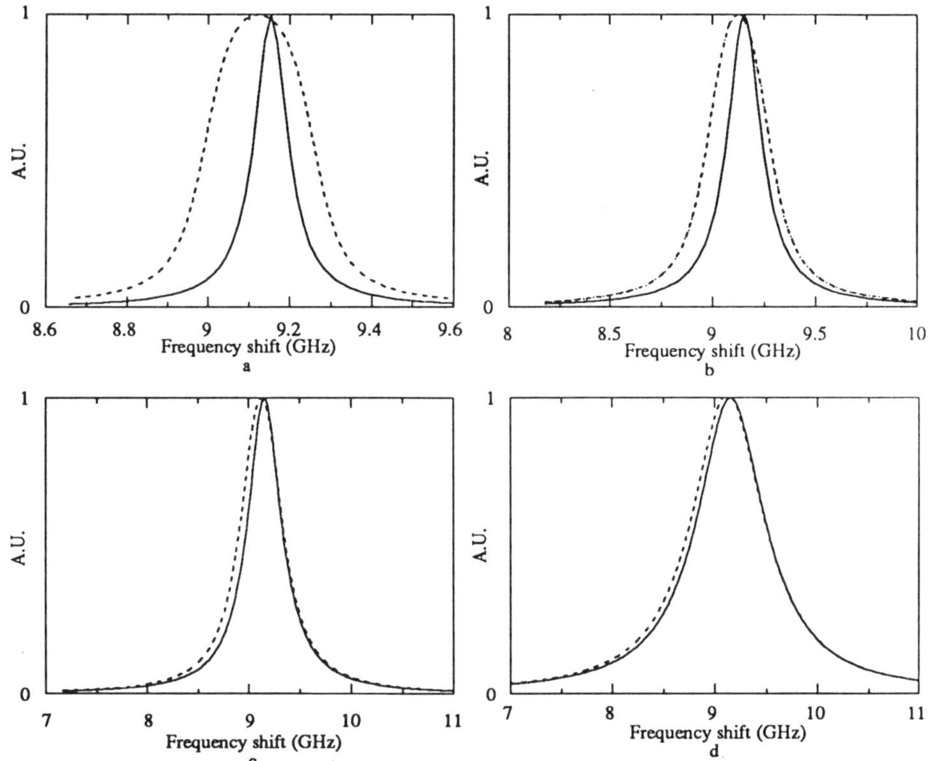

Figure 7.9. The surface phonon line shape calculated by Eq. (38) with a vertical slit compared with the intrinsic Lorentzian curve (v_R = 2700 m s^{-1}, angle of incidence 60°); (a) Γ = 100 MHz, (b) Γ = 200 MHz, (c) Γ = 400 MHz, (d) Γ = 800 MHz. (from Ref. 45)

frequencies by Brillouin scattering, since there is no need to excite a monochromatic acoustic mode directly. On the other hand, thermal surface acoustic modes are not coherently excited, which results in the low intensity of the Brillouin signals. This causes longer acquisition time for SBS experiments (typically 1 hour per spectrum) the measurement time for a $V(z)$ curve (typically less than 1 minute).

In addition to detect a Brillouin signal with an acceptable S/N ratio, it is often necessary to collect scattered radiation over a finite collection angle. This implies that acoustic modes over a range of wave vectors contribute to the signal with a consequent error in determining surface phonon propagation parameters. The resolution when determining SAW velocity that can be obtained with quantitative acoustic microscopy is therefore superior to using Brillouin-scattering spectroscopy. Brillouin scattering may be used to detect both SH and sagittal

Table 7.2. A Comparison between Surface Brillouin Scattering and Quantitative Acoustic Microscopy

Main experimental characteristics	Quantitative acoustic microscopy	Brillouin-scattering spectroscopy
Frequency range	From a few megahertz to 1–2 GHz depending on the acoustic lens. Given the large attenuation of sound in the couplant, measurements in the gigahertz regime have a rather poor resolution for quantitative purposes.	For surface phonons, the highest frequency is limited by the maximum attainable angle of incidence ($\theta_i \approx 85°$). This is therefore ≈30 GHz for a SAW velocity of 8000 ms^{-1}
Probed acoustic modes	All those surface acoustic modes that can couple to the coupling fluid. These are SAWs with normal displacement components (RWs and pseudo-RWs in substrates, Sezawa modes in supported films) and lateral waves.	Thermal surface and bulk acoustic modes. Both transverse and sagittal surface modes can be detected via elasto-optic and ripple effects. Bulk phonons are visible only in transparent or partially opaque materials.
Acoustic wave velocity resolution	Ranges from 10^{-5} for samples possessing only one SAM to $\approx 10^{-2}$ when multiple acoustic modes with similar velocities are present.	The relative error can be on the order of 5×10^{-3} when a vertical slit is used in the collection.
Acoustic wave attenuation resolution	The relative error in attenuation of leaky SAWs is 1–2%. The major contribution to attenuation is due to the leakage of energy in the coupling fluid. Intrinsic acoustic attenuation inside the material is usually too small to be detected.	The minimum phonon lifetime that can be measured is limited by the parasitic geometrical and instrumental broadening of the spectral line shape. When the vertical slit is used, this correspond to $\Gamma \approx 200$–300 MHz.
Compatibility with solid materials	The fluid used as a coupling medium sometimes reacts with the specimen surface. With water-coupled acoustic microscopy, problems can be encountered due to corrosion of the surfaces of some classes of metals and semiconductors, such as GaAs.	The incident laser power (110–200 mW) focused on the sample surface can produce local heating in the specimen. Sometimes a local disruption of the sample occurs when low-melting-temperature materials are used (Fuellerenes, low-conductivity polymers).

Source: From Ref. 45.

surface acoustic modes and also to give information about bulk acoustic phonons. A comparison with quantitative acoustic microscopy is summarized in Table 7.2.

7.5. Experimental Results

7.5.1. Observations of Surface Waves, Pseudosurface Waves, and Guided Waves

Results in this section primarily illustrate the discussion about semi-infinite materials and layered structures in Section 7.2.

7.5.1.1. Semi-Infinite Media

The simplest case in which a SAW can exist is the plane-free surface of a semi-infinite homogeneous medium. A medium behaves as though semi-infinite when its thickness is much larger than the penetration depth of the SAW displacement field. The penetration depth is on the order of the surface acoustic wavelength $\sim q_x^{-1}$, e.g., in light scattering, on the order of the wavelength of the incident light. In a Brillouin-scattering experiment using visible light, a medium whose thickness exceeds a few microns thus behaves as though semi-infinite.

Detecting SAWs using light scattering was made possible by the development of tandem multipass Fabry–Perot interferometry (see Section 7.3). Using this high-contrast spectrometer, the RW and the continuum of waves above the transverse threshold (*Lamb shoulder*)[5] were detected in semi-infinite samples of semiconductors,[43–45] insulators,[46] and metals.[47,48] The Brillouin spectrum of a GaAs(001) sample is presented in Fig. 7.10.[45] The RW is the only *true* surface wave, since its acoustic Poynting vector is parallel to the surface, and the particle displacement field decays exponentially in the medium. The continuous spectrum, as discussed in Section 7.3.3 consists of acoustic waves composed of propagating bulk transverse waves and evanescent longitudinal waves. For frequencies below the longitudinal threshold $\omega_l = v_l q_x$, their displacement field is somewhat localized at the surface, thus conferring some surface character on them. The continuum of waves can be considered as the Lamb modes of a free-standing slab having infinite thickness; e.g., in the limit, $hq_x \to \infty$ (where h is the sample thickness).

Pseudosurface waves are SAWs whose Poynting vector is not parallel to the surface, but it has a perpendicular component that radiates energy into the bulk of the medium.[19] Energy is thus leaked to bulk waves, and pseudosurface waves are attenuated as they propagate. In several cases, attenuation is however small enough so that it is possible to detect these modes as surface waves with a finite lifetime. Energy transfer to bulk waves is possible, since the phase velocities of pseudosurface waves always lie above the transverse wave velocity.

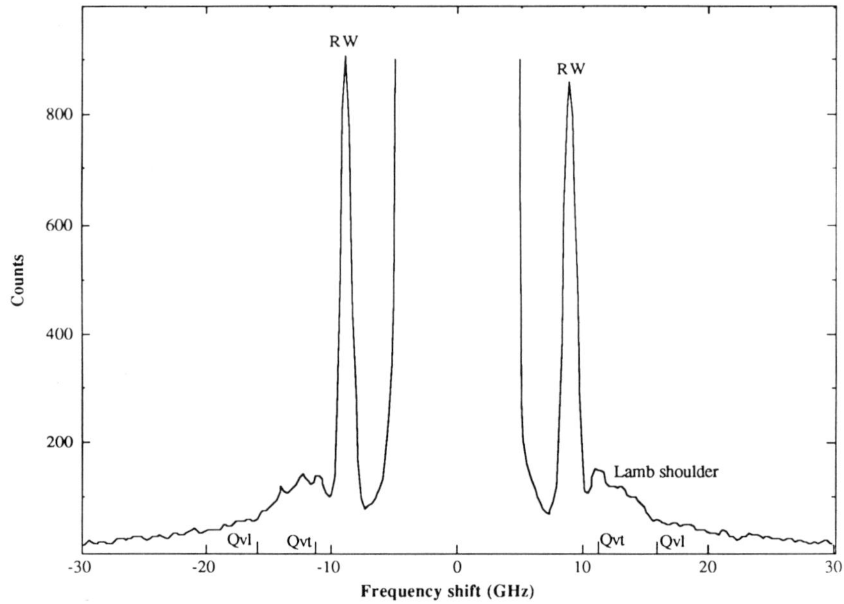

Figure 7.10. The Brillouin spectrum of GaAs(001) along ⟨110⟩. The angle of incidence was 60° and the *FSR* 100 GHz; RW Rayleigh wave peak. The transverse (Ω_t) and longitudinal (Ω_l) thresholds are shown. (from Ref. 45)

Two types of pseudosurface waves can be observed in a semi-infinite medium: the so-called *pseudo-SAW* (PSAW)[(49,50)] and the HFPSW (high frequency pseudosurface wave).[(43,44)] While the PSAW can exist only around specific directions in some anisotropic substrates, the HFPSW exists along any direction for anisotropic and isotropic materials.

The PSAW in cubic materials, as predicted by Lim and Farnell,[(50)] was observed by Brillouin scattering for directions close to the [110] on the (001) surface of Si and GaAs.[(15)] Along these directions, the Rayleigh wave approaches degeneracy with the bulk shear horizontal mode, and it is not observed by Brillouin scattering. The velocity of the PSAW lies between velocities of the shear horizontal and shear vertical bulk waves.

The HFPSW, also called *longitudinal resonance* (LR) due to its strong longitudinal polarization, was observed in isotropic materials[(21)] and cubic crystals[(43,44)]; its velocity, as discussed in Ref. 22, lies between the velocity of the transverse and longitudinal bulk waves. In materials having a Poisson ratio $\sigma < 1/3$, the HFPSW velocity coincides with the longitudinal velocity. The case where the RW, PSAW, and HFPSW all coexist was studied by Carlotti and others[(44)] in GaAs($1\bar{1}1$) for propagation directions scanning from [110] to [121]. Along [121] the PSAW disappears, and the RW has pure sagittal components. In the same work, attenuation of the pseudosurface modes was also analyzed.

Interesting modifications of the spectrum of acoustic waves at the surface of a semi-infinite medium appear when the surface is modulated by a periodic corrugation. This was recently observed and explained when the ratio of the height of the corrugation and its periodicity is much smaller than unity (*small roughness limit*).[51,52] The corrugation periodicity introduces a reduced Brillouin zone, and it causes phonon branches to fold; observed phenomena are similar to those occurring in the vibrational properties of superlattices. In the latter case, only the Brillouin zone of bulk phonons is involved, while the surface corrugation implies the folding of both the discrete spectrum of surface modes (RW) and the continuous spectrum of bulk modes. This is seen in Fig. 7.11, which shows the perturbation of the surface phonon spectrum due to a sinusoidal corrugation of periodicity Λ. Periodicity introduces a surface zone boundary at $Q_B = \pi/\Lambda$, and the dispersion curve branches fold at the zone boundary. At the intersection of folded branches, hybridizations occur, which open forbidden gaps in the dispersion curves. For the RW, the intersection and the gap occur at the surface zone boundary. Another intersection occurs between the RW branch and the folded branch of the longitudinal resonance. Near this intersection, hybridization occurs between the RW folded branch and the longitudinal resonance folded branch. This hybridization rotates the longitudinal polarization of the LR, makes it observable in the Brillouin spectra, and opens a gap at the intersection of the dispersion branches (see Fig. 7.12).

7.5.1.2. Films on a Substrate

The simplest case of a layered medium is that of an isotropic film deposited on a substrate. When the transverse velocity of the film v_t^F is smaller than that

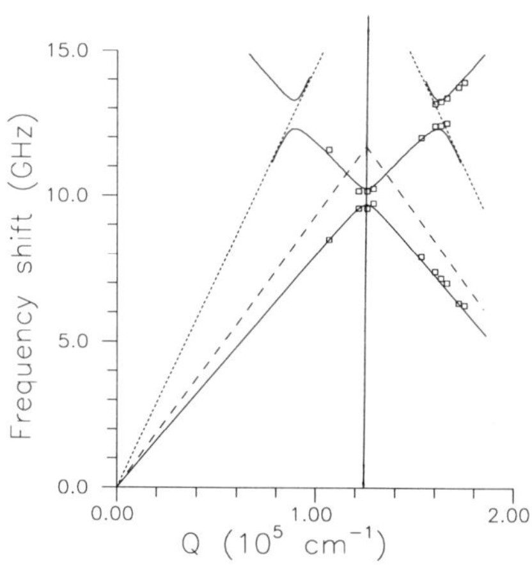

Figure 7.11. Dispersion relations for surface and pseudosurface waves (*solid lines*) for a Si(001) grating (Λ = 2500 Å). The surface zone boundary (Q_B = 1.26 × 10^5 cm^{-1}) is marked by a vertical line. The slanting straight lines represent the transverse threshold (*dashed line*) and the longitudinal threshold (*dotted line*). (from Ref. 52)

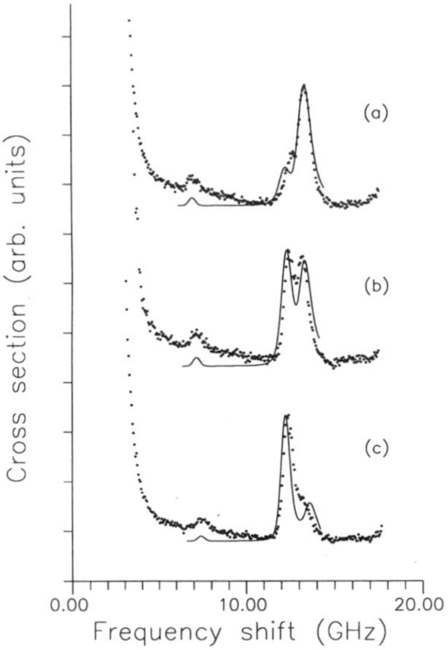

Figure 7.12. Calculated (*solid lines*) and experimental (*dots*) Brillouin spectra for the Si(001) grating. The transferred parallel wave vector is (a) 1.58×10^5 cm^{-1}, (b) 1.61×10^5 cm^{-1}, and (c) 1.64×10^5 cm^{-1}. (from Ref. 52)

of the substrate v_t^S, in addition to the RW, guided waves can exist whose particle displacement is localized either in the film or at the interface. These are the *generalized Lamb waves* or *Sezawa waves,* localized in the film and polarized in the sagittal plane[(29)]; the *generalized Love waves,* localized in the film and polarized parallel to the surface[(29)]; and the *Stoneley wave,* localized at the interface.[(18)]

As discussed in Section 7.3.3, an acoustic wave can scatter an incident optical wave by two different mechanisms: the elasto-optic effect and the ripple effect. The ripple effect is active only for acoustic waves having a shear vertical component, while the elasto-optical mechanism (provided the relevant elasto-optic coefficients are not too small) is active for waves of any polarization, including longitudinal and shear horizontal modes. If the film is transparent, scattering can occur by the ripple effect at the external surface and at the interface, and by the elasto-optic mechanism in both the film and the substrate. The Brillouin-scattering cross section is given by the combination of these four contributions, and interference effects between them can be relevant in determining the intensity of peaks.[(27)] Strong interference effects were observed for sagittal modes in a film of silica grown on a silicon substrate (see Fig. 7.13).[(53)] Peaks corresponding to Sezawa and pseudo-Sezawa waves, e.g., having a phase velocity larger than the transverse velocity of the substrate, were observed. Intensities of the peaks could be explained only by taking into account the interference between the four contributions in the cross section. In the case of CaF_2 films epitaxially

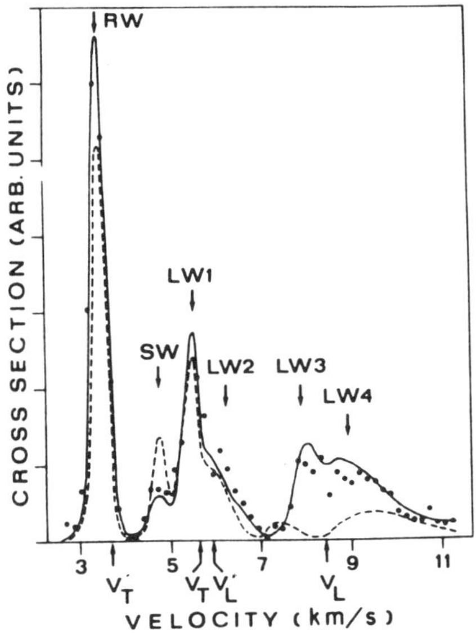

Figure 7.13. Measured Stokes spectrum (*dots*) and calculated cross section (*p-p*) for a 2250-Å film of SiO_2 on Si(001). *Solid line:* calculated cross section with ripple and elastooptic contributions fitted to the amplitude of the experimental RWave peak. *Dashed line:* calculated cross section with ripple terms only. The RW denotes the Rayleigh wave, SW, the Sezawa wave, LW1, etc., higher order Sezawa waves; v_T and v_L are the transverse and longitudinal velocities, respectively, of the film, and v'_T and v'_L are those for the substrate. (from Ref. 53)

grown on a silicon substrate,[54] the elasto-optical effect in the film is small, and the ripple mechanism at the film surface and the interface accounts for most of the scattered light. The theoretical analysis of interference between the two contributions shows a dependence of the peak intensity on the CaF_2 film thickness. Cancellation of some peaks, corresponding to destructive interference for specific values of thickness and incidence angle, is observed. In the case of opaque films, scattering occurs only by the ripple mechanism at the external surface; in such cases interference effects do not arise in the cross section. The RW, Sezawa waves, and a continuum of pseudo-Sezawa waves were observed by Brillouin scattering in a 2000-Å film of aluminium on silicon (see Fig. 7.14). Calculating the ripple effect at the Al surface reproduced experimental results very well.[55]

By using SBS, more exotic SAWs, not observable with other acoustic techniques, such as acoustic microscopy, have also been successfully investigated.

Stoneley waves at a solid interface were observed for the first time in *p-p* scattering by Bell and others[57] in Mo-sputtered films on a glass substrate. The interface was accessible thanks to the transparency of the substrate performing the SBS experiment at the Mo–glass interface (see Fig. 7.15). The Stoneley wave was also measured by Jorna and others[56] in Ni films on glass.

Generalized Love waves were first observed in Cu/Nb metallic superlattices[58] grown on a sapphire substrate. Generalized Love waves were identified according to selection rules for scattering from shear horizontal modes (see Section 7.3.2) using *p*-polarized incident light and for analyzing the polarization

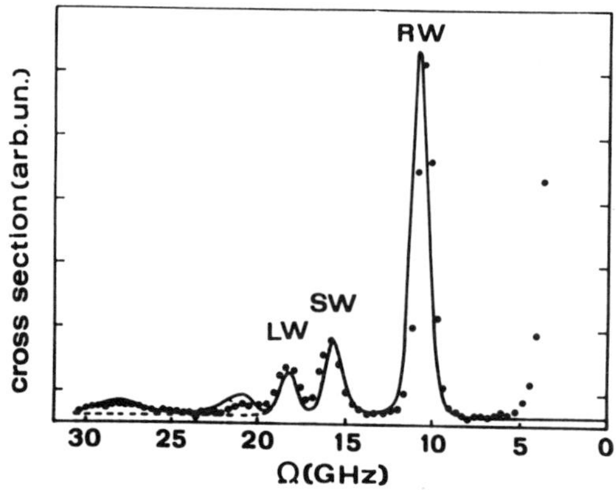

Figure 7.14. Measured (*dots*) and calculated (*solid line*) Stokes spectrum of a 2000-Å film of policrystalline Al on Si(001). The RW denotes the Rayleigh wave, SW, the Sezawa wave, LW, the second-order Sezawa wave. (from Ref. 55)

of scattered light (see Fig. 7.16). The *p-p* spectrum is dominated by the ripple mechanism at the superlattice–sapphire interface. The *p-p* spectrum is due to scattering from sagittal modes, while the *p-s* spectrum contains peaks due to generalized Love waves. Generalized Love waves can be detected by the elastooptical mechanism from the tail of the particle displacement field in the sapphire substrate, performing the experiment from the side of the sapphire. Peaks associated with *shear horizontal* (SH) waves were also observed in samples of Si/SiO_2/Si structures prepared by *separation by implantation of oxygen* (SIMOX).[59] In these experiments, SH modes in the continuum part of the spectrum were observed for the first time. The samples consisted of a Si(001) buffer with a 110-nm-thick SiO_2 layer, buried 350 nm below the surface. In the discrete part of the spectrum, analysis of SH acoustic modes in a bilayered structure shows the existence of a

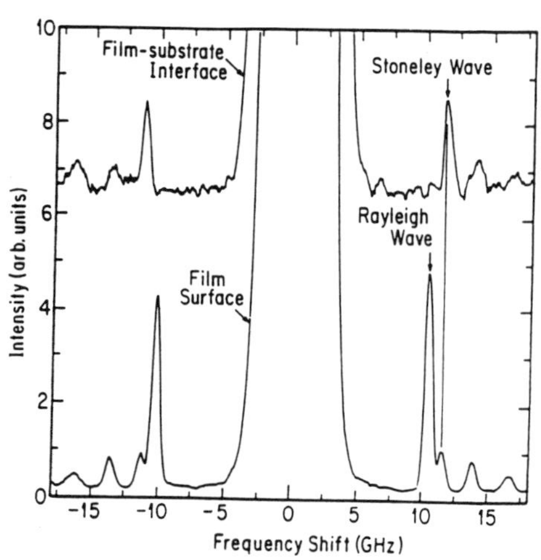

Figure 7.15. Detecting the Stoneley wave. Measured Brillouin spectra (*p-p* scattering) for a 4000-Å film of Mo on glass (incident angle 59°, scattering angle 70°). Spectra were acquired at the Mo-glass interface (*upper trace*) and the Mo-free surface (*lower trace*). (from Ref. 57)

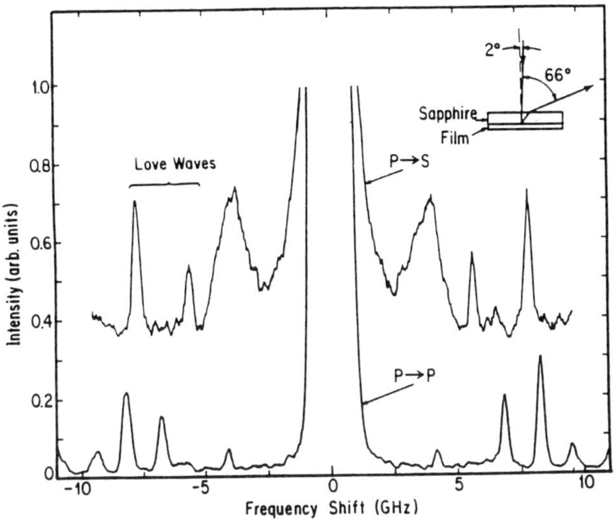

Figure 7.16. Detecting the generalized Love wave. Measured Brillouin spectra for a 3180-Å-thick Cu/Nb superlattice on sapphire. Spectra were acquired at the superlattice-sapphire interface. The inset shows the scattering and sample geometry. (from Ref. 58)

genuine SH-guided wave localized in the silica layer (a sort of *SH-guided mode*). Beyond the silicon transverse threshold, the theory predicts the existence of a pseudo-SH wave with a high degree of confinement in the surface silicon layer. Both these waves were experimentally observed, and their dispersion relation was measured (see Fig. 7.17).

Surface Brillouin scattering led to the observation and identification of a new class of long-wavelength acoustic excitations,[60] called *longitudinal guided modes* (LGM). These modes exist when the velocity of the longitudinal acoustic wave in the film v_l^F is smaller than the corresponding velocity in the substrate v_l^S. These modes were observed for a ZnSe film on a GaAs substrate.[60] The displacement field of the LGM is confined in the film, and its longitudinal component is overwhelming. The LGM couples with light solely through the elastooptic mechanism. The propagation velocity of the LGM falls in the interval between v_l^F and v_l^S, but it is much larger than the transverse velocity of the substrate. The LGM is therefore a pseudosurface mode that radiates energy in the substrate, and it therefore has a finite line width. In experiments the line width was found to be small (typically 0.5% of mode frequency), indicating a very weak radiative coupling in the substrate, which allows a clear detection of the mode, as shown in Fig. 7.18. The measured intensity of the LGM peak depended strongly on the thickness of the film. The authors explained this as a resonance condition for the existence of the LGM, implying constructive interfer-

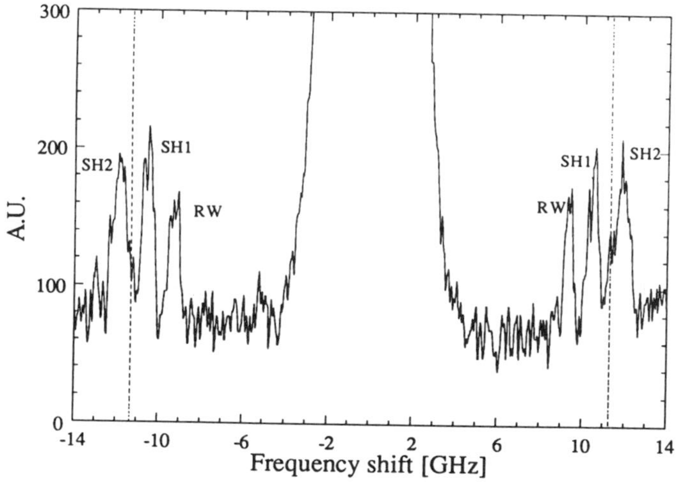

Figure 7.17. Detecting SH waves in the continuum. Measured Brillouin spectrum along ⟨100⟩ at $\theta = 30°$ for a Si(001)/SiO_2/Si(001) structure (*p*-polarized incident field, no analysis of the scattered field polarization). The transverse threshold of the substrate is marked by a vertical dashed line. Key: RW-Rayleigh wave, SH1-shear horizontal wave in the discrete (generalized Love wave), SH2-shear horizontal wave in the continuum (generalized pseudo-Love wave). (from Ref. 59)

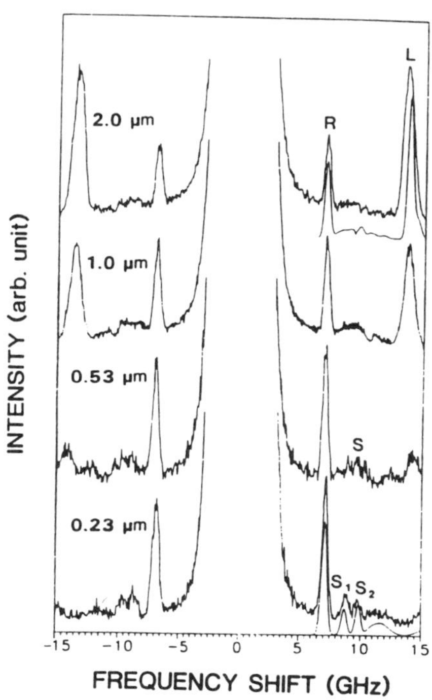

Figure 7.18. Measured Brillouin spectra of ZnSe(001) films grown on GaAs(001) along ⟨100⟩ at $\theta = 60°$. Film thicknesses h are indicated in the figure. In two cases ($h = 2$ and 0.23 μm), a comparison is made with calculated intensities (*thin lines*). Key: R-Rayleigh wave; S_1, S_2-Sezawa waves; L-longitudinal guided mode. (from Ref. 61)

ence of partial waves following two reflections at the external surface and the internal surface.

Surface Brillouin scattering was also used to study the properties of SAWs in a supported film at the *cluster-to-layer* transition. Brillouin spectra were measured by Hillebrands and others[61] in the case of Au films deposited on an NaCl(001) surface at several stages in the deposition process. Initially the film has the structure of isolated *islands* or *clusters;* as the amount of deposited gold increases, these islands grow and coalesce until at the percolation threshold, an incomplete continuous film is present. Film coverage becomes complete with a further deposition of gold atoms. A continuous decrease of the velocity of the RW with the film coverage is observed, as well as the appearance of a new surface mode below the layer percolation threshold. This localized mode was attributed to oscillation modes of the isolated clusters of gold that were coupled by the substrate.[28]

In the results just reported, the velocity of the transverse acoustic wave in the film v_t^F is smaller than the corresponding velocity in the substrate v_t^S. These structures are usually called the *slow-on-fast* type. Using SBS it was possible to observe the transformation of the RW in CaF_2/GaAs($1\bar{1}1$) heterostructures of the *fast-on-slow* or *accelerating* type.[62] Performing measurements by varying the thickness of the film demonstrated for the first time that in supported films of this type, the RW disappears at a critical film thickness. For propagation along $\langle 121 \rangle$ at the disappearance of the RW, a pseudosurface wave whose velocity tends to the RW of CaF_2($1\bar{1}1$) at a large value of film thickness is detected in the spectrum. Because of this behavior, the authors identified this pseudosurface wave as the *pre-Rayleigh* mode of the film. When propagation is along $\langle 110 \rangle$, the role of the *pre-Rayleigh* mode of the film is taken by the pseudosurface acoustic mode (see Fig. 7.19). The theoretical discussion was performed by comparing the Brillouin spectra with the surface projected density of states for shear vertical phonons calculated for CaF_2/GaAs($1\bar{1}1$) heterostructures.

7.5.2. Applications to Materials Science

In this section we summarize the most relevant applications of surface Brillouin scattering to the study of material properties. In particular we focus on determining elastic constants of substrates, thin films, and superlattices; and studying structural modifications caused by surface treatments, such as polishing damage and ion implantation.

7.5.2.1. Elastic Constants of Substrates, Thin Films, and Superlattices

The main advantage of Brillouin scattering for measuring material properties of substrates is its ability to sample elastic properties of regions 200–1000 nm

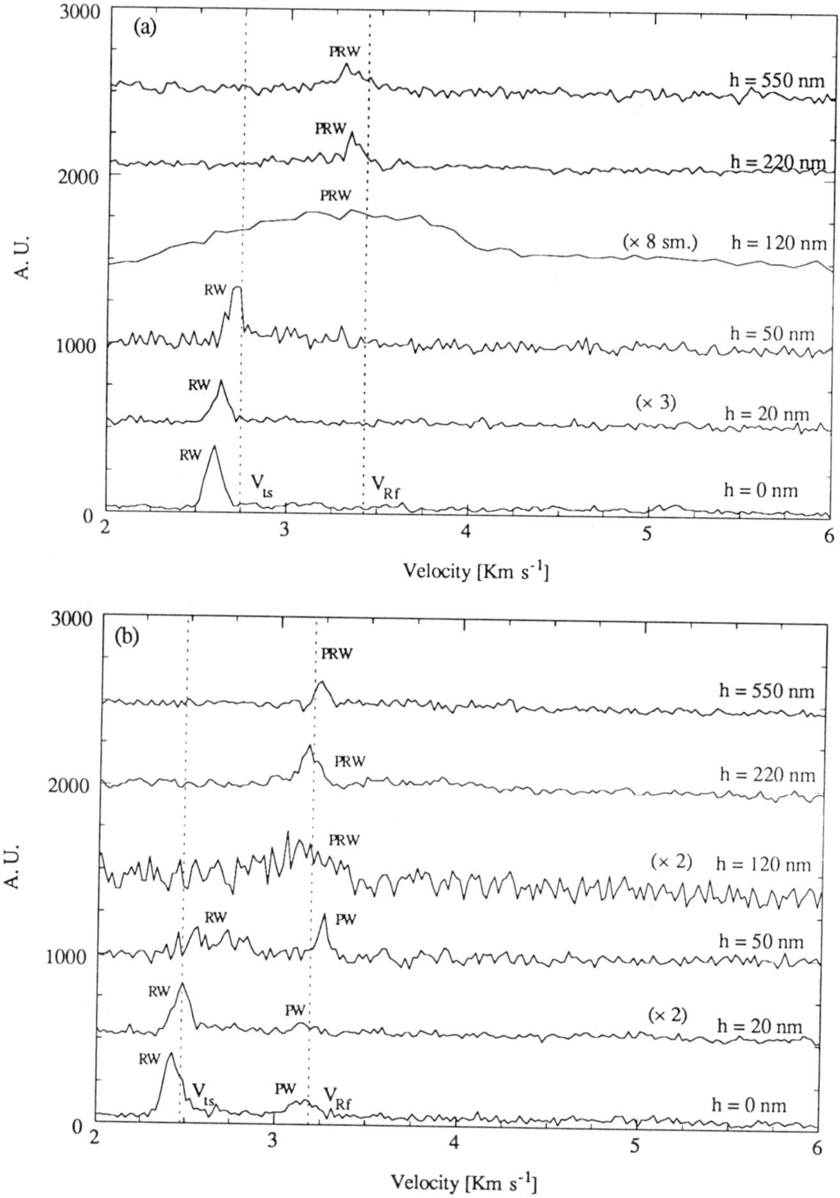

Figure 7.19. The SAW in a *fast-on-slow* structure. Experimental Brillouin spectra of CaF_2/GaAs(1$\bar{1}$1) heterostructures with varying values of film thickness *h* indicated in the figure; (a) propagation along ⟨121⟩, (b) propagation along ⟨110⟩. The transverse wave velocity in the GaAs substrate (V_{ts}) and the RW velocity in the CaF_2 layer (V_{Rl}) are shown as vertical dashed lines. Key: RW-Rayleigh wave, PW-pseudosurface wave, PRW-pre-Rayleigh wave. (from Ref. 62)

below the surface. Very small samples can therefore be investigated. In addition in transparent materials, information about elastic properties can be obtained from the combined measurements of SAWs and bulk acoustic waves. Brillouin scattering was used to study small $YBa_2Cu_3O_{7-\delta}$ (YBCO) single crystals of 0.1-mm thickness.[63,64] The values of C_{11}, C_{33}, and C_{44} were determined by the combined measurement of the LR, which propagates with a velocity $\nu_{LR} \approx \nu_l$ and the RW. A systematic work on the elastic constants of $Bi_2Sr_2CaCu_2O_{8+\delta}$ (BISCO) samples was performed by Boekholdt and others[65] making independent measurements in (001) and (010) planes. In these experiments, the position of the peaks corresponding to the RW, LR, and bulk transverse acoustic modes were measured. The independent determination of C_{11} and C_{33}, performed from the ν_{LR} data made it possible to estimate all elastic constants from the fitting of dispersion relations of the RW.

Determining independent elastic constants of a supported film using SWs were discussed in detail in Chapters 5 and 6. It was shown there that by fitting the measured dispersion relation of sagittal SAWs, namely, RWs and Sezawa waves, it is possible to determine the elastic constants of supported films. This can also be done using Brillouin scattering. As shown in Section 7.3.2, various surface acoustic modes of sagittal character can be detected with this method. Difficulties are encountered due to the large experimental errors that occur in determining experimental velocities (see Section 7.4.2). These can cause a relatively high change in the fitted parameters for some of the elastic constants and hence a high variance of the extracted value. Problems can also arise due to the loss of accuracy in SAW measurements when large collection apertures are used (see Section 7.4.3).

Nizzoli and others determined the elastic constants of films of polycrystalline aluminium having hexagonal symmetry on NaCl(001) using SBS.[66] The fit of the dispersion relation of the sagittal discrete modes (RW and Sezawa waves) on the measured velocity values and the fit of the theoretical cross section on the experimental spectra were both performed. The precision of the measured elastic constants was rather poor. This was because when varying values of elastic constants by as much as 20%, a negligible variation was found in both dispersion relations of discrete modes and in the computed cross section.[66] An interesting result of this work was that the width of the pseudo-Sezawa peaks predicted by the theory was broader than the measured one. The authors explained this result by assuming imperfect adhesion of the film, so that the leakage of the pseudo-Sezawa modes in the substrate was less than in the case of perfect adhesion. If their interpretation is correct, it may be feasible to use SBS to study adhesion problems. Other examples of determining elastic constants of supported films using SBS are also found in Ref. 67 for TiN films on Si(001) and in Ref. 54 for CaF_2 films on silicon. Elastic properties of Langmuir–Blodgett films were also determined by Brillouin scattering.[68,69]

Several studies have been conducted to measure effective elastic properties of metallic superlattices that are alternating layers of different metals having thickness d_1 and d_1. The great interest in this subject lies in observed elastic anomalies in the behavior of effective elastic constants as a function of modulation wavelength Λ; i.e., the bilayer thickness $\Lambda = d_1 + d_2$. This experimental finding, not predicted by linear elasticity theory, was first observed using SBS by Kueny and others[(70)] in Nb/Cu superlattices as a decrease in the RW velocity and hence a softening of C_{44}, for $20 \leq \Lambda \leq 40$ Å. Similar results were also found in the same system by Schuller and Grimsditch[(71)] and Bell and others[(72)] (see Fig. 7.20), while no elastic anomaly was observed for shear horizontal motion.[(58)] The decrease in C_{44} was also observed in other systems, such as Mo/Ni,[(73)] V/Ni,[(74)] and Mo/Ta.[(75)] An enhancement of the RW velocity with the modulation wavelength in the range of 50–100 Å was found in Au/Cr superlattices.[(76)] The enhancement of an elastic modulus, called *super modulus effect,* is however now generally believed to be erroneous.[(77)] On the other hand, due to the overwhelming experimental evidence, there are no doubts that a softening of effective elastic constants as a function of the modulation wavelength exists. Several models have been proposed to predict anomalous elastic behavior. These are based on charge transfer between two metal films,[(78)] band structure effects,[(79)] or interface effects.[(80)] The influence of the quality of the interface on the elastic properties of superlattices was studied by means of SBS in Co/Au[(81)] and in a more recent paper on Ag/Ni superlattices.[(82)]

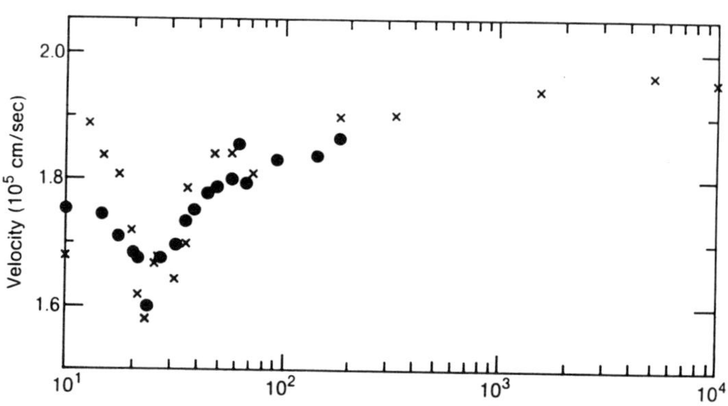

Figure 7.20. Elastic anomalies in the Nb/Cu superlattice. The measured Λ dependence of the C_{44} and C_{11} elastic constants for a series of Nb/Cu superlattice films. (from Ref. 72)

7.5.2.2. Residual Machining Damage

A great potential interest exists in the nondestructive detection of residual damage created by lapping and polishing such brittle materials as semiconductors, ceramics, and glasses. Remnant mechanical damage has a detrimental effect on the performance of electronic devices, or it can cause the catastrophic failure of ceramic components. Surface acoustic waves can probe the mechanical properties of regions near the surface of both isotropic and anisotropic samples. They are therefore very strong candidates for investigating residual machining damage. Among the various SAW-based techniques, quantitative acoustic microscopy and Brillouin scattering have reached a greater stage of instrumental evolution, and these techniques are therefore more suitable for such applications. While the great advantage of quantitative acoustic microscopy lies in the relatively high precision in determining SAW velocity, the frequency is limited to a few hundred megahertz for accurate $V(z)$ measurements. Brillouin scattering can probe thermal RWs having frequencies on the order of 10–20 GHz; this enables us to sample subsurface regions closer to the surface.

Sensitivity to mechanical machining damage of scanning acoustic microscopy (SAM) and SBS was recently explored in two studies by Mendik and others[83] and Mutti.[45] In the former study, nickel single crystals of {100}, {110}, and {111} surface orientation were subjected to mechanical polishing, and the angular dispersion of the RW were measured by means of Brillouin scattering and SAM. The bulk elastic constants were determined by standard ultrasound techniques. The elastic constants measured by Brillouin scattering were smaller than those measured by SAM, which in turn were smaller than those determined by ultrasound techniques (see Fig. 7.21). The softening of elastic constants at the surface was explained by the presence of residual machining damage. The difference in velocity values observed by SBS and SAM was attributed to the different penetration depth of the RW measured by these techniques (approximately 0.3 μm for SBS and 10 μm for SAM). The SBS data showed a lower anisotropy in the angular dispersion relation than that measured by SAM. This was justified by a greater disorder in the near-surface damaged region. After specimens were annealed by sputtering, differences between values obtained by the two methods and the bulk elastic constants measured by ultrasound were less.

Rayleigh wave propagation was used to characterize structural damage in chemomechanically polished GaAs wafers in Ref. 45. For this purpose, line-focus beam (LFB) acoustic microscopy and Brillouin-scattering spectroscopy were used to evaluate velocity and attenuation of the RW. To correlate RW propagation results with the nature of residual polishing damage, a comparative examination of a set of samples, prepared with a different amount of subsurface damage, was performed using destructive and nondestructive techniques. In par-

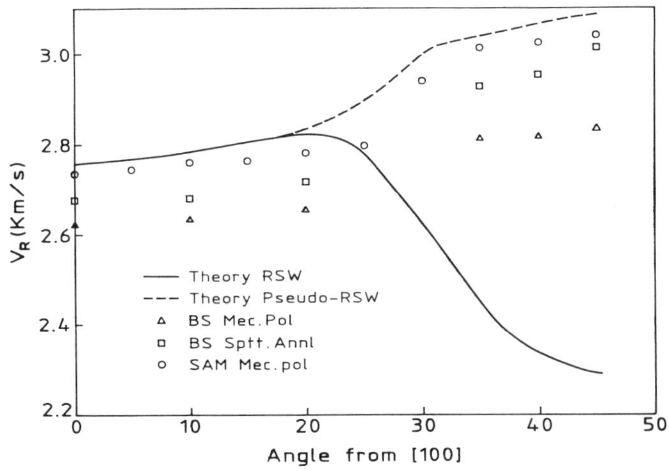

Figure 7.21. Mechanically polished damage in nickel. Measured SAW dispersion on a polished Ni (001) surface using Brillouin scattering (BS) and SAM prior and after annealing by sputtering. (from Ref. 83).

ticular the nature of defects was observed by cross-sectional transmission electron microscopy (TEM), and the presence of lattice strain and disorder was checked by X-ray diffractometry. Different polishing conditions in GaAs(001) surfaces were attained by varying the grit size. Three samples were polished with alumina powder in a solution of sodium hypochlorite using alumina powder of 0.05, 0.3, and 1 μm (*chemomechanical polishing*). A fourth sample was chemically polished using a solution of peroxide alkaline. An X-ray analysis revealed that the chemomechanically polished samples presented a *mosaic* structure with small regions differing slightly in relative crystal orientation. Boundaries between relative misoriented blocks were attributed to regions with a higher concentration of such defects as dislocations and short slip bands, although not many defects were detected by TEM. While using LFB, no significant difference was noticed among the samples; variations in the RW propagation parameters were indeed revealed with Brillouin-scattering spectroscopy. The RW velocity for propagation along ⟨100⟩ and ⟨110⟩ was less in samples that had more residual polishing damage (see Fig. 7.22). The variation was slightly anisotropic and larger for ⟨100⟩ propagation. A considerable attenuation was found for the sample polished with 1-μm grit size. These results are summarized in Table 7.3. For this sample, a velocity decrease in the bulk acoustic phonons was also observed.

From these results, we conclude that SAWs with wavelengths smaller than some hundreds of nanometres can successfully be used to detect residual mechanical damage. Nevertheless some comments are worth making on the possibility of investigating the exact nature of mechanical damage. When high-frequency

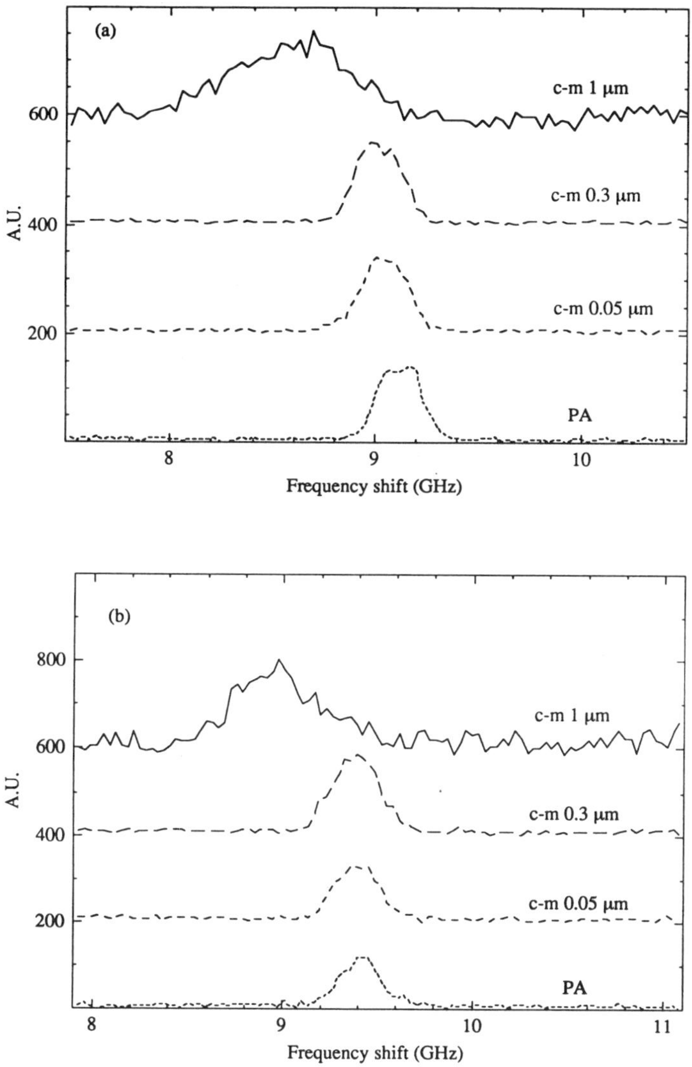

Figure 7.22. Chemomechanical polishing damage in semiconductors. Brillouin-scattering spectra of chemomechanically polished GaAs(001) samples (angle of incidence 60°); (a) ⟨100⟩ propagation, (b) ⟨110⟩ propagation. Key: PA-chemically polished in a peroxide alkaline solution, c-m-chemomechanically polished using 0.05-, 0.3-, and 1-μm alumina powder. (from Ref. 45)

Table 7.3. The RW Velocity Measured by LFB Acoustic Microscopy and SBS in GaAs(001) Samples Chemically Polished and Chemo mechanically (c-m) Polished using Different Abrasive Grain Sizes

Polishing conditions	LFB velocity ⟨100⟩ (m s^{-1})	SBS velocity ⟨100⟩ (m s^{-1})	LFB velocity ⟨110⟩ (m s^{-1})	SBS velocity ⟨110⟩ (m s^{-1})
c-m (1.00 μm)	2728	2549	2873	2670
c-m (0.30 μm)	2728	2681	2873	2797
c-m (0.05 μm)	2724	2695	2871	2797
Chemical	2725	2716	2872	2800

Source: From Ref. 45.

SAWs propagate in samples with residual polishing damage, two aspects must be considered. The first regards the influence of surface roughness on SAW propagation; the second concerns the modification in near-surface mechanical properties of the sample induced by mechanical damage (residual strain, micro-cracking, mosaic structures, etc.). As shown theoretically in Ref. 84, surface roughness causes the RW to become dispersive, inducing velocity decrease and attenuation. On the other hand, softening in the elastic properties of the material can be generated by residual lattice damage, and acoustic wave attenuation can be induced by the subsurface defects introduced during polishing. Both surface roughness and subsurface defects can contribute simultaneously to vary velocity and attenuation in SAWs and separating these contributions can be a problem. To estimate the effect of surface roughness on SAW velocity, we must know the *transverse correlation length a* of the surface profile (e.g., loosely speaking the average distance between peaks and valleys) and the *roughness h* (i.e., the average depth of the corrugated profile). If a random rough surface with a Gaussian cross-correlation function is considered, the strength of the perturbation depends on the $q_x a$ product (q_x SAW vector), and it is proportional to the square of the h/a ratio.[84] An estimate of the effect of surface roughness on the SAW propagation parameters is given in Ref. 45. In that work, according to TEM observations, the transverse correlation length of polished GaAs surfaces was taken equal to 10 nm and the h/a ratio on the order of unity. Using theoretical predictions in Ref. 84, a relative velocity shift of approximately 5% and a normalized attenuation on the order of 10^{-5} was obtained. These results gave the correct order of magnitude for the velocity variation measured in the more heavily damaged specimen, but they were inconsistent with attenuation measurements. This was probably due to the *extra* attenuation that acoustic waves suffered due to the presence of subsurface defects induced by mechanical damage. This was confirmed by observing a decrease in bulk acoustic phonon velocities, not affected by surface roughness for more seriously damaged samples.

7.5.2.3. Ion Implantation

Surface Brillouin scattering was successfully used to characterize the elastic properties of irradiated materials. In a series of studies conducted at the Argonne National Laboratory,[85–87] the variation of the RW velocity as a function of ion dose up to amorphization was measured to characterize the *crystalline-to-amorphous* (c-a) transformation in intermetallic compounds. The implantation was performed at mega-electron-volt energy to produce implanted layers several hundred nanometres thick. Probing only the first 350 nm below the surface, Brillouin scattering was able to provide information on the elastic properties solely of the implanted material. In the case of Nb_3Ir and Zr_3Al,[86,87] a change in RW velocity preceded the onset of amorphization. These results indicated that amorphization was triggered by an elastic instability similarly to the case of melting.[88] Similar experiments were conducted with semiconductors, Si[89] and GaAs,[90] but in those cases, the RW velocity decrease was observed after amorphization. The formation of growing amorphous islands that tend to coalesce with an increasing ion dose was hypothesized to interpret the results.

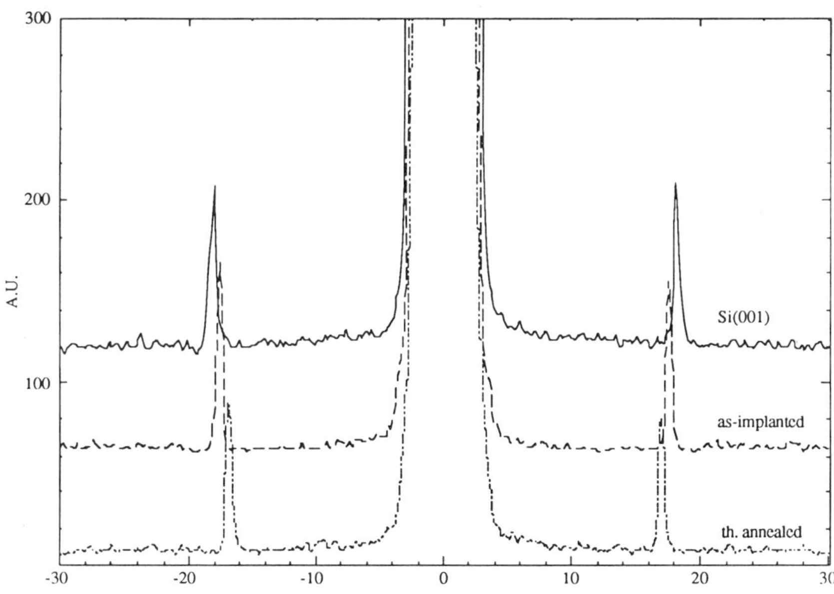

Figure 7.23. Brillouin scattering from a RW oxygen-implanted Si(001). Spectra were taken in backscattering at 70° incidence for propagation along ⟨100⟩. (from Ref. 45)

Table 7.4. The RW Velocity Measured by SBS in High-Dose Oxygen-Implanted Si(001) Samples in As-Implanted Conditions and after Thermal Treatment[a]

Sample	Incidence angle	⟨100⟩ v (m s^{-1})	⟨100⟩ $\Delta v/v$ (%)	⟨110⟩ v (m s^{-1})	⟨110⟩ $\Delta v/v$ (%)
As-implanted	70°	4810	2.6	4923	2.3
Annealed	70°	4635	6.1	4720	6.3
Si(001)	(theor.)	4920	//	5080	//
	70°	4932	//	5035	//

Source: From Ref. 45.
[a]The relative velocity decrease compared to the RW velocity measured in Si(001) is also shown.

High-dose oxygen-implanted silicon samples were studied by Mutti[(45)] in as-implanted conditions and after annealing at high temperature (see Fig. 7.23). Cross-sectional TEM was used to ascertain the nature of irradiation damage. In oxygen-implanted samples, a decrease in the Rayleigh phonon frequency shift with respect to Si(001) was observed; the variation was larger for the thermal annealed sample. Measured RW velocities are summarized in Table 7.4 and compared with the RW velocity for Si(001). The main reason for the velocity decrease in oxygen-implanted specimens was traced to the variation of elastic constants of the implanted material due to the presence of oxide precipitates in the silicon matrix. The density variation was neglected, since crystalline silicon and silicon dioxide have very similar density. According to TEM observations,[(91)] precipitates were present in the as-implanted condition in a region 300 nm thick with an average concentration $n_p \approx 0.4 \times 10^{21}$ cm^{-3} and average diameter $\langle d_p \rangle \approx 0.6$ nm. Since dimensions of the scatterers were much smaller than the acoustic wavelength, elastic constants of the implanted material were calculated using an *effective medium* approximation.[(92)] In the theory of the effective medium, elastic properties of the composite material depend on the volume fraction f occupied by scattering inhomogeneities. This was estimated, assuming spherical precipitates, as $f = (4/3)\pi(d_p/2)^3 n_p \approx 4.5\%$. Since the volume fraction of oxide precipitates was rather low ($f \ll 1$) Eshelby's model of effective elastic properties of an isotropic medium with spherical inclusions was used to calculate the elastic constants of the implanted material.[(93)] Assuming the variation of RW velocity was dominated solely by the decrease in the shear modulus and using the density of the precipitates estimated by Cerofolini and others,[(91)] the authors found an average diameter of the precipitates $\langle d_p \rangle$ on the order of 0.7 nm. This value was in very good agreement with the previous estimate. High-temperature annealing caused a redistribution of oxygen, which results in the dissolution of small precipitates and the growth of large ones.

7.6. Conclusions

Chapter 7 demonstrates the potential of SBS for investigating material properties.

Several applications were presented where Brillouin scattering was particularly successful. These involved determining elastic properties of small substrates and thin films, studying elastic anomalies of metal superlattices, investigating residual damage, and studying structural properties of implanted materials.

The most powerful aspect of SBS in these applications is the possibility of probing the population of thermal acoustic phonons at the surface of the sample. This implies that there is no need to excite a monochromatic acoustic wave, so that thermally generated SAW with frequencies of 10–20 GHz can be observed. Another advantage is that all of the population of surface excitations allowed by the sample geometry can be observed via the ripple or elastooptic contribution. Various types of SAWs not observable by acoustic microscopy can thus be investigated by Brillouin scattering, and more information can be obtained on the elastic properties of surfaces. The main drawback of Brillouin scattering is the limited resolution for SAW velocity determination, which is on the order of 0.5%. Brillouin-scattering experiments also require a long acquisition time, ranging from 1–2 hours per measurement and in the case of RW in semiconductors, up to 6 or 7 hours for some low-intensity modes in supported layers. In transparent material, the surface phonon contribution to the Brillouin cross section can be very low, and measurements can take even longer. The lack of sensitivity to surface phonon lifetimes in Brillouin scattering, caused mainly by the geometrical aperture effect, poses limitations for SAW attenuation measurements. Regarding compatibility with solid materials, the only problem that may arise is laser heating of the surface where an increase in the local temperature of 10–20 °C at the center of the laser spot can occur.

For the sake of completeness, we also mention other very interesting applications of Brillouin scattering to the study of spin waves,[94] diffuse excitations,[5] and phase transitions.[95]

Acknowledgments

The authors thank Dr. L. Giovannini for providing Figs. 11 and 12. Four of us (P. M., C. E. B., G. G., and M. B.) acknowledge funding from the "Progetto finalizzato materiali speciali per tecnologie avanzate" of the Consiglio Nazionale delle Ricerche.

References

1. Brillouin, L. (1922). Propagation of light in a dispersive medium. *Ann. Phys.* **17**, 88–96.
2. Mandelshtam, L. J. (1926). Thermal scattering by crystals. *Zh. Russ. Fiz. Khim. Ova.* **58**, 381–394.
3. Benedek, G. B. and Fritsch, K. (1966). Brillouin scattering in crystals. *Phys. Rev.* **149**, 647–662.
4. Vacher, R. and Boyer, L. (1972). Brillouin scattering: a tool for the measurement of elastic and photoelastic constants. *Phys. Rev. B* **6**, 639–673.
5. Sandercock, J. R. (1982). In: *Light Scattering in Solids III, Topics in Applied Physics* vol. 51 (M. Cardona and G. Güntherodt, eds.), pp. 173–206. Springer, Berlin.
6. Born, M. and Wolf, E. (1964). *Principles of Optics,* Pergamon, Oxford, UK, pp. 100–104.
7. Landau, L. and Lifshitz, E. M. (1958). *Electrodynamics of Continuous Media,* Pergamon, New York, chap. 14.
8. Mills, D. L., Maradudin, A. A., Burnstein, E. (1970). Theory of Raman effect in metals. *Ann. Phys.* **56**, 504–524.
9. Loudon, R. (1978). Theory of lineshapes for normal incidence Brillouin scattering by acoustic phonons. *Phys. Rev. Lett.* **40**, 581–583.
10. Loudon, R. (1978). Theory of surface ripple Brillouin scattering by solids. *J. Phys. C.* **11**, 403–417.
11. Subbaswamy, K. R. and Maradudin, A. A. (1978). Theory of inelastic light scattering by acoustic phonons in an opaque crystal. *Phys. Rev. B* **18**, 4181–99.
12. Rowell, N. L. and Stegeman, G. I. (1978). Brillouin scattering from isotropic metals. *Phys. Rev. Lett.* **41**, 970–73.
13. Rowell, N. L. and Stegeman, G. I. (1978). Brillouin scattering from surface phonons in thin films. *Phys. Rev. B* **18**, 2598–2615.
14. Marvin, A. M., Bortolani, V., Nizzoli, F. (1980). Surface Brillouin scattering from acoustic phonons: I. General theory. *J. Phys. C* **13**, 299–317.
15. Marvin, A. M., Bortolani, V., Nizzoli, F., and Santoro, G. (1980). Surface Brillouin scattering from acoustic phonons: Applications to semiconductors. *J. Phys. C* **13,** 1607.
16. Velasco, V. R. and Garcia-Moliner, F. (1980). Brillouin scattering from surface waves. *Sol. St. Comm.* **33**, 1–5.
17. Dervisch, A. and Loudon, R. (1976). Theory of the Brillouin scattering lineshape in an opaque material. *J. Phys. C* **9**, L669–L673.
18. Auld, B. A. (1973). *Acoustic Waves and Fields in Solids,* vols. 1 & 2, Wiley, New York.
19. Farnell, G. W. (1970). In: *Physical Acoustics,* vol. 6 (W. P. Mason and R. W. Thurston, eds.), pp. 109–160. Academic Press, New York.
20. Bortolani, V., Nizzoli, F., Santoro, G. (1978). Surface density of acoustic phonons in GaAs. *Phys. Rev. Lett.* **41**, 39–42.
21. Glass, N. E. and Maradudin, A. A. (1983). Leaky surface elastic waves on both flat and strongly corrugated surfaces for isotropic, nondissipative media. *J. Appl. Phys.* **54**, 796–805.
22. Camley, R. E. and Nizzoli, F. (1985). Longitudinal resonance in surface acoustic phonons. *J. Phys. C* **18**, 4795–4804.
23. Marvin, A. M., Toigo, F., Celli, V. (1975). Light scattering from rough surfaces: general incident angle and polarization. *Phys. Rev. B* **11**, 1392–1415.
24. Albuquerque, E. L. (1980). Theory of Brillouin scattering by Stoneley waves. *J. Phys. C* **13**, 2623–39.
25. Albuquerque, E. L., Loudon, R., Tilley, D. R. (1980). Theory of Brillouin scattering by Love waves. *J. Phys. C* **13**, 1775–89.
26. Marvin, A. M., Bortolani, V., Nizzoli, F., Santoro, G., Celli, V. (1982). Theory of the three wave mixing of volume EM, surface polaritons and surface acoustic waves. *J. Phys. C* **15**, 3273–80.

27. Bortolani, V., Marvin, A. M., Nizzoli, F., Santoro, G. (1983). Theory of Brillouin scattering from surface acoustic phonons in supported films. *J. Phys. C* **16**, 219–30.
28. Albuquerque, E. L., Oliveros, M. C., Tilley, D. R. (1984). Brillouin scattering from an elastic film. *J. Phys. C* **17**, 1451–63.
29. Farnell, G. W. and Adler, E. L. (1972). In: *Physical Acoustics,* vol. 9 (W. P. Mason and R. W. Thurston, eds.), pp. 35–121. Academic Press, New York.
30. Rowell, N. L. and Stegeman, G. I. (1982). Brillouin scattering in thin film optical waveguides: I. Theory of the phonon modes. *Can J. Phys.* **60**, 1788–1803.
31. Rowell, N. L. and Stegeman, G. I. (1982). Brillouin scattering in thin film optical waveguides: II. Scattering theory. *Can. J. Phys.* **60**, 1804–20.
32. Vaughan, J. M. (1989). *The Fabry–Perot interferometer,* Adam Hilger, Bristol U, UK.
33. Gustaffson, S. E., Khan, M. N., van Eijk, J. (1979). Wavelength scanning and mirror alignment of a new multipass Fabry–Perot interferometer. *J. Phys. E* **12**, 1100–02.
34. Sandercock, J. R. (1976). Simple stabilization scheme for maintenance of mirror alignment in a scanning Fabry–Perot interferometer. *J. Phys. E.* **9**, 566–69.
35. Sandercock, J. R. (1971). "The design and use of a stabilized multipass interferometer of high contrast ratio." In: *Proc. 2nd international conference on light scattering in solids* (M. Balkanski, ed.), pp. 6–12. Flammarion, Paris.
36. Roychoudhury, C. and Hercher, M. (1977). Stable multipass Fabry–Perot interferometer: design and analysis. *Appl. Opt.* **16**, 2514–20.
37. Lindsay, S. M., Anderson, M. W., Sandercock, J. R. (1981). Construction and alignment of a high performance multipass vernier Fabry–Perot interferometer. *Rev. Sci. Instr.* **52**, 1478–89.
38. Mock, R., Hillebrands, B., Sandercock, J. R. (1987). Construction and performance of a Brillouin scattering set up using a triple pass tandem Fabry–Perot interferometer. *J. Phys. E* **20**, 656–59.
39. Cannell, D. S. and Benedek, G. B. (1970). Brillouin scattering from xenon near its critical point. *Phys. Rev. Lett.* **25**, 1157–1161.
40. Dil, J. G. and Brody, E. M. (1976). Brillouin scattering from isotropic metals. *Phys. Rev. B* **14**, 5218–27.
41. Sandercock, J. R. (1980). US Patent 4 225 236.
42. Briggs, G. A. D. (1993). *Acoustic Microscopy,* Clarendon Press, Oxford, UK.
43. Loudon, R. and Sandercock, J. R. (1980). Analysis of the light scattering cross section for surface ripples on solids. *J. Phys. C.* **13**, 2609–22.
44. Carlotti, G., Fioretto, D., Giovannini, L., Nizzoli, F., Socino, G., Verdini, L. (1992). Brillouin scattering by pseudosurface acoustic modes in $(1\bar{1}1)$ GaAs. *J. Phys. Cond.* **4**, 257–62.
45. Mutti, P. (1993). Surface Acoustic Waves for Semiconductor Characterization. D. Phil. Thesis, Oxford University.
46. Güntherodt, G., Blumenröder, S., Hillebrands, B., Mock, R., Zirngrieb, E. (1985). Light scattering in metallic compounds and thin supported layers. *Z. Phys. B* **60**, 423–32.
47. Sandercock, J. R. (1978). Light scattering from surface acoustic phonons in metals and semiconductors. *Sol. St. Comm.* **26**, 543–51.
48. Bassoli, L., Nizzoli, F., Sandercock, J. R. (1986). Surface Brillouin scattering from polycrystalline gold. *Phys. Rev. B* **34**, 1296–99.
49. Lim, T. C. and Farnell, G. W. (1968). Elastic surface wave propagation in crystals. *J. Appl. Phys.* **39**, 4319–21.
50. Lim, T. C. and Farnell, G. W. (1969). Character of pseudo surface waves in anisotropic crystals. *J. Acoust. Soc. Am.* **4**, 845–48.
51. Dutcher, J. R., Lee, S., Hillebrands, B., McLaughlin, G. J., Nickel, B. G., Stegeman, G. I. (1992). Surface-grating-induced zone folding and hybridization of surface acoustic modes. *Phys. Rev. Lett.* **68**, 2464–66.
52. Giovannini, L., Nizzoli, F., Marvin, A. M. (1992). Theory of surface acoustic phonon normal

modes and light scattering cross section in a periodically corrugated surface. *Phys. Rev. Lett.* **69**, 1572–78.

53. Bortolani, V., Nizzoli, F., Santoro, G., Sandercock, J. R. (1982). Strong interference effects in surface Brillouin scattering from a supported transparent film. *Phys. Rev. B* **25**, 3442–45.
54. Karanikas, J. M., Sooryakumarand, R., Phillips, J. M. (1989). Brillouin cross section and localized phonons in CaF_2/Si heterostructures. *Phys. Rev. B* **39**, 1388–91.
55. Bortolani, V., Nizzoli, F., Santoro, G., Marvin, A., Sandercock, J. R. (1979). Brillouin scattering from surface phonons in Al-coated semiconductors. *Phys. Rev. Lett.* **43**, 224–26.
56. Jorna, R., Visser, D., Bortolani, V., Nizzoli, F. (1989). Elastic and vibrational properties of nickel films measured by surface Brillouin scattering. *J. Appl. Phys.* **65**, 718–825.
57. Bell, J. A., Zanoni, R. J., Seaton, C. T., Stegeman, G. I., Mokous, J., Falco, C. M. (1988). Elastic constants of, and Stoneley waves in molybdenum films measured by Brillouin scattering. *Appl. Phys. Lett.* **52**, 610–12.
58. Bell, J. A., Zanoni, R. J., Seaton, C. T., Stegeman, G. I., Bennet, W. R., Falco, C. M. (1987). Brillouin scattering from Love waves in Cu/Nb metallic superlattices. *Appl. Phys. Lett.* **51**, 652–54.
59. Bottani, C. E., Ghislotti, G., Mutti, P. (1994). Brillouin scattering from surface horizontal surface phonons in buried SiO_2 layers. *J. Phys. Cond.* **6**, L85.
60. Hillebrands, B., Lee, S., Stegeman, G. I., Cheng, H., Potts, J. E., Nizzoli, F. (1988). Evidence for the existence of guided longitudinal acoustic phonons in ZnSe films on GaAs. *Phys. Rev. Lett.* **60**, 832–35.
61. Hillebrands, B., Mock, R., Güntherodt, G., Bechthold, P. S., Herres, N. (1986). Dispersion of localized elastic modes in thin supported gold layers measured by Brillouin scattering. *Solid State Comm.* **60**, 649–52.
62. Aleksandrov, V., Bottani, C. E., Caglioti, G., Ghislotti, G., Marinoni, C., Mutti, P., Yakovlev, N. L., Sokolof, N. S. (1994). Experimental evidence of the transformation of the Rayleigh surface phonon in GaAs($1\bar{1}1$) heterostructures of the accelerating type. *J. Phys. Cond.* **6**, 1947–54.
63. Baumgart, P., Blumenroder, S., Erl, A., Hillebrands, B., Güntherodt, G., Schmidt, H. (1989). Sound velocities of $YBa_2Cu_3O_{7-\delta}$ single crystals measured by Brillouin spectroscopy. *Sol. St. Comm.* **69**, 1135–37.
64. Baumgart, P., Blumenroder, S., Erle, A., Hillebrands, B., Splittgerber, P., Güntherodt, G., Schmidt, H. (1989). Sound velocities of $YBa_2Cu_3O_{7-\delta}$ and $Bi_2Sr_2CaCu_2O_x$ single crystals measured by Brillouin spectroscopy. *Physica C* **171**, 173–82.
65. Boekholt, M., Harzer, J. V., Hillebrands, B., Guntherodt, G. (1991). Determination of the sound velocities and the complete set of elastic constants of $Bi_2Sr_2CaCu_2O_{8+\delta}$ single crystals using Brillouin light scattering. *Physica C* **179**, 101–106.
66. Nizzoli, F., Bhadra, R., de Lima, O. F., Brodsky, M. B., Grimsditch, M. (1988). Problems with the determination of elastic constants from higher order surface waves: results for Al on NaCl. *Phys. Rev. B* **37**, 1007–10.
67. Bottani, C. E., Elena, M., Beghi, M., Ghislotti, G., Guzman, L., Miotello, A., Ossi, P. M. (1993). Elastic properties of thin TiN films. *Mat. Res. Soc. Symp. Proc.* **308**, 95–100.
68. Zanoni, R., Naselli, C., Bell, J. A., Stegeman, G. I., Seaton, C. T. (1986). Brillouin spectroscopy of Langmuir–Blodgett films. *Phys. Rev. Lett.* **57**, 2838–40.
69. Nizzoli, F., Hillebrands, B., Lee, S., Stegeman, G. I., Duda, G., Werner, G., Knoll, W. (1989). Determination of the whole set of elastic constants of a polimeric Langmuir–Blodgett films by Brillouin spectroscopy. *Phys. Rev. B* **40**, 3323–28.
70. Kueny, A., Grimsditch, M., Miyano, K., Banerjee, I., Falco, C. M., Schuller, I. K. (1982). Anomalous behaviour of surface acoustic waves in Cu/Nb superlattices. *Phys. Rev. Lett.* **48**, 166–69.

71. Schuller, I. K. and Grimsditch, M. (1986). Effect of strain on the elastic properties of superlattices. *J. Vac. Sci. and Technol. B* **4**, 1444–46.
72. Bell, J. A., Bennet, W. R., Zanoni, R., Stegeman, G. I., Falco, C. M., Seaton, C. T. (1987). Elastic constants of Cu/Nb superlattices. *Sol. St. Commun.* **64**, 1339–42.
73. Khan, M. R., Chun, C. S. L., Felcher, P., Grimsditch, M., Kueny, A., Falco, C. M., Schuller, I. K. (1983). Structural, elastic and transport anomalies in molybdenum/nickel superlattices. *Phys. Rev. B* **27**, 7186–93.
74. Danner, R., Huebener, R. P., Chun, C. S. L., Grimsditch, M., Schuller, I. K. (1986). Surface acoustic waves in Ni/V superlattices. *Phys. Rev. B* **33**, 3696–3701.
75. Bell, J. A., Bennet, W. R., Zanoni, R., Stegeman, G. I., Falco, C. M., Nizzoli, F. (1987). Elastic constants of Mo/Ta superlattices measured by Brillouin scattering. *Phys. Rev. B* **343**, 9004–07.
76. Bisanti, P., Brodsky, M. B., Felcher, G. P., Grimsditch, M., Sill, L. R. (1987). Surface waves in Au/Cr superlattices. *Phys. Rev. B* **35**, 7813–19.
77. Moreau, A., Ketterson, J. B., Davis, B. M. (1990). A new ultrasonic method for measuring elastic moduli in unsupported films: application to Cu/Pd superlattices. *J. Appl. Phys.* **68**, 1622–1627.
78. Huberman, M. L., and Grimsditch, M. (1989). Lattice expansion and contraction in metallic superlattices. *Phys. Rev. Lett.* **62**, 1403–06.
79. Cammarata, R. C. (1986). The supermodulus effect in compositionally modulated superlattices. *Scr. Metall.* **20**, 479–81.
80. Wolf, D. and Lutsko, J. F. (1988). Structurally induced supermodulus effect in superlattices. *Phys. Rev. Lett.* **60**, 1170–73.
81. Hillebrands, B., Krams, P., Sporl, K., Weller, D. (1991). Influence of the interference quality on the elastic properties of Co/Au multilayers investigated by Brillouin scattering. *J. Appl. Phys.* **69**, 938–42.
82. Carlotti, G., Fioretto, D., Socino, G., Rodmacq, B., Pelosin, V. (1992). Interface effects and elastic constants of Ag/Ni superlattices studied by Brillouin scattering. *J. Appl. Phys.* **71**, 4897–902.
83. Mendik, M., Sathish, S., Kulik, A., Gremaud, G., Wacther, P. (1992). Surface acoustic wave studies on single crystal nickel using Brillouin scattering and scanning acoustic microscopy. *J. Appl. Phys.* **71**, 2830–34.
84. Eguiluz, E. G. and Maradudin, A. A. (1983). Frequency shift and attenuation length of Rayleigh wave due to surface roughness. *Phys. Rev. B* **28**, 728–747.
85. Rehn, L. R., Okamoto, P., Pearson, J., Bhadra, R., Grimsditch, M. (1987). Solid-state amorphization of Zr_3Al: evidence of an elastic instability and first order transformation. *Phys. Rev. Lett.* **59**, 2987–90.
86. Grimsditch, M., Gray, K. E., Bhadra, R., Kampwirth, R. T., Rehn, L. R. (1987). Brillouin scattering study of lattice-stiffness changes due to ion irradiation: dramatic softening in Nb_3Ir. *Phys. Rev. B* **35**, 883–85.
87. Okamoto, P., Rehn, L. R., Pearson, J., Bhadra, R., Grimsditch, M. (1988). Brillouin scattering and transmission electron microscopy studies of radiation-induced elastic softening, disordering and amorphization of intermetallic compounds. *J. Less Common Met.* **140**, 231–44.
88. Koike, J. (1993). Elastic instability of crystals caused by static atom displacement: a mechanism for solid state amorphization. *Phys. Rev. B* **47**, 7700–04.
89. Bhadra, R., Pearson, J., Okamoto, P., Rehn, L. R., Grimsditch, M. (1988). Elastic properties of Si during amorphization. *Phys. Rev. B* **38**, 12656–60.
90. Sharma, R. P., Bhadra, R., Rehn, L. E., Baldo, P. M., Grimsditch, M. (1989). Crystalline to amorphous transformation in GaAs during Kr ion bombardment: a study of elastic behavior. *J. Appl. Phys.* **66**, 152–55.
91. Cerofolini, G. F., Bertoni, S., Fumagalli, P., Meda, L., Spaggiari, C. (1993). SiO_2 precipitation in highly supersaturated oxygen implanted single-crystal silicon. *Phys. Rev. B* **47**, 1174–82.

92. Elliott, R. J., Krumhansl, J. A., Leath, P. L. (1974). The theory and properties of randomly disordered crystals and related physical systems. *Rev. Mod. Phys.* **46**, 465–543.
93. Eshelby, J. D. (1959). The elastic field outside an ellipsoidal inclusion. *Proc. R. Soc. A* **252**, 561–80.
94. Nizzoli, F. and Mills, D. L. (1991). In: *Nonlinear Surface Electromagnetic Phenomena* (H.-E. Ponath and G. I. Stegeman, eds.), Chap. 7, North Holland, Amsterdam.
95. Cummins, H. Z. (1983). In: *Light Scattering Near Phase Transitions* (H. Z. Cummins and A. P. Levanyuk, eds.), pp. 1031–65, North Holland, Amsterdam.

8

New Approaches in Acoustic Microscopy for Noncontact Measurement and Ultra High Resolution

Kazushi Yamanaka

8.1. Introduction

As an imaging method of elastic properties and subsurface features in microscopic scale, the scanning acoustic microscope (SAM) provides spatial resolution comparable or superior to that of optical microscopes.[1] Nondestructive evaluation methods of defects and elastic properties in microscopic scale were developed by using the SAM,[2–4] and they have been widely applied to various fields in science and technology.[4–7]

In the SAM however, coupling fluids pose some limitations. For example operation at high temperature or in a vacuum is impossible, and measured SAW data are affected by the perturbation of coupling fluids. As an approach to removing limitations imposed by coupling fluids, a noncontact SAW measurement method using the photoacoustic effect of lasers is being developed; it is specially

---

K. YAMANAKA • Nanotechnology Division, Mechanical Engineering Laboratory, Ministry of International Trade and Industry, Namiki 1–2, Tsukaba, Ibaraki, 305 Japan.

Advances in Acoustic Microscopy, Volume 1, edited by Andrew Briggs.
Plenum Press, New York, 1995.

adapted for precise SAW velocity measurement at frequencies above 100 MHz. This method is described in Section 8.2.

Another major challenge in acoustic microscopy is its resolution. The best resolution of SAM with water as the coupling fluid has been 240 nm at the frequency of 4.4 GHZ (See Ref. 4 Chap 3). At most conveniently realized frequency of 1 GHz, the resolution is about 1 μm. Therefore the resolution of SAM is not always sufficient for nanoscopic defects and advanced micro-nanodevices. In Section 8.3 a new approach to imaging the mechanical property of materials on the nanometer scale is described.

8.2. Phase Velocity Scanning Method

In this section, we describe a novel method for noncontact generation of the surface acoustic waves (SAW) using the phase velocity scanning (PVS) method,[8,9] in which a laser beam is scanned on a opaque solid surface with a velocity equal to the phase velocity of acoustic waves. At ultrasonic frequencies above 100 MHz, scanning interference fringes (SIF) are employed as a thermo-elastic source of the SAW generation.[10,11] The SIF is produced by intersecting two laser beams with a frequency difference, an extension of the bulk acoustic wave generation using electrostrictive mixing by laser beams.[12] For material characterization, SAW velocity within the laser beam spot is measured as the ratio of observed frequency and wave number of the SIF of the SAW.[13] The frequency measurement can be quite precisely performed because of the large number of generated SAW carriers and the amplitude enhancement effect of the PVS method. This principle was successfully applied to anisotropic velocity measurement of SAW and pseudo-SAW on Si(001) surface.[14] This measurement is free from the water-loading effect accompanying leaky SAW measurements.

8.2.1. Principle of the PVS Method

Noncontact excitation of acoustic waves by means of pulsed lasers is expected to provide a nondestructive evaluation (NDE) method of materials in circumstances where fluid or gel coupler cannot be used.[15] However the main limitation of this technique is its low sensitivity compared with the standard piezoelectric transducer method, and the amplitude of acoustic waves that can be generated without damage to the tested object is not large. If we use high-peak power laser to generate strong acoustic waves, ablation damage of the surface results as illustrated in Fig. 8.1(a). Furthermore both SAW and bulk acoustic waves (BAWs) are generated, which sometimes interfere with each other, and directionality cannot be easily controlled. At frequencies higher than

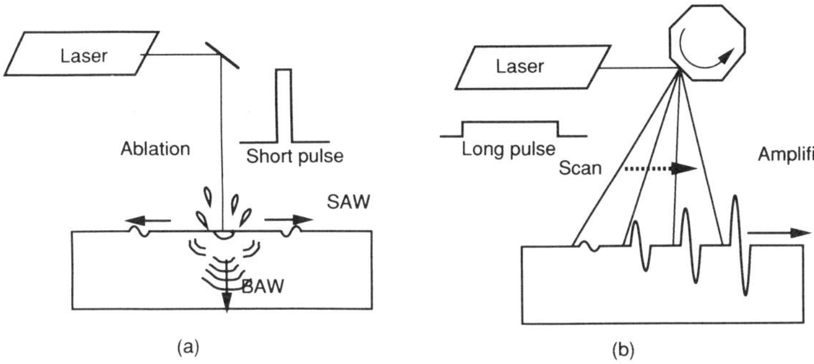

Figure 8.1. Two methods for generating SAWs by using the photoacoustic effect. (a) The laser ultrasonic method; SAW and BAWs are generated together. Sometimes ablation damage is induced. (b) The phase velocity scanning method; only the amplified and unidirectional SAW is generated with much less possibility of ablation damage.

100 MHz, the requirement of the laser is also rather high, and mode-locked picosecond lasers are usually required.[16]

To solve these problems, we developed a novel method for generating the SAW, phase velocity scanning (PVS) method.[8,9] First we scanned a single laser beam (SSB) at the phase velocity of SAW in the 1–3 MHz frequency range. After proving the advantages of the PVS method, another approach was developed to scan interference fringes formed by laser beams[10,11] to achieve SAW generation with frequencies higher than 100 MHz.

The principle of the PVS method is illustrated in Fig. 8.1(b). If a laser beam is scanned along the surface of a solid, the absorbed energy of the laser beam at a specific point forms a heat pulse. The effective length of this heat pulse and the corresponding frequency is given by $\tau = \boldsymbol{d} / \boldsymbol{V}$ and $\boldsymbol{F} = \boldsymbol{V} / \boldsymbol{d}$, where $\boldsymbol{d}$ is the width of the laser beam and $\boldsymbol{V}$ is the scanning velocity. This heat pulse is the thermoelastic source for generating SAW. For example the frequency $\boldsymbol{F}$ = 3 MHz if $\boldsymbol{v}$ = 3000 m/s and $\boldsymbol{d}$ = 1 mm. The generated SAW is coherently amplified if the scanning velocity $\boldsymbol{V}$ is equal to the phase velocity of the SAW.

8.2.2. Observation of Rayleigh and Lamb Waves by the SSB Approach

To verify the amplitude enhancement effect, a broad-band pulse of Rayleigh wave was observed on a 6-mm-thick aluminum plate.[9,13] Figure 8.2 is a waterfall plot of observed waveforms as the scanning velocity $\boldsymbol{V}$ was varied from 2701–2980 m/s across a Rayleigh wave velocity $\boldsymbol{v_R}$ of 2906 m/s. The time delay of the oscilloscope was adjusted depending on the scanning velocity, so that the center of the Rayleigh wave pulse is always located at the same position in the time window.

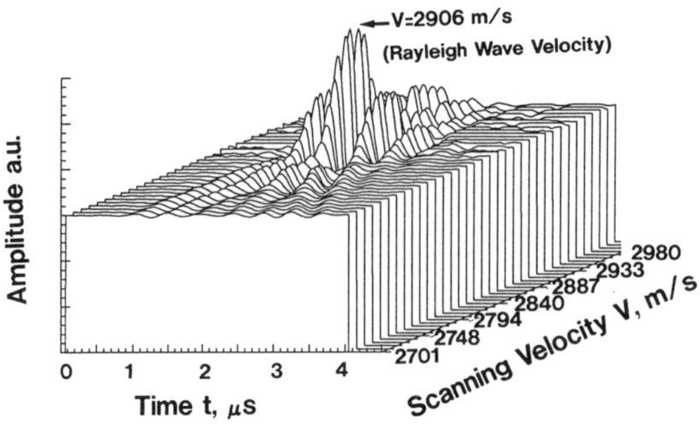

Figure 8.2. A waterfall plot of SAW wave forms on an aluminum sample generated by the SSB with the scanning velocity near the Rayleigh wave velocity.

The amplitude of generated waveforms increased as the scanning velocity V approached v_R, and a significant amplitude enhancement of waveforms took place at $v = v_R$. The amplitude of the Rayleigh wave decreased rapidly when the scanning velocity V was further increased.[9] This result proved that scanning at the SAW velocity is required to generate a large amplitude SAW.

Since in dispersive media, the phase velocity and the group velocity are different, the next problem was to find the velocity at which to scan the laser beam. To solve this problem, we studied the dispersive Lamb waves on an aluminum plate. When the scanning velocity was varied from 2608–3204 m/s on a 1.2-mm-thick plate, acoustic waves with different amplitude, duration, and frequency were observed (see Fig. 8.3). Observed waveforms were clearly separated into two modes depending on whether the scanning velocity were larger or smaller than the Rayleigh wave velocity v_R. Similar results were obtained for thicknesses of 1.5 and 1.9 mm.

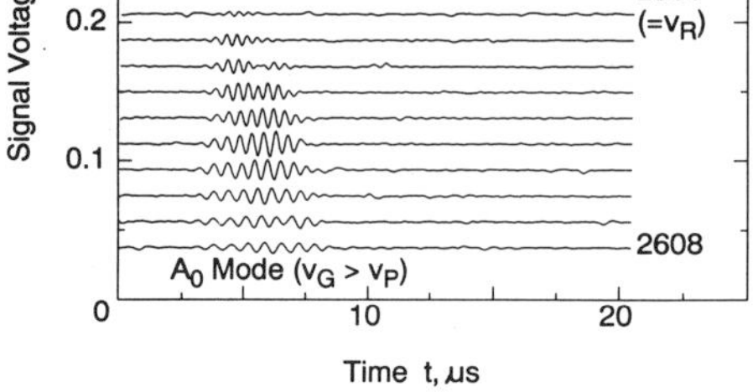

Figure 8.3. Lamb waves on an 1.2-mm-thick aluminum plate generated by the SSB with the scanning velocity near the Rayleigh wave velocity v_R. Waveforms with a scanning velocity below v_R belong to the A_0 mode, and waveforms with a scanning velocity above v_R belong to the S_0 mode.

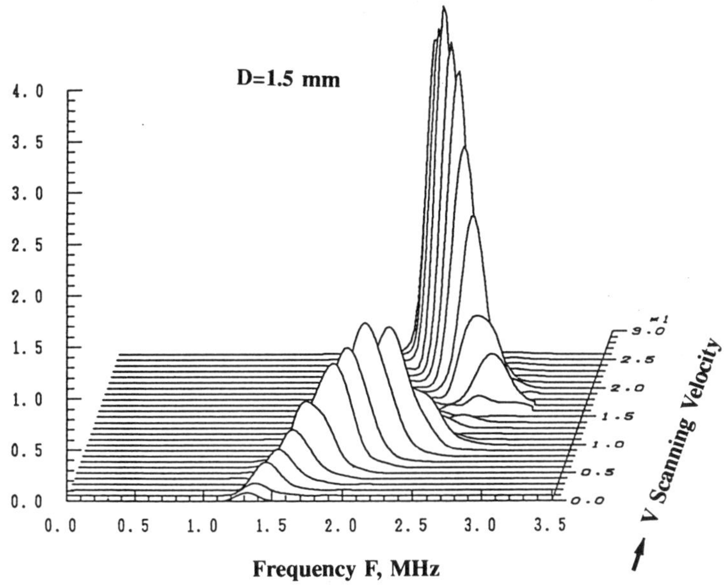

Figure 8.4. Power spectra of Lamb waveforms on a 1.5-mm-thick aluminum plate.

Power spectra of typical Lamb waves ($\boldsymbol{D}$ = 1.5 mm) are plotted in Fig. 8.4 for different scanning velocities $\boldsymbol{v}$. The peak is wider for the small velocity mode and narrower for the larger velocity mode. The frequency $\boldsymbol{F}$ of each waveform was evaluated from the peak position of the power spectra. The value of product $\boldsymbol{FD}$ is plotted in Fig. 8.5 as a function of the scanning velocity $\boldsymbol{V}$ at the ordinate. Note that all data of $\boldsymbol{D}$ = 1.2, 1.5, and 1.9 mm fell on the same curve.

In Fig. 8.5 the solid curve is the calculated phase velocity of the lowest asymmetric A_0 mode, and the dotted curve is the lowest symmetric S_0 mode as a function of the $\boldsymbol{FD}$ value. Obviously the experimentally observed relation between $\boldsymbol{V}$ and $\boldsymbol{FD}$ falls on the calculated phase velocity dispersion curves, and the agreement between them is excellent. The other two curves are the group velocity dispersion of the corresponding modes, which clearly differ from experimental data. This result indicates that the laser-scanning velocity should not be identical to the group velocity but to the phase velocity. This is the origin of the term PVS method.

Interestingly enough the duration of the observed waveform is expressed by a simple equation:

$$\Delta t \cong L \left| \frac{1}{v_G} - \frac{1}{v_P} \right| \tag{1}$$

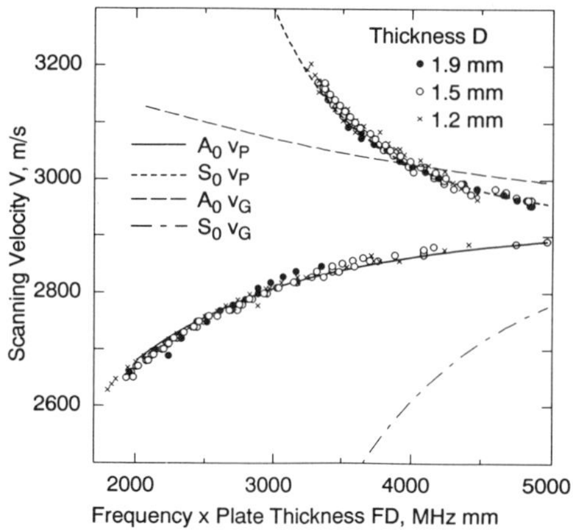

Figure 8.5. The relation between the frequency F of generated Lamb waves multiplied by the plate thickness D and the laser-scanning velocity V (*closed circles*). The solid curve and dotted curve are calculated phase velocities of the fundamental symmetric (S0) and asymmetric (A0) Lamb wave, respectively. The broken curve and chained curve are calculated group velocities of the two modes.

where L is the scanning length and v_P and v_G are the phase and group velocities, respectively.[8] This equation was derived from a consideration that the width of a generated wave should be the difference between a propagation distance of phase (laser beam) and a propagation distance of acoustic wave energy. Then, for waves of large dispersion, this width is large. For waves without dispersion, the width vanishes, and the actual waveform is determined only by the frequency bandwidth, or the laser beam width. In fact in the A_0 mode below 2700 m/s and in the S_0 mode above 3100 m/s, the difference between the phase and group velocities is large, as seen in Fig. 8.5. Correspondingly the width of an observed waveform is large in Fig. 8.3. Note that though this feature looks like a quite common wave phenomenon, it was not described in the literature before. This feature can also be derived by solving thermal and elastic wave equations.[17] Usually narrow-band power spectra are preferable for precise frequency measurements. Since bandwidth is related to the width of tone bursts, Eq. (1) may be useful in designing testing conditions for plates using the PVS method. These observations show that phase velocity plays a dominant role, whereas group velocity determines the pulse width of generated waves. The PVS method can thus provide a precise measurement of phase velocity dispersion related to the thickness or elastic properties of plates and coatings.

8.2.3. The Scanning Interference Fringe (SIF)

Though the SSB approach is useful for generating SAWs with a frequency below 10 MHz and for establishing the principle of the PVS method, another approach is required to achieve SAW generation with frequencies higher than 100 MHz. For this purpose, SIF can be used.

The SIF is regarded as a sequence of scanning laser beams, and each fringe is essentially similarly to the SSB. Since fringe spacing can easily be made less than 30 μm, a typical wavelength of a 100-MHz SAW, the SIF is more suited to generating SAW in the frequency range above 100 MHz. Furthermore the number of fringes can be made quite large, and the bandwidth of generated SAWs can be very narrow. The SIF also has the advantage of selectively excitating BAWs by realizing the phase-matching condition at the surface with BAW propagating in a particular direction.[19]

The SIF is produced by intersecting two laser beams with a frequency difference, as shown schematically in Fig. 8.6. The amplitude I of the laser beams on the surface of the specimen is expressed as,[10]

$$\boldsymbol{I} = \boldsymbol{I}_1 \,\mathbf{exp}\boldsymbol{i}[(-\,\boldsymbol{k}_1 \,\mathbf{sin}\theta)\boldsymbol{x} - \omega_1\boldsymbol{t}] + \boldsymbol{I}_2 \,\mathbf{exp}\boldsymbol{i}[(\boldsymbol{k}_2 \,\mathbf{sin}\theta)\boldsymbol{x} - \omega_2\boldsymbol{t}] \tag{2}$$

where $\boldsymbol{k}_1$, $\boldsymbol{k}_2$, $\boldsymbol{I}_1$, $\boldsymbol{I}_2$ are the magnitudes of the wave number and the amplitudes of the two laser beams and θ is the incident angle of the laser beams. Thus the intensity of the interference fringes is

$$\boldsymbol{I} \cdot \boldsymbol{I}^* = \boldsymbol{I}_1^2 + \boldsymbol{I}_2^2 + 2\boldsymbol{I}_1\boldsymbol{I}_2 \,\mathbf{cos}[(2\boldsymbol{K}\,\mathbf{sin}\theta)\boldsymbol{x} - \omega_a\boldsymbol{t})] \tag{3}$$

where $\boldsymbol{K} \cong \boldsymbol{k}_1, \boldsymbol{k}_2$ is the approximate magnitude of the wave number of the laser beams and $\omega_a = \omega_2 - \omega_1$ is the frequency difference between the two laser beams. Equation (3) shows that interference fringes are scanned along the x-axis at a velocity

$$\boldsymbol{v}_f = \frac{\omega_a}{2\boldsymbol{K}\,\mathbf{sin}\theta} \tag{4}$$

and the wave number is

$$\boldsymbol{k}_f = 2\boldsymbol{K}\,\mathbf{sin}\theta \tag{5}$$

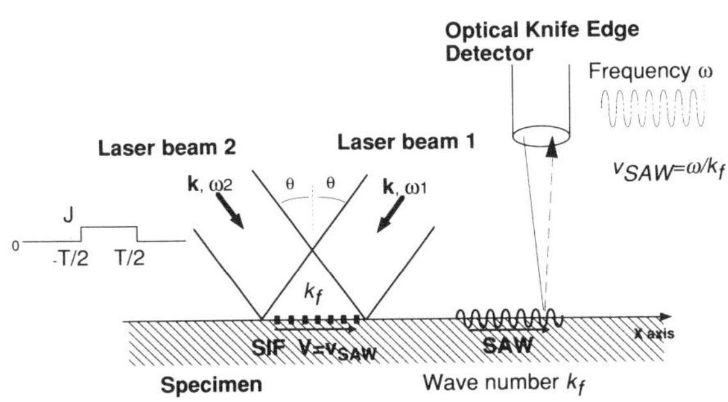

Figure 8.6. The principle of the SIF approach of the PVS method for SAW generation and optical knife edge detection. T, pulse width; J, power density.

When the SIF is formed on optically absorbing surface, a spatially and temporally periodic thermal expansion pattern is introduced. The frequency difference can be introduced by a Bragg cell frequency shifter or a single laser with two frequency oscillation. When the laser pulse width is smaller than the propagation time of the SAW along the laser beam spot, the displacement $u(x,t)$ of the SAW generated by the SIF is[13]

$$\boldsymbol{u(x,t) = AJC_G T}\mathbf{sinc}\left[\frac{(\boldsymbol{k_f v_R} - \omega_a)\boldsymbol{T}}{2}\right]\mathbf{exp}[\boldsymbol{ik_f(x - v_R t)}] \tag{6}$$

where A, J, T, k_f, v_R, and ω_a are absorption coefficient, laser power density, laser pulse width, wave number of SIF, SAW velocity, and the frequency difference of two laser beams, respectively. The coefficient C_G is a material constant for SAW generation, represented by

$$\boldsymbol{C_G} = \frac{\boldsymbol{iK}\alpha}{2\rho^2 \boldsymbol{c_l^2 C X_R}} \frac{\boldsymbol{c_R^2}}{\boldsymbol{c_t^2}} \left(\frac{1}{\boldsymbol{c_t^2}} - \frac{2}{\boldsymbol{c_R^2}}\right) \tag{7}$$

and

$$\boldsymbol{X_R} = \frac{8 - [(\boldsymbol{v_R^2/c_t^2}) - 2]\,[(\boldsymbol{v_R^4/c_t^4}) - 2\,(\boldsymbol{v_R^2/c_t^2}) - 4]\,(\boldsymbol{c_l^2 c_t^2})}{\boldsymbol{c_l^2}[\boldsymbol{c_l^2/c_t^2} - 2]} \tag{8}$$

where K, α, ρ, C, c_l, and c_t are bulk modulus, the volumetric thermal expansion coefficient, density, thermal capacity, longitudinal wave velocity, and transverse wave velocity, respectively. The exponential part of Eq. (6) indicates carrier signals of the SAW. The amplitude of the SAW follows a sinc function dependence that becomes a maximum when the scanning velocity of the SIF equals the phase velocity of the SAW. The amplitude of the SAW is proportional to the laser pulse width, showing the amplitude enhancement effect of the PVS method. For an aluminum specimen, the SAW amplitude is estimated to be 1.0 nm from Eq. (6) when $A = 0.2$, $J = 0.71$ MW/cm^2, $T = 100$ ns, $C = 0.90$ J/gK, $\alpha = 69.6{*}10^{-6}\text{K}^{-1}$, $\rho = 2.7{*}10^6$ g/m^3, $v_R = 2840$ m/s, $c_l = 6420$ m/s, and $c_t = 3040$ m/s. These parameters are employed in the experiment described later.

Typical calculated waveforms are given in Fig. 8.7. In this example, the frequency difference of laser beams is 110 MHz, the laser pulse width is 140 ns; and laser beam spot size is 0.5 mm; the upper curve is at resonance, and the lower curve is out of resonance. For the initial and last part of the waveform, other solutions exist,[13] and the initial growing part of the resonance reflects the amplitude enhancement effect of the PVS method.

From Eq. (6) we know that the frequency of the generated SAW is given as $\omega = \boldsymbol{v_R K_f}$. Note that this frequency ω is not always equal to the frequency difference ω_a of the two laser beams. Consequently the SIF is a variable frequency source of the SAW, with the efficiency decreasing as the mismatch of the two

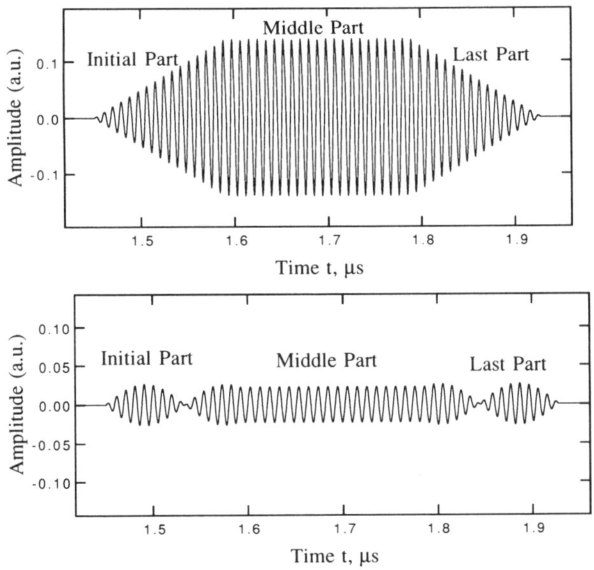

Figure 8.7. The calculated SAW waveform for the short-pulse regime of the SIF approach. A frequency difference between laser beams is assumed to be 110 MHz; the laser pulse width is 140 ns; and the laser beam spot size is 0.5 mm. The upper curve shows the resonance condition (0.565°), and the lower curve shows the out-of resonance condition (0.525°) with a beating signal.

frequencies increases, as seen from the argument of the sinc function in Eq. (6). Therefore a useful equation is obtained for the SAW velocity measurement

$$v_R = \frac{\omega}{k_f} \tag{9}$$

Equation 9 indicates that the SAW velocity is calculated from the measured SAW frequency and the wave number of the SIF, as illustrated in Fig. 8.6. The wave number is determined from the laser beam incident angle, and the accuracy in setting this angle dominates the measurement accuracy of the SAW velocity. It was applied to precise SAW velocity measurements of Si(100) surfaces as described in the next section.

8.2.4. Experimental Setup

An experiment was carried out for generating SAW and pseudo-SAW (PSAW) on a silicon (001) surface (see Fig. 8.8). The specimen was a 3-in. diameter silicon (001) wafer. The pulse of the Q-switched, injection-seeded Nd:YAG laser was used as a thermoelastic source of SAW generation. The wave

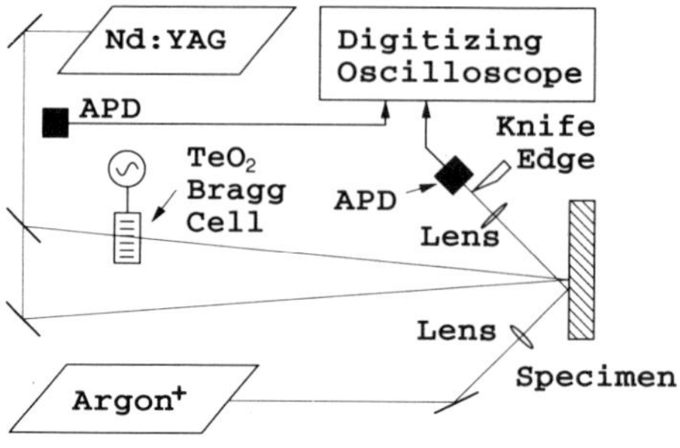

Figure 8.8. An experimental setup of the noncontact QNDE system implementing the SIF approach of the PVS method.

length, energy density, pulse width, and beam diameter of the laser were 532 nm, 0.5 MW/cm^2, 90–120 ns, and 3 mm, respectively. Such a long pulse in a Q-switched laser was realized by a long cavity of 2-m length and a earlier or later Q-switch triggering than the optimum timing. A TeO_2 Bragg cell was used as a frequency shifter of the laser beam, because of its high efficiency. The durability of the Bragg cell under high-power laser was acceptable. The frequency shift of the laser ω_a was set to 110 MHz.

The SAW or the PSAW was detected by a ZnO piezoelectric transducer with a water coupler oriented at the Rayleigh critical angle or by the optical knife edge method using an avalanche photodiode (APD) optical sensor.[18] The propagation length of the SAW and PSAW between the detection and generation positions was about 5 or 10 mm. Scattered light from the laser, which was received by another APD, was employed as a trigger signal in the digitizing oscilloscope.

8.2.5. Measuring Anisotropic SAW Velocity

Experiments to measure SAW velocity for each direction of silicon (001) surface were carried out. Figure 8.9(a) shows the APD signal of the optical knife edge detector, for a typical waveform of SAW generated by the SIF propagating along the [100] direction on a Si (001) surface. The signal was averaged 20 times to give a clear waveform, and a tone burst with many carriers was obtained. Even without averaging, a satisfactory signal was obtained, thanks to the amplitude enhancement effect of the PVS method.

In a separate experiment on an aluminum sample, an estimation of the SAW amplitude was attempted by calibrating the optical knife edge detector using the SAW generated by interdigital transducers (IDTs). Hence we found that a SAW amplitude of 1.4 nm was obtained with a laser power density of 0.71 MW/cm^2 at 532 nm,[18] which is only 4% of the ablation threshold of aluminum.

The power spectrum of the waveform in Fig. 8.9(a) is shown in Fig. 8.9(b). From the peak frequency of a generated SAW or PSAW for each direction and wave number of the SIF, the velocity of an SAW or PSAW was obtained using

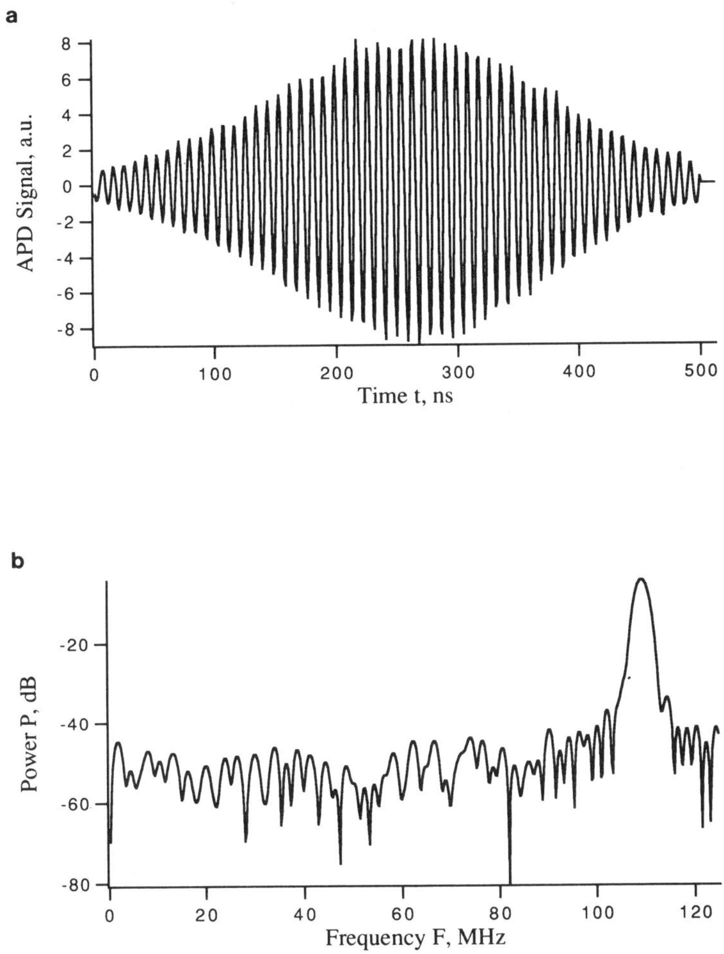

Figure 8.9. A SAW waveform detected by the SIF approach of the PVS method; (a) a typical waveform of a generated SAW on Si (100) surface, averaged 20 times; (b) the power spectrum (center frequency 110 MHz) of the waveform in (a).

Eq. (9). In Fig. 8.10 closed squares indicate experimental results of velocity measurements using the SIF and a piezoelectric transducer detection.[14] Solid and dotted curves show calculated values of SAW and PSAW velocity, respectively, propagating at the free surface of Si (100). Experimental data were in fairly good agreement with calculated curves. When the optical detection was employed, similar results were obtained.

Two closed circles also indicate experimental data of velocity measurements. These two experimental data were obtained from an observed waveform shown

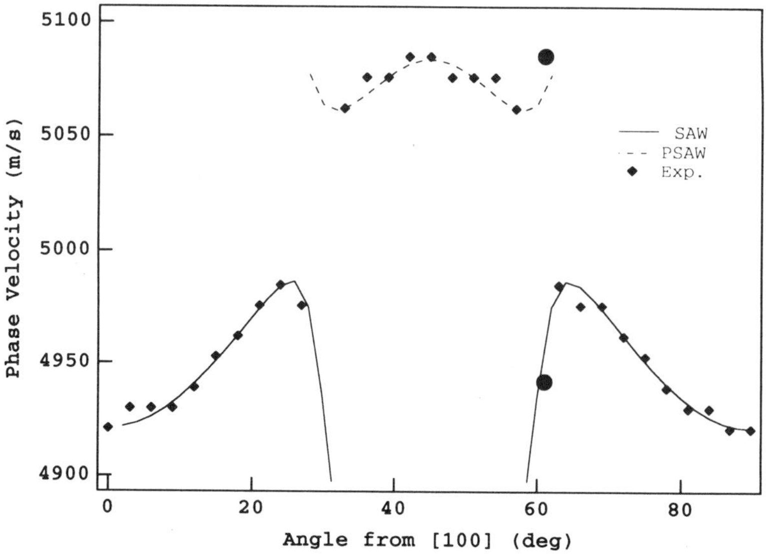

Figure 8.10. The SAW and PSAW velocities of an Si (100) surface as a function of the angle from the [100] direction. Closed squares and circles indicate experimental results in two attempts. Solid and dotted curves indicate calculated SAW and PSAW velocities, respectively.

in Fig. 8.11(a). A beat signal was seen from the interference of two frequency components. The fast fouńer transform (FFT) spectrum of the wave form showed two peaks, as shown in Fig. 8.11(b). From these peaks, two velocities were obtained, and these velocities turned out to correspond to the SAW and PSAW velocities. This example shows that whether the piezoelectric or optical detection is employed, the SIF approach gives precise SAW and PSAW velocities at free surfaces.

Anisotropic SAW velocity measurements of crystals are carried out most conveniently and precisely by using the line-focus-beam acoustic microscopy (Ref. 3; Chapter 12 of Ref. 4). In this method, the measured quantity is the leaky SAW (LSAW) or leaky PSAW (LPSAW) velocity, which are slightly modified from the SAW (PSAW) velocity due to perturbation of the coupling fluid. Excellent agreement is obtained between measurement and theory if the LSAW or LPSAW velocity is calculated.(3)

Accurately calculating LSAW and LPSAW velocities is not always a simple task, especially when they show complex transition behavior or velocity dispersion exists due to a coating or layered structure. Thus SAW and PSAW measurements by the PVS method is favorable in some cases to avoid perturbation from the coupling fluid. In standard SAW velocity measurement for SAW devices and/or materials, interdigital transducers are commonly employed to generate an

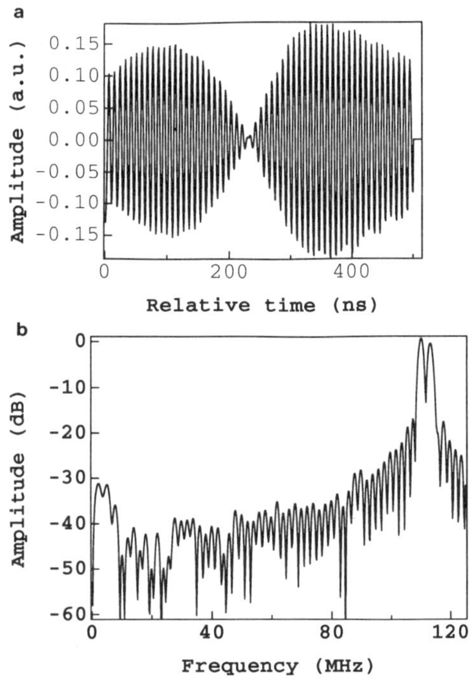

Figure 8.11. The SAW waveform generated by the SIF approach of the PVS method. (a) Observed waveform of a generated SAW on an Si (100) surface at the propagation direction in the transition region between the SAW and PSAW. A beating was seen from the interference of two frequencies components; Detected by a ZnO transducer and averaged 20 times. (b) Power spectrum of the waveform in (a) with two peaks corresponding to SAW and PSAW velocities.

SAW. However the deposition of electrodes or piezoelectric films are required in that method, therefore other methods applicable to any material are still required. Since the present method meets this requirement, it has many applications in evaluating crystals, coatings, and layered structures.

8.2.6. Generating Bulk Waves

The role of BAWs is quite important for elastic and subsurface imaging. Fortunately BAWs can also be generated by the PVS method when the velocity of an SIF is larger than the SAW velocity.[19] In this case, phase matching is required between the BAW wave front at the surface and the SIF. In contrast to bulk wave generation in optically transparent media,[12] the present method is capable of controlling the beam steering. The propagation direction of the BAW measured from the normal to the surface is given by

$$\phi(\theta,\omega_a) = \sin^{-1}\left(\frac{c}{v_f}\right) = \sin^{-1}\left(\frac{2cK\sin\theta}{\omega_a}\right) \tag{10}$$

where c is the BAW velocity. When v_f is larger than v, BAWs are excited in the direction determined by Eq. (10), as schematically shown in Fig. 8.12.

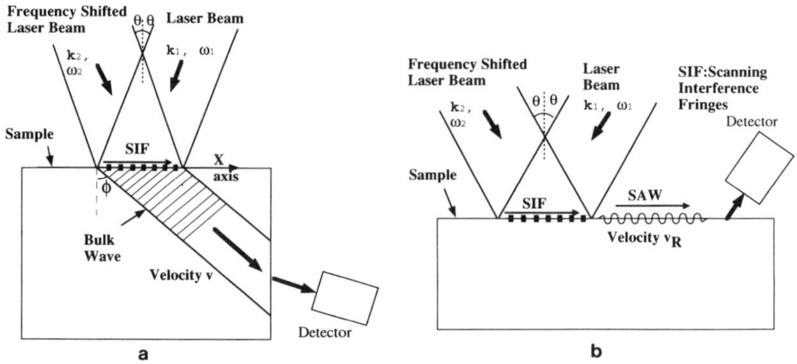

Figure 8.12. The principle of selectively generating (a) BAWs and (b) SAWs.

This principle was verified by selectively exciting SAWs and BAWs in a aluminum plate 120 × 80 × 20 mm thick (v_R = 2960 m/s, c_t = 3040 m/s, c_l = 6420 m/s). The SAW or the PSAW was detected with a 2 mm × 2 mm ZnO piezoelectric transducer with a center frequency of 120 MHz and water coupling. The propagation length of the SAW and BAW between the transducer and the laser beam spot was about 5 mm.

When θ = 0.570°, v_f is calculated to be 2941 m/s from Eq. (4). A clear tone burst was detected by the ZnO transducer placed on the irradiated side of the plate coupled by water at the Rayleigh critical angle. It has been assigned to the SAW. The upper trace of Fig. 8.13(a) shows a center part of the SAW tone burst with approximately 1-ms duration. When the angle θ was increased to 0.585°, an SAW was still observed with slightly reduced amplitude, as shown by the center trace.

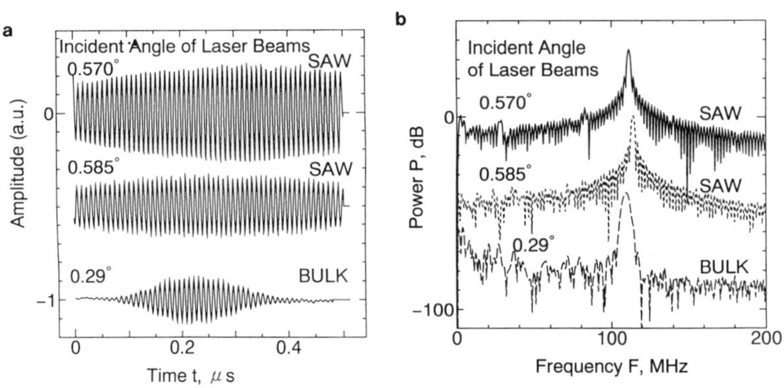

Figure 8.13. The experimental verification of selectively generated (a) BAWs and (b) SAWs; (a) waveforms, (b) power spectra.

When θ was more than 0.63° or less than 0.50°, no SAW signal was detected. However when θ was reduced to 0.29°, a tone burst was detected on the side of the plate by a ZnO transducer, as shown in the bottom trace. In this case, v_f is calculated as 5852 m/s from Eq. (4), and $\phi = 31.3°$ from Eq. (10). Since this velocity is more than the shear wave velocity but less than the longitudinal wave velocity, we assigned this signal as a shear wave. The duration of the BAW tone burst was about 300 ns, much smaller than the SAW tone burst. The signal to noise (S/N) ratio was better than 50 dB for the SAW, whereas it was 30 dB in the BAW. The power spectra of the SAWs and BAWs are plotted in Fig. 8.13(b). Both SAWs and BAWs have a clear peak around 110 MHz. Although the center frequency of the SAW varied from 100–120 MHz, depending on the θ, the BAW peak was always located at 110 ± 1 MHz.

8.2.7. Conclusion

A novel method of excitation and velocity measurement of SAWs with frequencies higher than 100 MHz was described. The potential high accuracy of this method was experimentally verified by the anisotropic SAW and PSAW velocity measurements on silicon (001) surface. In the transition region of SAWs and PSAWs, the present method succeeded in measuring the velocity of the two modes. It provides a powerful tool for evaluating acoustic properties of micro-scaled devices without a coupling fluid.

8.3. Atomic Force Microscopy (AFM) with a Vibrating Sample

For materials characterization on a nanometer scale, some methods using vibration forces between a sample and a probing tip have been developed by extending scanning tunneling microscopy (STM)(20) and atomic force microscopy (AFM) or scanning force microscopy (SFM).(21) In the tunneling acoustic microscopy (TAM), the tip is vibrated,(22) and in the force modulation mode (FMM), the sample is vibrated(23) in the frequency range from 100 Hz to hundreds of kHz. Measured response to the vibrating force is used to image ion–implanted layers, embedded wires, carbon fiber/epoxy composites, and Langmuir-Blodgett films.

Detecting MHz-range acoustic waves was also tried in a few ways, and among these, a method was developed where the nonlinear (rectifying) properties of the tunneling behavior allowed detection of the envelope of the acoustic burst in STM.(24) Other methods were proposed to detect the envelope of amplitude-modulated acoustic waves in SFM using the nonlinear force interaction between tip and surface.(25,26)

Based on those methods, we have developed two novel imaging methods employing the vibration of an AFM sample. The methods developed are lateral force modulation AFM (LM-AFM),[27,28] and ultrasonic force microscopy (UFM).[29,30] In the LM-AFM mode, the AFM sample is vibrated laterally (parallel to the sample surface) to obtain enhanced images of surface steps and friction forces. The UFM employs two imaging modes; in one mode, amplitude modulated ultrasonic frequency vibration (UFV) is applied to the AFM sample, and the cantilever deflection at the modulation frequency is measured. In another mode of UFM, the UFV is simultaneously applied to the LM-AFM, and the cantilever torsion is measured, with enhanced contrast by the shear elasticity. Combining the two modes of UFM, selective imaging of different types of subsurface features was achieved even where different features overlapped each other.

After showing the implementation of AFM/UFM and the lateral FMM we explain the principle of UFM using a spring model. The expected image contrast due to contact elasticity was compared between the UFM and the FMM using an approximate analytic expression for the $z_a(\boldsymbol{a})$ characteristics and the additional cantilever deflection z_a as a function of the UFV amplitude $\boldsymbol{a}$. After showing images of highly oriented pyrolytic graphite (HOPG) and DNA, a more detailed theoretical explanation is introduced, and quantitative measurements of the tip sample interaction are demonstrated. Finally, additional images are shown, such as a polymer blend and the floppy disk surface.

8.3.1. Implementing Ultrasonic Force Microscopy (UFM)

Figure 8.14 is a schematic illustration of the AFM/UFM system. For the LM-AFM, the sample scanner is used to apply the low-frequency (0.7–20 kHz) vibration. For UFM at frequencies between 1–10 MHz, a thickness mode PZT piezoelectric transducer bonded on a sample stage is used to excite UFV. For the lateral UFM, low-frequency lateral vibration is used with the UFV. The cantilever deflection and torsion are monitored by a four-segment photo diode. The deflection signal is used for the constant-force mode operation of AFM: The vibrating component, measured by a lock-in amplifier, is used for the FMM and UFM.

An advantage of the sample vibration over a tip vibration scheme, such as TAM[22] is the flexibility in choosing the mode, frequency, and direction of the vibration. Vertical or lateral vibration can be applied independently or with a certain phase relation, using a combination of piezoelectric transducers. A concave transducer was used for the UFM operation at frequencies above 100 MHz.[31] At the bottom of the sample stage, a ZnO piezoelectric film concave transducer was sputter–deposited as shown in Fig. 8.15(a). It was employed to emit a continuous wave or a tone burst of focused ultrasonic waves within the sample

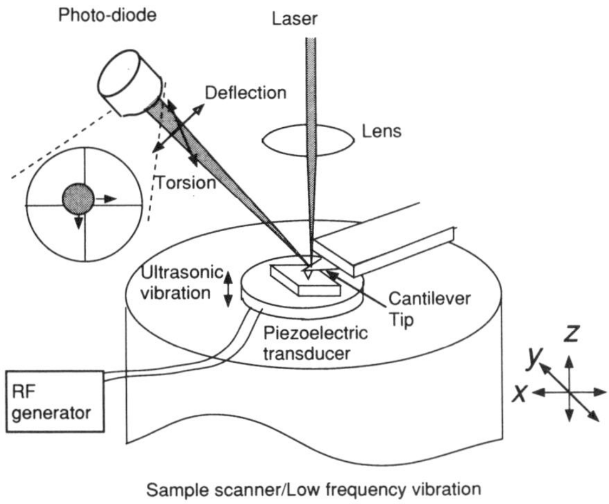

Figure 8.14. A schematic illustration of an AFM/UFM. A thickness mode PZT piezoelectric transducer is bonded on a sample stage to excite UFV with frequencies 1–10 MHz.

stage. By virtue of the focusing and the resonance effect, the central region of the sample stage with an approximate diameter of 200 μm was easily vibrated at a 1-nm amplitude with electric input power of 1 mW. The sawtooth modulated radio frequency (RF) signal of 114 MHz and the cantilever deflection at the modulation frequency of 1 kHz are displayed in Fig. 8.15(b).

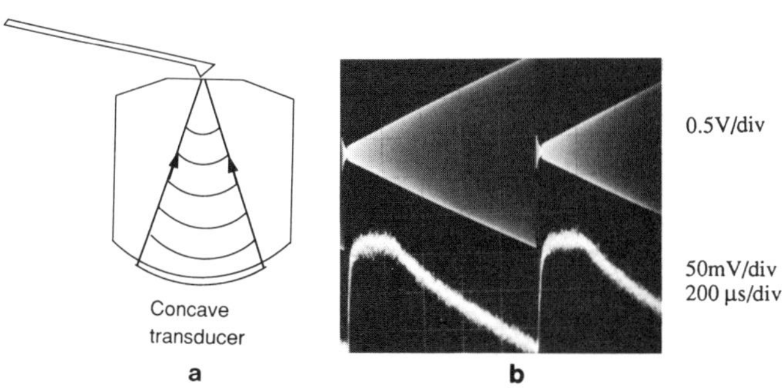

Figure 8.15. (a) An ultrasonic transducer and (b) observed signal of a UFM operated at a frequency above 100 MHz. (a) A fused quartz sample stage with a ZnO piezoelectric film concave transducer deposited on its back surface. It is used to excite the high-amplitude UFV of frequencies above 100 MHz. (b) Amplitude modulated RF signal for a sample vibration 0.5V/div (*upper trace*) and the cantilever vibration at the modulation frequency 50m V/div 200 ms/div (*lower trace*).

8.3.2. LM-AFM for Friction Imaging

The friction force microscope (FFM)[32] was developed as an extension of the AFM and enabled to measure friction force distribution on the nanometer scale and sometimes reaching the atomic scale. However on rough surfaces, the contrast due to the local gradient usually obscures the friction contrast, and interpreting FFM images becomes difficult. If we vibrate the AFM sample in a lateral direction and measure the amplitude of the torsion vibration of the cantilever, LM-AFM microscope images are obtained. Figure 8.16(a) illustrates an AFM cantilever and tip. When the sample surface comes into contact with the AFM tip, the normal force F_N acting between the sample and the tip deflects the cantilever by an amount z_c. If the sample surface is tilted by an angle θ, the direction of the normal force is also tilted, which induces cantilever torsion even when there is no friction. This is the origin of the contrast due to the local

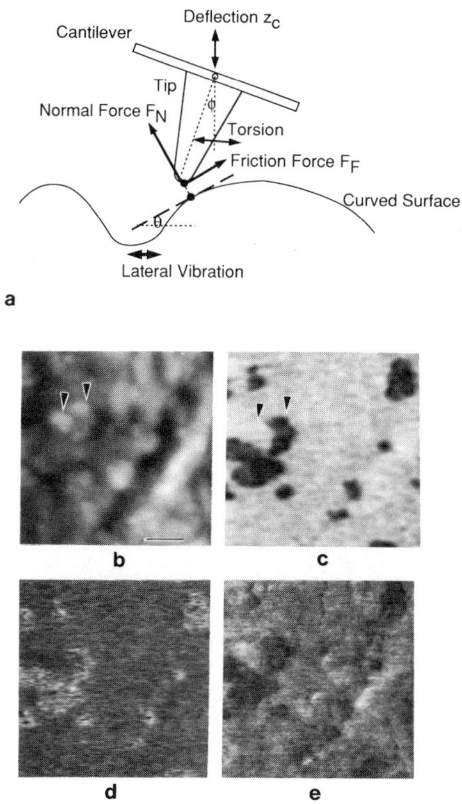

Figure 8.16. The principle and images of lateral FMM. (a) The principle of the lateral FMM showing the cantilever torsion due to contact with a tilted sample surface. The friction force changes sign when the sample displacement is inverted, whereas the normal force does not. (b) Topography of a gold film surface. The height difference between the brightest and darkest area is 14 nm; field of view 500 nm, scale bar 100 nm. (c) Lateral FMM (amplitude) image of the same area; modulation frequency 16 kHz. (d) Lateral FMM (phase) image. (e) FFM image.

gradient in FFM images. If a friction force $\boldsymbol{F}_F$ is present, it reduces or increases the torsion, depending on the scanning direction.

When the sample is laterally vibrated, both the normal force and the friction force are modulated. However the variation of the friction force $\boldsymbol{F}_F$ is very large because the direction of $\boldsymbol{F}_F$ is inverted during one cycle of the lateral vibration. On the other hand, the variation of the normal force $\boldsymbol{F}_N$ is small when the lateral vibration amplitude is small.[27,28] Therefore the cantilever torsion vibration is dominated by the friction, and it can be used for selectively imaging the friction force distribution. From the phase of the cantilever torsion vibration, the amount of slip and/or energy dissipation due to the friction can also be estimated.

For an experimental verification of this idea, a 100-nm-thick gold thin film deposited on a glass substrate was imaged in air. The sample was left in ambient air for a few weeks. Figure 8.16(b) shows a topography image taken at a constant repulsive force of 1.2 nN in a view area of 500 $\times$ 500 nm. The maximum height difference was 14 nm. Besides a groove running from the upper right-hand corner to a lower central position, a large number of hills of 2–10 nm height was observed. From this image, the maximum local gradient was estimated to be about 30°.

Figure 8.16(c) shows LM-AFM images obtained from the amplitude of the cantilever torsion vibration at 16 kHz with an amplitude of 5 nm. Here clearly distinguished dark spots were observed in the surrounding area of almost uniform brightness. On the dark spots, the friction force was about 50–70% of the surrounding area. Dark spots were assigned to contamination, since they disappeared after ultrasonic cleaning in organic solvent. By comparing topography (b) and the LM-AFM image (c), we notice that only some of asperities in (b) correspond to dark spots in (c). For example although the two asperities indicated by arrows in (b) look similar, the one on the right corresponds to a dark spot in (c), whereas the one on the left does not correspond to a dark spot. In other words, at least two different kinds of asperities are distinguished only in the LM-AFM image. This is a good example showing that the LM-AFM image is free from the topography and selectively displays the friction force.

In the phase image in Fig. 8.16(d), some spots look brighter than the surrounding area, indicating smaller phase delay. This is interpreted as a result of a larger amount of slip or smaller energy dissipation caused by a smaller friction force. Since spots in the amplitude and phase images corresponded to each other, it was unambiguously shown that spots were places of smaller friction force than in other areas. However since some asperities in the topography showed no frictional force anomaly, there are at least two different kind of asperities on the observed surface of gold film. Figure 8.16(e) shows an FFM image of the same location recorded in a left-to-right scan. On this image, spots of smaller friction were not distinguished from other asperities. This indicates that discriminating the friction force from the local gradient is not easy on rough surfaces.

A method to separate the topography and friction contribution in AFM has been proposed where two images obtained from different scanning direction is numerically subtracted.[23] However, this method requires exact pixel to pixel matching of the two images, which is not easy due to the hysteresis of the piezoelectric scanner. In contrast, the subtraction is automatically performed in real time in the LM-AFM. Moreover the phase sensitive detection enhances the signal to noise ratio, and the phase delay provide the slip and energy dissipation information. Therefore, the LM-AFM has some unique value in nanoscopic materials evaluation. A further improvement with the addition of UFV will be described in sections 8.3.3.4 and 8.3.7.2.

8.3.3. The Principle of UFM

8.3.3.1. Low-Frequency Limit

In this section, we explain the operation of UFM by a spring and mass model of a sample tip cantilever system, as illustrated in Fig. 8.17. As in the constant force mode of AFM, the cantilever is deflected by an amount z_c from

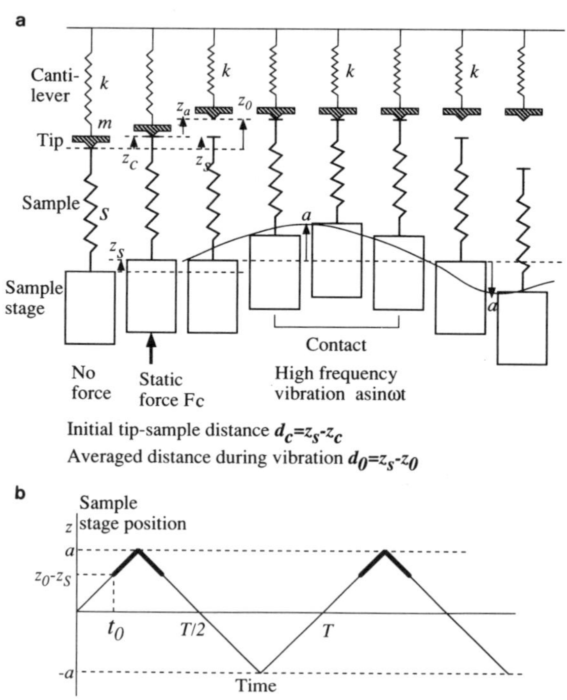

Figure 8.17. (a) A spring model for the operation of AFM with sample vibration; (b) assumed sample vibration waveform of triangular function time dependence.

its free position due to a static repulsive force $\boldsymbol{F_c}$. First we consider the situation encountered in the FMM. When the sample is vibrated at a frequency $\boldsymbol{f}$ lower than the cantilever resonant frequency $\boldsymbol{f_0}$, the cantilever is vibrated following the sample vibration. This vibration is influenced by the tip sample contact stiffness $\boldsymbol{s}$, expressed as the slope of the force and tip sample distance relation. If we introduce a linear spring approximation, the tip sample contact force is expressed as

$$\boldsymbol{F(d) = sd} \tag{11}$$

where $\boldsymbol{d}$ is tip sample distance with a plus sign for increasing the indentation. If the frequency is low enough to allow quasistatic force balance, the peak-to-peak cantilever deflection vibration amplitude is given by

$$\boldsymbol{V} = 2z_c \frac{\boldsymbol{a} / z_c}{1 + (\boldsymbol{k} / \boldsymbol{s})} \tag{12}$$

where $\boldsymbol{a}$ is the sample vibration amplitude and $\boldsymbol{k}$ is the cantilever spring constant. The amplitude $\boldsymbol{V}$ does not significantly depend on the relative sample stiffness $\boldsymbol{K} = \boldsymbol{k} / \boldsymbol{s}$ when $\boldsymbol{K}$ is in the range of 10^{-1}–10^{-4} (see solid and dashed lines in Fig. 8.18 labeled $\boldsymbol{f} << f_0$). No significant image contrast is expected between two domains with $\boldsymbol{s} = 20\boldsymbol{k}$ and $\boldsymbol{s} = 10^4\boldsymbol{k}$. There is almost no tip sample indentation, as illustrated in Fig. 8.19(a)

8.3.3.2. High-Frequency Limit

Next we consider the case of UFM when the sample is vibrated at ultrasonic frequencies much higher than the cantilever resonant frequency ($\boldsymbol{f} >> f_0$). In contrast to the low-frequency limit, the cantilever cannot follow the vibration due to its inertia. When the vibration amplitude exceeds the initial sample compression ($z_s - z_c$), i.e., $\boldsymbol{a} > z_s - z_c = (\boldsymbol{k} / \boldsymbol{s})z_c$, the tip is detached from the sample for a certain period within one vibration cycle (see Fig. 8.17). During contact a repulsive

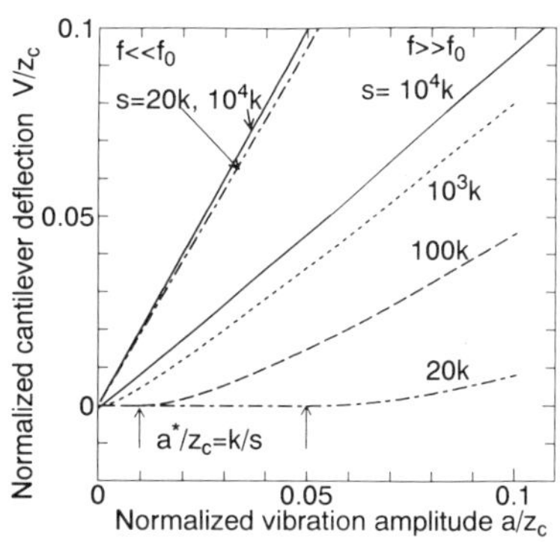

Figure 8.18. The calculated cantilever vibration amplitude in the low-frequency force modulation mode (FMM), ($f << f_0$) and the additional cantilever deflection due to the UFV in the vertical UFM mode ($f >> f_0$).

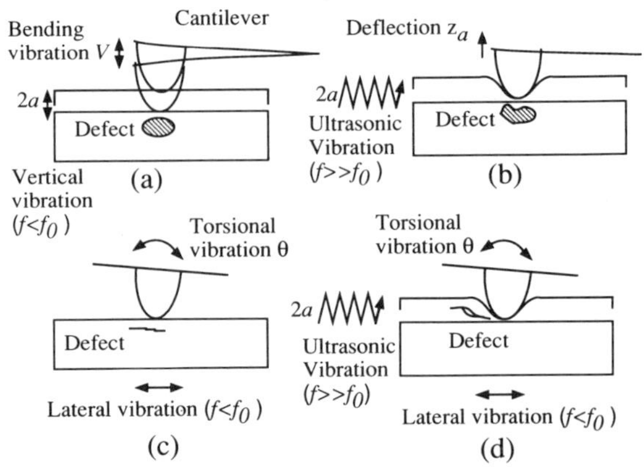

Figure 8.19. Imaging schemes of the force modulation modes in AFM and UFM.[30] (a) Low-frequency vertical force modulation mode; (b) vertical UFM mode; (c) low-frequency lateral force modulation mode; (d) lateral UFM mode.

force acts, so that the tip is indented into the sample even when the sample is much more rigid than the cantilever. To investigate the performance of a UFM with simple analytic expressions, we assume that the sample stage is vibrated around its initial position z_s with the triangular function time dependence illustrated in Fig. 8.17(b), where $T = 1/f$ is the period of vibration. Under this assumption, the sample contacts the tip at time

$$t_0 = \frac{z_0 - z_s}{4a} T,$$

where z_0 is the equilibrium cantilever deflection, for a period

$$\frac{T}{2} - 2t_0.$$

Therefore the tip sample repulsive force averaged for one cycle is

$$F_m = \frac{s}{4a}\left(\frac{k}{s}z_c + a - z_a\right)^2 \tag{13}$$

where $z_a = z_0 - z_c$ is the additional cantilever deflection due to the vibration (see Fig. 8.17). Since this force is balanced with the cantilever restoring force, i.e., $F_m = k(z_c + z_a)$, z_a can be solved as

$$z_a = z_c\left[\frac{k}{s} + \frac{a}{z_c} + 2\frac{ka}{sz_c} - 2\sqrt{\frac{ka}{sz_c}\left(\frac{k}{s}+1\right)\left(\frac{a}{z_c}+1\right)}\right] \quad a > (k/s)z_c \tag{14a}$$

On the other hand, when the UFV amplitude is less than the initial sample compression ($z_S - z_c$), i.e., $a < z_S - z_c = (k / s)z_c$, the tip cannot be detached from the sample. Because of the linearity of the force distance relation in Eq. (11), the averaged force for one cycle becomes zero, and hence the additional cantilever deflection is

$$z_a = 0 \qquad a \leq (k / s)z_c \tag{14b}$$

8.3.3.3. Vertical UFM

Now we are ready to describe a new elastic and subsurface-imaging method, UFM, based on the spring model in Fig. 8.17 and Eq. (14). In principle we can measure the cantilever deflection z_a by switching on and off the vibration while keeping z_c constant with the vibration switched off.[30] However it is more practical to amplitude modulate the UFV and measure the cantilever deflection vibration at the modulation frequency.[25,26,30] If the ultrasonic vibration frequency f is much above the cantilever resonant frequency f_0, the cantilever is not vibrated at f. However when the UFV is amplitude modulated at a frequency below f_0, the cantilever is vibrated at the modulation frequency, and this signal is used in UFM.

If the modulation is 100% with no z-feedback from the sample, the peak-to-peak cantilever vibration amplitude V equals z_a. Thus we obtain images representing the contact elasticity in the vertical direction, which we call vertical UFM [see Fig. 8.19(b)]. Another mode, lateral UFM, is described later, where the sample is also vibrated in the lateral direction. Sometimes the sample is z-feedback-controlled to suppress the variation of cantilever deflection at frequencies much lower than the modulation frequency. This enables us to obtain a simultaneous approximate topography image and avoid tip crashing to the sample during the scanning.

From Eq. (14a, b), we obtain useful information about the UFM. Most importantly the performance of the UFM is expressed by only three parameters: the normalized cantilever deflection z_a/z_c, the normalized UFV amplitude a / z_c, and the normalized cantilever stiffness k/s relative to that of the tip sample contact. By virtue of this remarkable simplicity, the performance of UFM is illustrated in a compact form in Fig. 8.18 ($f >> f_0$). In this figure, note that the contrast in UFM images significantly depends on k/s. A change in k/s from 10^{-4} to 10^{-3} is easily detected. It is also useful to consider the most effective operating condition of UFM to give a good contrast for any particular sample. Furthermore if we keep the static cantilever deflection z_c to a small level, output is expected to be large. This trend is confirmed by more detailed numerical analysis described later in Section 8.3.6 and Fig. 8.23.

From Eq. (14a) the quantity within square brackets equals zero at an amplitude threshold $a^* = (k / s)z_c$, below which additional cantilever deflection vanishes [see Eq. (14b)]. Examples of the threshold are indicated by arrows in Fig. 8.18 when $s = 20k$ ($a^* = 0.05z_C$) and $s = 100k$ ($a^* = 0.01z_C$). The threshold is proportional to the initial force bias ($a^* \propto z_c$) and inversely proportional to the tip sample contact stiffness ($a^* \propto s^{-1}$). Therefore the dynamic elasticity can be measured from the shift in the threshold in the measured z_a (a) characteristic due to the change in z_C[33] as far as the threshold can be clearly identified.

Now we estimate an expected UFM contrast of HOPG. The contact stiffness s is obtained from a relation $s = 2AE^*$, where A is the contact radius given from Hertzian contact theory as $A = (3FR/4E^*)^{1/3}$ and E^* is the effective elastic modulus, defined as

$$E^* = \left[\frac{(1 - \nu_1^2)}{E_1} + \frac{(1 - \nu_2^2)}{E_2} \right]^{-1} \tag{15}$$

where F is a force, R is the tip radius, E_1, ν_1 and E_2, ν_2 are the Young's modulus and Poisson's ratio of the tip and sample, respectively. Since the effective elastic modulus of the Si_3N_4 tip and HOPG sample is around 28 GPa, the Hertzian contact rigidity s of HOPG with a tip radius of 20 nm and a force of 1 nN is estimated to be about 45 N/m. Hence the relative sample stiffness K lies between 10^{-4}–10^{-3} for a typical microfabricated cantilever spring constant k of 0.1 N/m. In Fig. 8.18 there is a substantial difference in V for $s = 10^3k$ ($K = 10^{-3}$) and $s = 10^4k$ ($K = 10^{-4}$). Therefore if some subsurface features of different elasticity modify K in this range, they can be detected in a UFM image.

8.3.3.4. Lateral UFM

When the sample is laterally vibrated at frequencies lower than the cantilever resonance, torsion vibration of the cantilever is excited by the surface friction force as already described in section 8.3.2 (LM-AFM). If additional vertical UFV of the sample is excited, the torsion torque of the cantilever is changed. This torque is not only sensitive to the surface friction but also to the subsurface shear rigidity, because it is generated while the tip is indented into the sample. Therefore subsurface features, such as a delamination or an edge dislocation, that modify the shear rigidity would be imaged by measuring the torsion vibration [see Fig. 8.19(d)]. We call this imaging mode lateral UFM.[30] In this mode, it is necessary to consider a combined directional force, so that modeling and analysis are more difficult than in the vertical UFM. However considering the maximum indentation force, and a quantitative evaluation of the effect of UFV on torsion vibration, provides more insight into the nature of this mode (see Sections 8.3.7.2 and 8.3.7.1).

8.3.4 UFM Images of Subsurface Defects in HOPG

To verify the proposed principles of UFM, we conducted the following experiment on a HOPG sample. A piezoelectric transducer was bonded onto a sample holder of an AFM, and the sample was glued on the transducer. A 400-mm-thick 100-μm-long Si_3N_4 cantilever and Si_3N_4 tip with a spring constant of 0.09 N/m and resonant frequency of 40 kHz were used. In the vertical UFM mode, a 5.6-MHz ultrasonic vibration of 0.5-nm amplitude was amplitude modulated at 10 kHz.

Figure 8.20(a) shows a topography image of a sample with a 500 × 500 nm field of view and 5.4-nm total height difference, cleaved in an ambient air prior to imaging. Several surface steps were observed. Figure 8.20(b) shows a low-frequency (10-kHz) FMM image. Though the contrast of some edges were enhanced, no significant difference from the topography image was noticed. In a vertical UFM image in Fig. 8.20(c), a distinct feature of bright bandlike structure, labeled α, was observed between two steps. Since this contrast was not observed in either the topography or in the vertical FMM image, this contrast seems to have a subsurface origin with different elasticity corresponding to vertical forces. It is not surprising that the FMM image did not show this contrast, because the FMM, using a soft cantilever, is not sensitive to the variation of elasticity in rigid samples, as described by Eq. (12). Note that the spatial resolution for this bright feature is better than 10 nm.

Figure 8.20(d) shows a LM-AFM image taken at 10 kHz, and the surface steps observed in topography (a) were enhanced. When a continuous UFV of 5.6 MHz was added, stringlike features different from the surface steps appeared [labeled β in the lateral UFM image in Fig. 8.20(e)]. Since this feature showed an asymmetric contrast consisting of a dark and a bright part, it probably accompanies two sides of small and large shear rigidities.[30] Perhaps it could be a subsurface edge dislocation with extra atomic planes on one side.

Some stringlike features were also slightly visible in the LM-AFM image (d) and other images, although they were much enhanced in (e). These features do not seem to be completely isolated from the surface but located near the surface. In contrast the bright bandlike feature in the vertical UFM image was completely invisible in other images. Therefore the bandlike feature probably lies more deeply in the subsurface than the stringlike features. It is also interesting that they were selectively imaged in the vertical (c) and lateral (e) UFM images even in the overlapped area indicated by an arrow in (e). Such an argument suggests a promising performance of the UFM for depth discrimination and defect characterization in subsurface imaging.

The basic features of HOPG images are thus consistently explained by the simple linear spring model. An important implication of Eqs. (12) and (14) is that if a linear spring is an appropriate model, no contrast reversal is expected

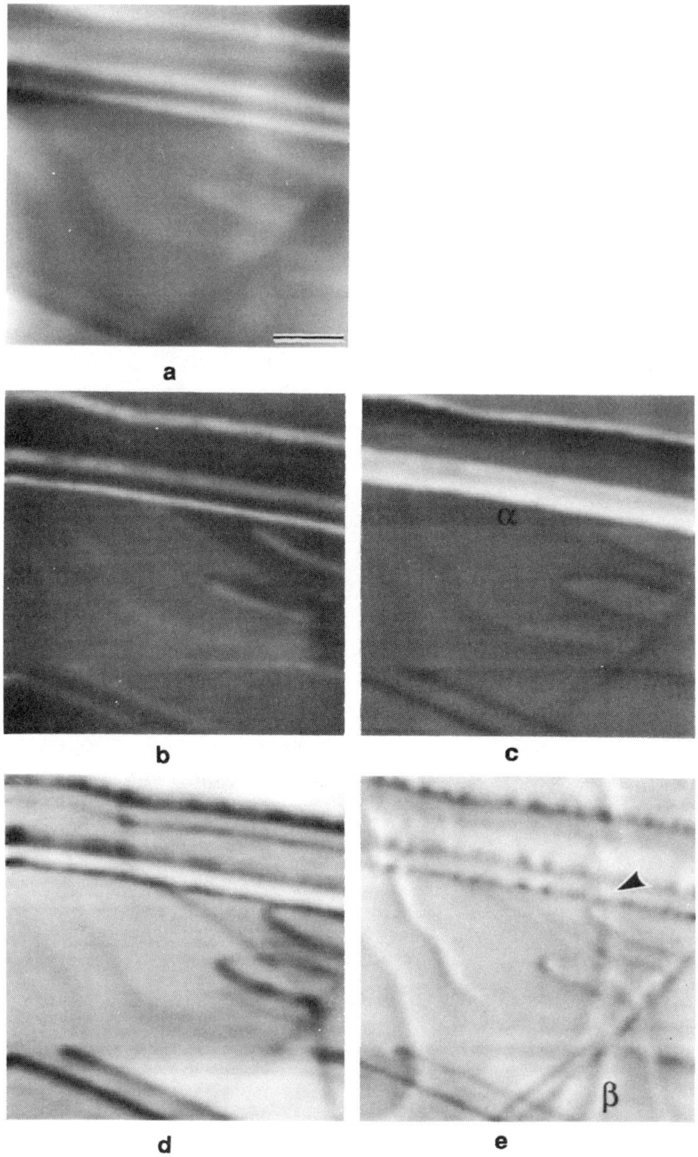

Figure 8.20. Images of HOPG; scale bar-100 nm. (a) Topography of 500 × 500 nm area with 5.4-nm total height difference; (b) low-frequency vertical force modulation image; (c) vertical UFM image; (d) low-frequency lateral force modulation image; (e) lateral UFM image.

between the FMM image and the UFM image, and a stiffer part should always look brighter than less stiff parts in both kinds of images. Even under the intense nonlinearity of a Hertzian contact, the situation is not changed. Therefore if a contrast reversal is experimentally observed for a certain feature between FFM and UFM, other models, such as capillary forces from condensed water or intrinsic nonlinear elasticity, would have to be considered. These points are discussed again in Section 8.3.6.

8.3.5. DNA

Ultrasonic wave measurements for investigating the dynamic elastic property of biopolymer molecules have so far been performed for bulk samples of aqueous solution. Apart from its high-frequency capability, the UFM offers unique improvement in nanoscopic spatial resolution.

A demonstration of imaging DNA molecules was performed. A droplet of an aqueous solution of Lambda DNA (4.4 μg/ml; Nippon gene, molecular weight of 31.5×10^6 daltons, 48,502 bps) was dried on the sample stage. In the topography image shown in Fig. 8.21(c), individual DNA molecules are imaged, indicated by an arrow. Figure 8.21(d) is the UFM image of the same area taken at the ultrasonic frequency of 114.03 MHz with a maximum cantilever vibration amplitude of 1.3 nm.[34] All molecules showed a clearly dark contrast corresponding to a 5–13% smaller cantilever vibration amplitude than the surrounding area. It is the first demonstration of a UFM image of biopolymer molecule.

The observed contrast is qualitatively explained by the smaller elasticity of the molecules, according to Fig. 8.18, but the signal difference seems to be rather modest, whereas the difference of elasticity between fused quartz and DNA molecules is quite large. A possible explanation is that the thickness of the molecule (1–2 nm) is not enough to eliminate the effect of rigid substrate. As another possibility, the dynamic elasticity of DNA at 114 MHz may be significantly higher than the static elasticity because the intramolecular motion is frozen at high frequencies due to relaxation. For investigating the relaxation behavior of individual molecules, the present UFM is particularly advantageous because the high-frequency range of 50–150 MHz and the nanometer range spatial resolution are simultaneously achieved. Although the advantage of the UFM over the FMM is established for rigid objects, it is not obvious for soft objects. These issues have to be explored to establish the application of UFM to biopolymer molecules with a nanometer scale.

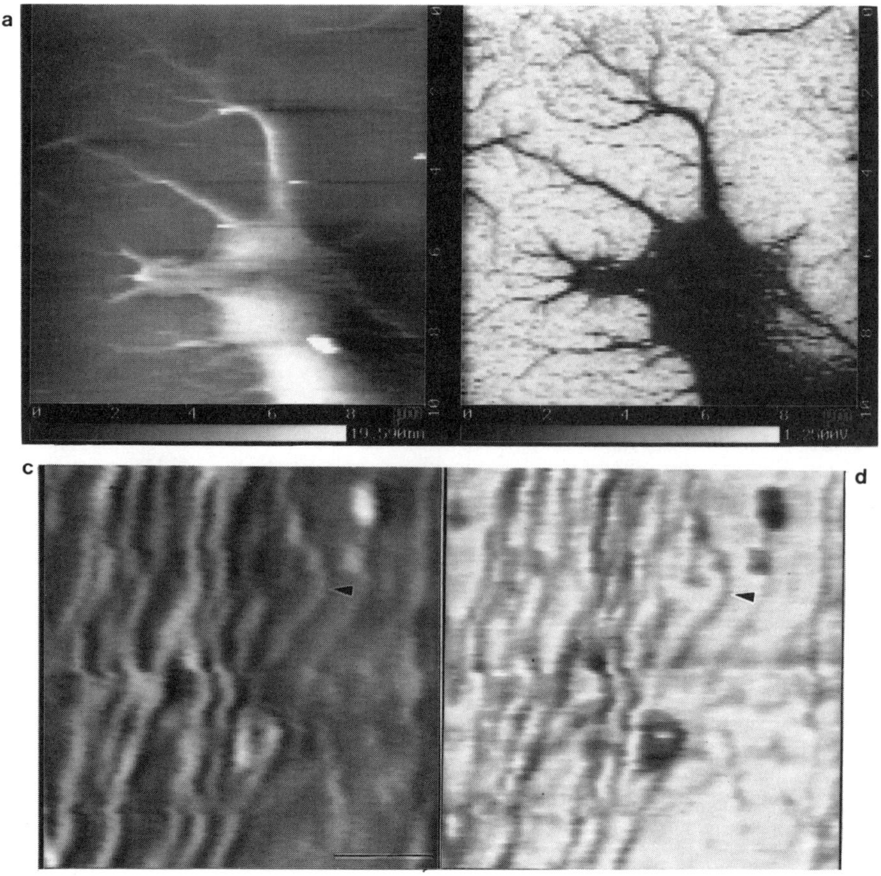

Figure 8.21. Images of λ phage DNA. (a) Topography of aggregated DNA molecules on Si with a bundle of a small number of DNA molecules with an image width of 10 μm. (b) UFM image of the same area as (a); ultrasonic frequency-3 MHz, modulation frequency-1 kHz. (c) Topography of individual DNA molecules on fused quartz with an image width of 400 nm; scale bar-100 nm. (d) UFM image of the same area as (c); ultrasonic frequency-114.03 MHz, modulation frequency-1 kHz.

8.3.6. Theoretical Analysis

8.3.6.1. Hertzian Model with Surface Energy

We already considered the Hertzian contact model in section 8.3.3.3. However it is well-known in the field of AFM, that the attractive force due to a water film condensed on a surface or the van der Waals force significantly affects the tip sample interaction. In this section, we consider the attractive force effect. Though a number of approaches for formulating this attractive force were presented, a simple analytical expression is not available. However a simplified semiempirical formula for a tip sample force was recently proposed.[35] In this model, the force is expressed by

$$F(d) = \frac{E^* A^3}{R} - \left(\frac{3\pi}{2} w E^* A^3\right)^{1/2} - F_{att} \tag{16}$$

or converting this equation to

$$F(d) = E^* (Rd^3)^{1/2} - \left(\frac{3}{2}\pi w E^*\right)^{1/2} (Rd)^{3/4} - F_{att} \tag{17}$$

using the Hertzian approximation $A^2 \approx Rd$. Here d is the tip sample distance with the plus sign for increasing indentation and w is the adhesion energy defined as the work per unit area of contact required to separate two solid surfaces or the change of surface energy per unit area of two solid surfaces due to the contact.[36]

The first term on the right-hand side of Eqs. (16) and (17) gives the Hertzian repulsion, the second term gives the transition from attraction to repulsion, and the third term F_{att} gives the long-range attraction force. Since we want to investigate the effect of the second term, we neglect F_{att} here. The force curve $F(d)$ for the effective elasticity of 50 GPa and 5 GPa, with the cantilever spring constant k of 0.2 N/m, tip radius R of 20 nm, and adhesion energy w of 360 mJ/m^2 are shown in Fig. 8.22.[31] The force curve $F(d)$ exhibits intense nonlinear behavior. It may be useful here to specify two kinds of nonlinearity in a force distance relation, namely, geometric nonlinearity and intrinsic nonlinearity. The former is due to the geometric effect of contact between sphere and flat surface. The stress–strain relation in the material is linear and described by linear elastic constants [e.g., Eq. (15)]. On the other hand, the latter is due to inelasticity or the nonlinear stress-strain relation of the material. It is dominant in a large deformation or for soft materials. So far there has been no rational evidence of

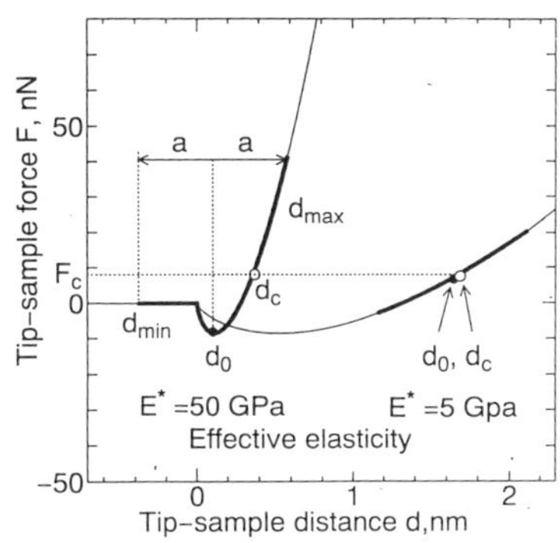

Figure 8.22. The tip sample force $F(d)$ as a function of the tip sample distance. The effective elasticity of 50 GPa and 5 GPa, with the cantilever spring constant k-0.2 N/m, tip radius R-20 nm, and adhesion energy w-360 mJ/m^2 are assumed. Thick curves indicate the variation of force and distance during the UFV.

the intrinsic inelasticity experimentally verified in UFM. To verify the intrinsic inelasticity, comparison of the FMM and UFM is at least required.[30] Therefore we assume geometric nonlinearity in this section.

Although there is a discontinuous change in the slope at $\boldsymbol{d} = 0$, it has no essential effect on the following analysis. This discontinuous change can be reduced or eliminated by assuming an appropriate attraction term $\boldsymbol{F}_{att}$ as a function of $\boldsymbol{d}$. Regardless of the simplicity of Eqs. (16) and (17), they reproduce essential features of a more elaborate theoretical treatment describing the transition from attraction to repulsion.[37]

8.3.6.2. Cantilever Deflection

Using Fig. 8.22 we now start to look at the cantilever deflection as well as the tip sample force and distance during the UFM operation. As previously described, the cantilever is deflected by z_C from its free position by applying a tip sample static force $\boldsymbol{F}_C$. Due to this force, the tip is indented in the sample, and the tip sample distance is determined from the force curves $\boldsymbol{d}_C = \boldsymbol{F}^{-1}(\boldsymbol{F}_C)$, shown by open circles in Fig. 8.22.[31]

When the UFV is applied, the cantilever cannot follow the sample vibration due to inertia, since the frequency of a UFV above 1 MHz is much higher than the cantilever resonant frequency. However due to the additional repulsive force caused by contact with the vibrated sample, cantilever deflection is increased from z_C to z_0. This increase is brought about by the nonlinear rectifying effect,[25] which is determined by solving an integral equation.[26]

$$\boldsymbol{k}z_0 = \frac{1}{\boldsymbol{T}}\int_0^T \boldsymbol{F}(z_s + \boldsymbol{a}\cos\omega\boldsymbol{t} - z_0)\boldsymbol{dt} \tag{18}$$

where z_S is the sample stage displacement required to lift the cantilever by z_C, determined by

$$z_s = z_c + \boldsymbol{F}^{-1}(\boldsymbol{k}z_c) \tag{19}.$$

Once z_C is assumed, z_0 is obtained by solving Eqs. (18) and (19), and the additional cantilever deflection $z_a = z_0 - z_C$ due to the UFV is calculated as a function of $\boldsymbol{a}$.

The $z_a(\boldsymbol{a})$ characteristic for $\boldsymbol{E}^* = 5$ GPa with four different static repulsive forces $\boldsymbol{F}_C$ of 4, 8, 16, and 32 nN are plotted in Fig. 8.23. In this figure, (a) shows the case with no adhesion energy, and (b) shows the case with adhesion energy $w = 360$ mJ/m^2. The cantilever spring constant $\boldsymbol{k}$ and the tip radius $\boldsymbol{R}$ were 0.2 N/m and 20 nm, respectively. These parameters were chosen to be close to those used in experiments in the order of magnitudes, although not exactly adjusted to be identical.

In Fig. 8.23 the deflection z_a increases with the increasing UFV amplitude $\boldsymbol{a}$. Though the slope $\delta z_a / \delta \boldsymbol{a}$ is small at smaller $\boldsymbol{a}$ and increases at larger $\boldsymbol{a}$, it

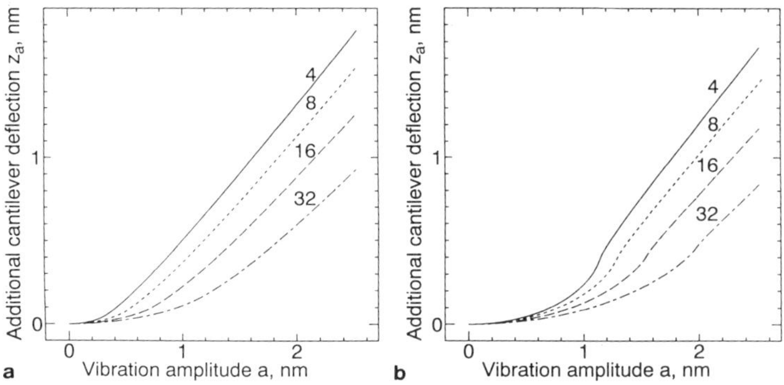

Figure 8.23. The calculated additional cantilever deflection as a function of UFV amplitude z_a (***a***) characteristics for different static repulsive forces. An effective elasticity-5 GPa, cantilever spring constant ***k***-0.2 N/m, and tip radius ***R***-20 nm are assumed; effective elasticity E*-SGPa; force F_c, nN; (a) no adhesion energy, (b) adhesion energy ***w***: 360 mJ/m^2.

is not easy to define a clear threshold. The deflection z_a also increases as the static repulsive force F_C decreases. Furthermore the adhesion energy reduces the cantilever deflection z_a, especially at small UFV amplitudes ***a,*** as seen by comparing (a) and (b), although at large amplitudes, this effect becomes less significant. Consequently a kink occurs around a certain range of UFV amplitudes ***a.*** The kink formation is particularly obvious for a smaller repulsive force F_C of 4 and 8 nN.

The z_a (***a***) characteristic for F_C = 8 nN with four different effective elastic constants ***E**** of 1, 5, 20, and 50 GPa are plotted in Fig. 8.24. In this figure, (a) shows the case with no adhesion energy, and (b) shows adhesion energy ***w*** of

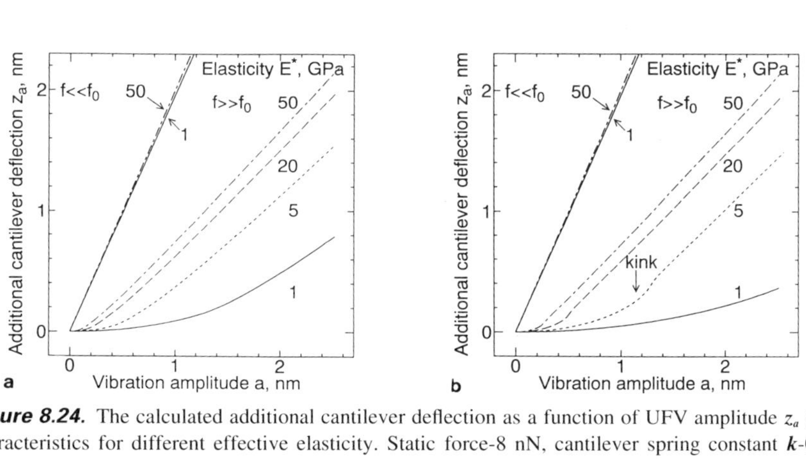

Figure 8.24. The calculated additional cantilever deflection as a function of UFV amplitude z_a (***a***) characteristics for different effective elasticity. Static force-8 nN, cantilever spring constant ***k***-0.2 N/m, and tip radius ***R***-20 nm are assumed; (a) no adhesion energy, (b) adhesion energy ***w***: 360 mJ/m^2.

360 mJ/m^2. Deflection is larger for a larger effective elastic modulus E^*. A kink formation is again observed, and the kink position is shifted to a smaller UFV amplitude as the effective elastic modulus E^* increases.

When the vibration frequency is much lower than the cantilever resonant frequency, we can assume that the cantilever stays at the equilibrium position for each sample stage displacement z_s. Then the peak-to-peak vibration amplitude of cantilever deflection z_{FMM} when the sample stage displacement is varied from $z_S - a$ to $z_S + a$ is given by

$$z_{FMM} = \frac{[F(d_2) - F(d_1)]}{k} \tag{20}$$

where d_1 and d_2 are solutions of equations

$$F(d_1) = k(z_S - a - d_1) \qquad F(d_2) = k(z_S + a - d_2) \tag{21}$$

respectively.

The amplitude z_{FMM} is also plotted in Fig. 8.24 for two effective elasticities of $E^* = 1$ GPa and 50 GPa. Since the difference between two curves is negligibly small, it is difficult to find a large image contrast in the FMM images for elasticity differences of 1 and 50 GPa. Although the modulus of carbon and industrial graphite ranges from 3.5 GPa–28 GPa, it is difficult to distinguish them by the FMM. This is in contrast to the UFM, where the difference for each E^* is large enough to show a large contrast. Qualitatively this trend agrees with the linear spring model. The experimentally observed large contrast in UFM and poor contrast in FMM of HOPG (see Section 8.3.4) and a floppy disk surface (see Section 8.3.8) are consistent with this prediction.

8.3.6.3. Tip Sample Force and Distance

During the UFV, the tip sample force F and the distance d (indentation depth) vary with time at the same ultrasonic frequency. The force and distance are important parameters for the physical process that takes place at the contact. The range of the tip sample distance d between the maximum $d_{\max}$ and minimum $d_{\min}$ for the case of the UFV amplitude a of 0.5 nm are shown as thick curves in Fig. 8.22. This UFV amplitude is above the kink for 50 GPa and below the kink for 5 GPa, as seen from Fig. 8.24(b).

The average tip sample distance during the UFV, $d_0 = z_S - z_0$, is also calculated from the solution to Eq. (18); it is also indicated in Fig. 8.22. Note that d_0 is much smaller than the initial indentation depth d_C for E^* of 50 GPa. In this force curve, the UFV amplitude is large enough to reach the nonlinear region. Then substantial repulsive force is generated by averaging for one cycle. As a result, d is shifted from the initial d_C to a more separated position d_0. For a given UFV amplitude, this tendency is less dominant in a small elasticity of $E^* = 5$ GPa, as shown in Fig. 8.22. and Table 8.1.

The maximum instantaneous tip sample force $F_{max} = F(z_S + a - z_0)$ is also a useful quantity in considering the deformation. As listed in Table 8.1, the maximum force tends to increase with the effective elasticity E^*. Such a tendency of the indentation and force has important implications in the interpretation of UFM images.

8.3.7. Detailed Experiments and Comparison with Theory

8.3.7.1. Dependence of Cantilever Deflection on the UFV Amplitude

In our initial trial to detect the UFV,[26] the deflection of the cantilever was measured on a Si (100) surface as the piezoelectric transducer was driven at a sawtooth-modulated signal with the carrier frequency of 3 MHz, peak-to-peak amplitude of 2 V, and modulation frequency of 700 Hz. A microfabricated silicon nitride cantilever with a spring constant of 0.024 N/m and a resonance frequency of 33 kHz was used. To diminish the effect of the z-feedback loop on the cantilever deflection, the integral gain of the lowest possible value was chosen. Thus the experimentally observed cantilever deflection against the UFV amplitude, the $z(a)$ curve, is a good approximation of the theoretically defined $z_a(a)$ characteristic. However it should be kept in mind that the $z(a)$ curve is distorted when a high gain is employed in the z-feedback loop.

The resultant $z(a)$ curves with a static force of 1.2, 0 and -1.2 nN are plotted in Fig. 8.25, where the downward direction corresponds to an increased cantilever deflection. At small UFV amplitudes, deflection was very small for small UFV amplitudes, then suddenly increased after a certain threshold. The magnitude of the threshold decreased as the static force changed from 1.2 nN (repulsive) to -1.2 nN (attractive). Also a decrease in the slope at higher UFV amplitude is visible.

Table 8.1. Numerical Example of Tip Sample Forces and Distance during the UFV of AFM Samples[a]

E^* (GPa)	F_{max} (nN)	d_{min} (nm)	d_0 (nm)	d_{max} (nm)	d_c (nm)
1	12.2	4.42	4.90	5.38	4.91
5	20.7	1.16	1.64	2.12	1.68
20	36.6	0.06	0.54	1.02	0.67
50	40.5	−0.38	0.10	0.58	0.36

[a]The constant E^* is the effective elasticity, F_{max} is the maximum tip sample force, and d_{min}, d_{max}, and d_0 are the minimum, maximum, and averaged tip sample distance during UFV, respectively. $a = 0.5$ nm, tip radius = 20 nm, and adhesion energy = 360 mJ/m^2 are assumed. Note the difference between d_c and d_0 grows as the effective elasticity increases.

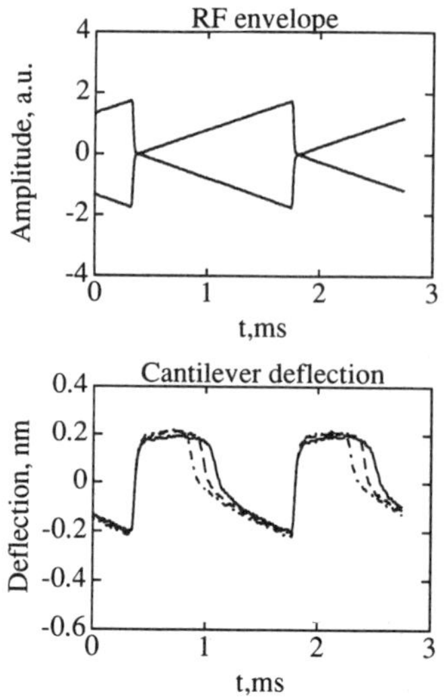

Figure 8.25. Waveforms in a vertical UFM; (*upper trace*) a sawtooth-modulated RF signal at 3 MHz for exciting the UFV on an Si(100) surface; (*lower traces*) a cantilever deflection vibration signal at the modulation frequency (0.7 kHz) showing the measured $z(\boldsymbol{a})$ curves at three different static forces—1.2, 0, and −1.2 nN. *Solid line,* F = 1.2 nN, *dashed line,* F = OnN; *dash-dot line,* F = −1.2nN.

The theoretically predicted features of the $z_a(\boldsymbol{a})$ characteristic seems to be generally consistent with those measurements. In particular the change of the slope of the $z(\boldsymbol{a})$ curve in Fig. 8.25 is similar to the theoretically predicted kink.

Another experiment was tried on a fused quartz surface at the very high UFV frequency of 114 MHz. A 400-nm-thick and 100-μm-long Si_3N_4 cantilever and Si_3N_4 tip with a spring constant of 0.09 N/m and resonant frequency of 40 kHz were used. Although the strong tip-sample interaction may enhance the highest possible vibration frequency of the cantilever-tip sample system, we can assume that all cantilever vibration modes are frozen at this frequency, since it is more than three orders of magnitude higher than the basic resonant frequency of independent cantilever-tip.

When the UFV at 114 MHz was amplitude modulated at 1.0 kHz (see Fig. 8.26, upper trace), the cantilever vibrated at the modulation frequency shown in lower traces. The deflection z_a remained at almost zero for a small amplitude of $\boldsymbol{a}$ and increased as the amplitude $\boldsymbol{a}$ increased over a threshold indicated by an arrow. This is consistent with previous experimental data and theoretical analysis. Moreover note that the slope of $z_a(\boldsymbol{a})$ first increased, then slightly decreased, and a kink was formed, though it was not so evident as in Fig. 8.25.

The magnitude of the kink changed on different samples, and even in different experiments on the same sample, and reproducibility was not very good. This may be due to changes in the surface condition and hence the adhesion energy. Therefore atmosphere control is essential for better reproducibility and a more detailed comparison with theory. Once this is achieved, the measurement

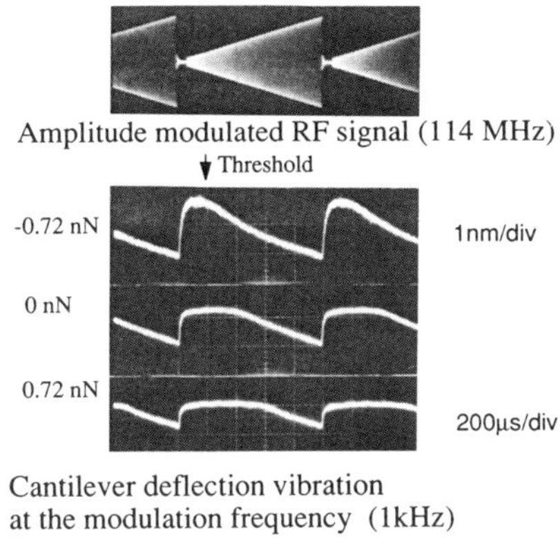

Figure 8.26. Waveforms in vertical UFM; (*upper trace*) a sawtooth-modulated RF signal at 114 MHz for exciting the UFV on fused quartz sample stage; (*lower traces*) a cantilever deflection vibration signal at the modulation frequency (1 kHz) showing the measured $z(a)$ curves at three different static forces—0.72, 0, and −0.72 nN.

described here would provide a useful tool for quantitative elasticity evaluation on a nanoscopic scale and at ultrasonic frequencies.

8.3.7.2. Reduction of Cantilever Torsion Vibration by the UFV

We investigated the effect of UFV amplitude on the cantilever torsion vibration on a fused quartz sample. The UFV amplitude was estimated as follows. First the additional cantilever deflection under the amplitude-modulated UFV was measured using the photodiode output voltage, which is calibrated from a DC displacement of the previously calibrated piezoelectric sample scanner. Then the UFV amplitude was approximated by the additional cantilever deflection, since these quantities are almost the same in materials with large elasticity according to Fig. 8.24 (for fused quartz; Young's modulus E = 73 GPa and Poisson's ratio $\nu = 0.17$).

When the sample was laterally vibrated at 1 kHz, cantilever torsion vibration was excited as in LM-AFM. The static repulsive force was 0.72 nN. When the continuous UFV at 114 MHz was applied with a lateral vibration of the sample, the amplitude of the cantilever torsion vibration was reduced, as shown in Fig. 8.27[31]. The torsion vibration became very small at UFV amplitudes above 1.8 nm. This damping of the torsion vibration was commonly observed on other samples, such as Si or HOPG at UFV frequencies of 1–10 MHz, though required UFV amplitude varied depending on experimental conditions.

This damping of the torsion vibration is rather surprising, since the torque for the cantilever torsion increases as the tip sample force increases and the UFV increases the maximum instantaneous as well as the average tip sample force, according to the theoretical prediction given in the preceding section. A possible

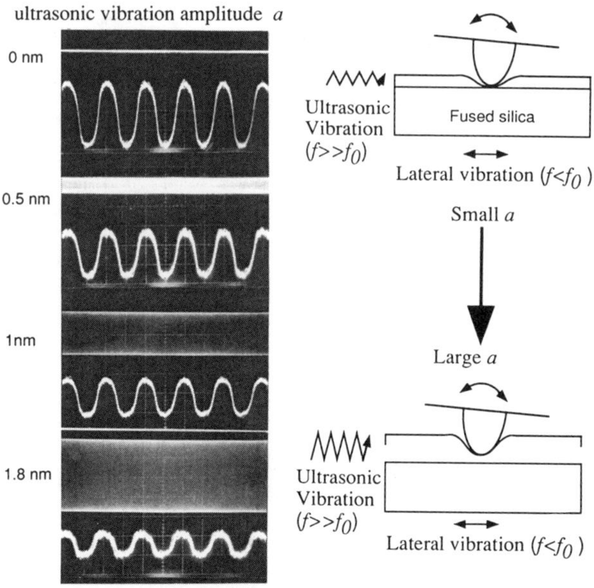

Figure 8.27. Waveforms in lateral UFM; (*upper traces*) a continuous-wave RF signal for exciting the UFV on a fused quartz sample stage at 114 MHz with the amplitude ranging from 0–1.8 nm; (*lower traces*) a cantilever torsion vibration caused by the lateral force modulation at 1 kHz. This vibration is damped by the UFV.

reason for the torsion damping is that the decrease in torque during the tip sample-detaching period of the UFV is more significant than the increase in torque during the tip sample contact period. More detailed measurements and analysis are required for a complete understanding of this phenomenon. This phenomenon may find an interesting application for evaluating nanoscopic local friction coefficient and shear elasticity, especially in tribology.

8.3.8. More Images

To compare the performance of UFM with other imaging modes of AFM, such as the topography and FFM images, we obtained a set of images by different techniques. In Fig. 8.28(a) the topography, (b) FFM, (c) FMM, and (d) UFM images of a floppy disk surface (Maxell Super RD II, 5.25™) are shown; image width is 400 nm.[33] A grain structure with a grain size of about 100 nm was observed on all images, whereas smaller grains of 50 nm were seen on FFM and FMM images. All grains in (b) and (c) had nearly equal contrast within each image. The FFM image resembled an FMM image with enhanced topographic features.

Figure 8.28. Images of a floppy disk surface (Maxell Super RD II, 5.25™ in.); (a) topography with an image width of 500 nm, (b) FFM image, (c) FMM image, (d) UFM image. (ultrasonic frequency 6 MHz, modulation frequency 2.4 kHz). (courtesy of Oleg Kolosov)

The UFM image of the same area was obtained by applying a 6-MHz sawtooth modulated UFV with 2.4-kHz modulation frequency. The contrast of grains was much higher than in the topography and FFM images. The resolution of the grain structure seems to be enhanced with a clearly visible fine grain structure of 10–20 nm width.

Moreover the contrast among different grains in the UFM image was much larger than in the topography and the FFM images. This may be due to different elastic modulus of ferrite particles and the polymer matrix of the floppy disk surface, though chemical analysis, such as EPMA or scanning Auger images, are required to present a definite interpretation. Once this is achieved, it will be another application of UFM to the local variation of elastic properties.

8.3.9. PS/PEO Polymer Blend

As the final image, we show a polymer blend film of polystyrene (PS) and poly(ethylene oxide) (PEO).[39] The morphology of polymer blends is usually investigated by using small-angle X-ray scattering, TEM, SEM, and so on. However because PS is very rigid and PEO is very soft, elastic properties of the blend would be quite interesting. It is expected to give a sensitive measure about the elastic morphology. If the distribution of the elastic property could be imaged on a nanometer scale, phase separation would be clearly imaged.

We observed a PS/PEO film spin-coated on a mica substrate, using AFM and UFM. Figure 8.29(a) shows a topography image with width of 10 μm. The image was composed with a few tall mountains 50–100 nm high and a flat plane. Small asperities within the plane are formed around dust particles. This topography is probably due to different surface tension of two phases with a different composition PS/PEO.

In the UFM image of the same area shown (b), with an ultrasonic frequency of 5 MHz and a modulation frequency of 2.4 kHz, mountains and hills looked darker, suggesting a softness. In contrast to the topography, the plane was composed of two phases with different softness. In the enlarged UFM image (c), with image width of 1.5 μm, more rigid phase of 200–400 nm width were identified. This is probably due to a finer phase separation. The nature and composition of this finer phase is not known, and will be investigated in the future.

8.3.10. Conclusions

As a subsurface and elastic-imaging method for materials characterization on the nanometer scale, we have developed two methods that employ vibration forces between the tip and the AFM sample. The lateral force modulation AFM (LM-AFM) gives images that show the distribution of friction forces, even on rough surface, by suppressing the effect of the local gradient.

Furthermore, the friction force measurement in LM-AFM is performed in real time, in contrast to the conventional FFM[32] where numerical subtraction of forward and backward scan data is required.[23] This advantage is attractive in investigating the dependence of friction force upon other parameters, such as the normal force and vibration frequency. In particular, implication of the phase signal recently elucidated in relation to slip and deformation will provide a new insight in the tip-sample interaction phenomenon.[40]

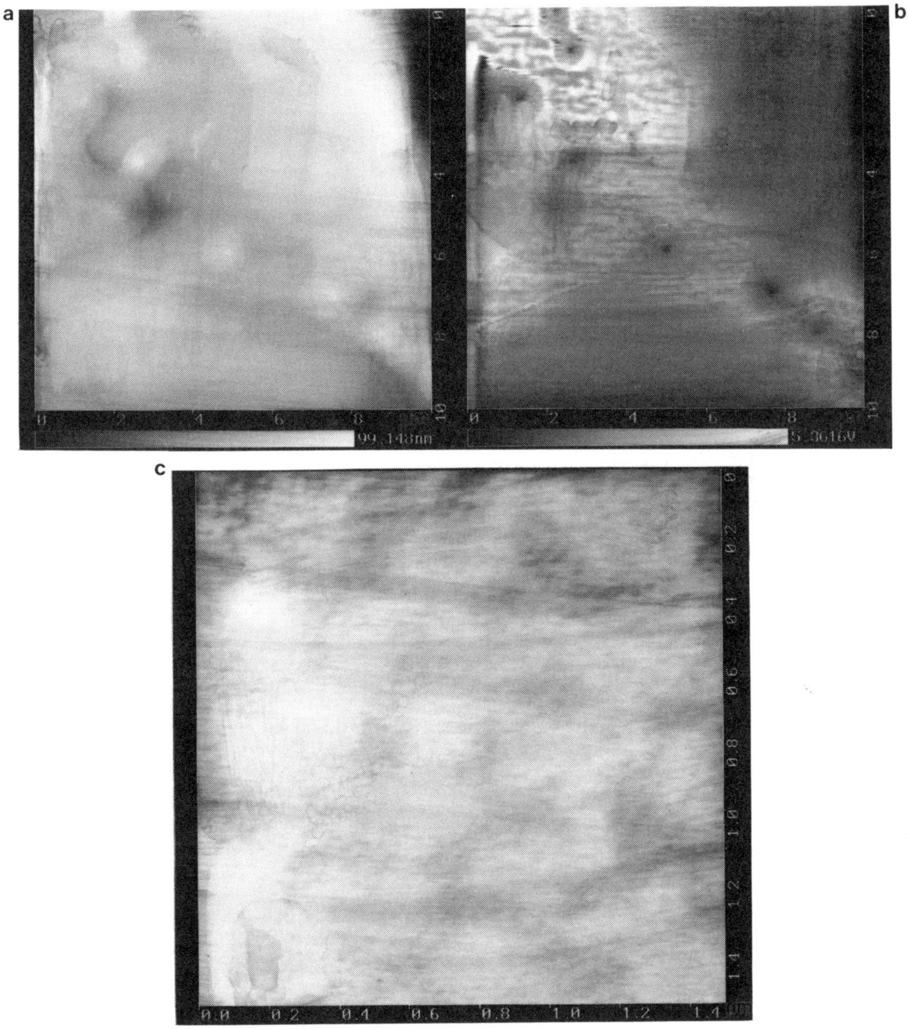

Figure 8.29. Images of PS/PEO polymer blend. (a) Topography with an image width of 10 μm. (b) UFM image of the same area as (a); ultrasonic frequency-5 MHz, modulation frequency-2.4 kHz. (c) Enlarged UFM image with an image width of 1.5 μm. (courtesy of Makoto Motomatsu)

In the UFM, the AFM sample is vibrated at ultrasonic frequencies, and the tip is elastically indented into the sample. Subsurface features of rigid objects are observed using a soft cantilever. When the direction of the tip sample vibration force is controlled and the cantilever torsion as well as the deflection are monitored, subsurface features sensitive to compression and shear are selectively imaged. Viscoelastic relaxation of biological and synthetic polymers at ultrasonic frequencies can also be evaluated on a nanometer scale. Although we have assumed in the present analysis that the cantilever does not vibrate at the ultrasonic frequencies, the strong tip-sample interaction force may enhance the highest possible vibration frequency up to the MHz range, as recently pointed out by Rabe and Arnold.[41] The influence of this effect on the UFM signal should be rigorously investigated in the future.

Acknowledgments

This work was done in collaboration with Hideo Nishino, Yoshihiko Nagata, Yusuke Tsukahara, Oleg Kolosov, Hisato Ogiso, Harumichi Sato, and Toshio Koda. It was part of the precision acoustic microscopy project supported by the Science and Technology Agency (STA), Japan. We acknowledge partial support of this work by a Science and Technology Agency (STA) fellowship and the National Institute for Advanced Interdisciplinary Research (NAIR) of Japan.

References

1. Quate, C. F., Atalar, A., Wickramasinghe, H. K. (1979). Acoustic microscopy with mechanical scanning. *Proc. IEEE* **67,** 1092–1114.
2. Yamanaka, K. and Enomoto, Y. (1982). Observation of surface cracks with the scanning acoustic microscope. *J. Appl. Phys.* **53,** 846–50.
3. Kushibiki, J. and Chubachi, N. (1985). Material characterization by line-focus beam acoustic microscope. *IEEE Trans.* **SU32,** 189–212.
4. Briggs, G. A. D. (1992). *Acoustic Microscopy,* Clarendon, Oxford, UK.
5. Kushibiki, J. (1990). *Acoustic Microscopy* in *Ultrasonic Spectroscopy, vol. 2* (N. Mikoshiba and A. Ikushima, eds.), pp. 147–87. (Baifu-kan, Tokyo) (in Japanese)
6. Yamanaka, K. (1992). In: *Materials and Ultrasonics* (R. Sakata, ed.), pp. 127–46. (Shokabo, Tokyo) (in Japanese)
7. Mizuhara, K., Taki, T., Yamanaka, K. (1993). Anomalous cracking of bearing balls under a liquid-butane environment, *Tribol. Int.* **26,** 135–42.
8. Yamanaka, K., Nagata, Y., Koda, T. (1991). Selective excitation of single-mode acoustic waves by phase velocity scanning of a laser beam. *Appl. Phys. Lett.* **58,** 1591–93
9. Yamanaka, K., Nagata, Y., Koda, T. (1992). Generation of dispersive acoustic waves by the phase velocity scanning of a laser beam. *Review of Progress in Quantitative Nondestructive Evaluation.* vol. 11, 633–40.
10. Nishino, H., Tsukahara, Y., Nagata, Y., Koda, T., Yamanaka, K. (1993). Excitation of high-frequency surface acoustic waves by phase velocity scanning of a laser interference fringe. *Appl. Phys. Lett.* **62,** 2036–38.
11. Nishino, H., Tsukahara, Y., Nagata, Y., Koda, T., Yamanaka, K. (1993). Generation of 100-MHz-band Rayleigh waves by phase velocity scanning of a laser interference fringe. *Jpn. J. Appl. Phys.* **32,** 2536–39.

12. Caddes, D. E., Quate, C. F., Wilkinson, C. D. W. (1966). Conversion of light to sound by electrostrictive mixing in solids. *Appl. Phys. Lett.* **8,** 309–11.
13. Yamanaka, K., Kolosov, O., Nagata, Y., Koda, T., Nishino, H., Tsukahara, Y. (1993). Analysis of excitation and coherent amplitude enhancement of surface acoustic waves by the phase velocity scanning method. *J. Appl. Phys.* **74,** 6511–22.
14. Nishino, H., Tsukahara, Y., Nagata, Y., Koda, T., Yamanaka, K. (1993). Surface acoustic wave generation by phase velocity scanning of laser interference fringes and its application to nondestructive materials evaluation. Submitted to Proceedings of 6th International Symposium on Nondestructive Characterization of Materials, 7–11 June.
15. Hutchins, D. A. and Tam, A. C. (1986). Pulsed photoacoustic materials characterization. *IEEE Trans. UFFC* **33,** 429–49.
16. Harata, A. and Sawada, T. (1993). Evaluation and imaging of materials using picosecond laser induced ultrasonics. *Jpn. J. Appl. Phys.* **32,** 2188–91.
17. Tsukahara, Y. (1991). Analysis of the elastic wave excitation in solid plates by phase velocity scanning of a laser beam. *Appl. Phys. Lett.* **59,** 2384–85.
18. Nishino, H., Tsukahara, Y., Nagata, Y., Koda, T., Yamanaka, K. (1994). Optical probe detection of high-frequency surface acoustic waves generated by phase velocity scanning of laser interference fringes. *Jpn. J. Appl. Phys.* **33,** 3260–64.
19. Yamanaka, K., Nishino, H., Tsukahara, Y., Nagata, Y., Koda, T. (1993). Selective excitation of bulk and surface acoustic waves by rapid scanning of a laser interference fringe. *Proc. Ultrasonics International* **93,** 807–10.
20. Binnig, G., Rohrer, H., Gerber, Ch., Weibel, E. (1982). Surface studies by scanning tunneling microscopy. *Phys. Rev. Lett.* **49,** 57–61.
21. Binnig, G., Quate, C. F., Gerber, Ch. (1986). Atomic force microscope. *Phys. Rev. Lett.* **56,** 930–33.
22. Takata, K., Hasegawa, T., Hosaka, S., Hosoki, S., Komada, T. (1989). Tunneling acoustic microscope. *Appl. Phys. Lett.* **55,** 1718–20.
23. Radmacher, M., Tillmann, R. W., Fritz, M., Gaub, H. E. (1992). From molecules to cells: Imaging soft samples with the atomic force microscope. *Science* **257,** 1900–05.
24. Moreau, A. and Ketterson, J. B. (1992). Detection of ultrasound using a tunneling microscope. *J. Appl. Phys.* **72,** 861–64.
25. Rohrbeck, W. and Chilla, E. (1992). Detection of surface acoustic waves by scanning force microscopy. *Phys. Stat. Sol. A* **131,** 69–71.
26. Kolosov, O. and Yamanaka, K. (1993). Nonlinear detection of ultrasonic vibrations in an atomic force microscope. *Jpn. J. Appl. Phys.* **32,** 1095–98.
27. Yamanaka, K., Kolosov, O., Ogiso, H., Sato, H., Koda, T. (1993). Atomic force microscope with lateral vibration of sample. *Proceeding of Japanese Acoustic Society.* Spring Meeting, pp. 889–90.
28. Yamanaka, K., Kolosov, O., Ogiso, Y., Sato, H., Koda, T. (1993). Atomic force microscope and sample observation methods in atomic force microscope. Japanese Patent Application Heisei 5- No. 135342 (1993.5.13).
29. Kolosov, O., Ogiso, H., Yamanaka, K. (1993). Ultrasonic force microscope (UFM)—a new approach to a NDE on nanometer scale. Abstract for *Internal Research Presentation Meeting of Mechanical Engineering Laboratory,* May 12–13, 1993. Tsukuba, Japan, pp. 19–20.
30. Yamanaka, K. Ogiso, H., Kolosov, O. (1994). Ultrasonic force microscopy for nanometer resolution subsurface imaging. *Appl. Phys. Lett.* **64,** 178–80.
31. Yamanaka, K., Ogiso, H., Kolosov, O. (1994). Analysis of subsurface imaging and effect of contact elasticity in the ultrasonic force microscope. *Jpn. J. Appl. Phys.* **33,** 3197–3203.
32. Mate, C. M. (1992). Atomic force microscope study of polymer lubricants of silicon surfaces. *Phys. Rev. Lett.* **68,** 3323–26.

33. Kolosov, O., Ogiso, H., Yamanaka, K. (1993). Ultrasonic force microscope (UFM)—a nondestructive evaluation of elastic properties at nanoscale. Proceeding of the Third Japan SAMPE Symposium Dec. 8–10, 1993, Tokyo Japan (Society for advancement of materials and process engineering), pp. 2196–2201.
34. Yamanaka, K., Kolosov, O., Ogiso, H. (1994). Ultrasonic force microscopy of biopolymers at frequencies above 100 MHz. Proceeding of JRDC International Symposium on Nanostructures and Quantum Effects (Research Development Corporation of Japan, Tokyo, 1993), pp. 349–53.
35. Burnham, N. A., Colton, R. J., Pollock, H. M. (1993). Interpretaton of force curves in force microscopy. *Nanotechnology* **4,** 64–80.
36. Johnson, K. L., Kendall, K., Roberts, A. D. (1971). Surface energy and the contact of elastic solids. *Proc. R. Soc. London, A* **324,** 301–13.
37. Pashley, M. D. (1984). Further consideration of the DMT model for elastic contact, *Colloids and Surf.* **12,** 69–77.
38. Kolosov, O., Ogiso, H., Tokumoto, H., Yamanaka, K. (1994). Elastic imaging with nanoscale and atomic resolution by ultrasonic force microscopy (UFM). Proceeding of JRDC International Symposium on Nanostructures and Quantum Effects (Research Devevelopment Corporation of Japan, Tokyo, 1993), pp. 345–48.
39. Motomatsu, M., Mizutani, W., Nie H.-Y., Kolosov, O., Yamanaka, K., Tokumoto, T. (unpublished).
40. Yamanaka, K., and Tomita, E. (1995). Lateral force modulation AFM for selective imaging of friction forces. *Japan. J. Appl. Phys.* **34,** to be published.
41. Rabe, U., and Arnold, W. (1994). Acoustic microscope by the atomic force microscopy. *Appl. Phys. Lett.* **64,** 1493–1495.

Index